Microcomputers and Modern Control Engineering

CASSELL

MICROCOMPUTERS AND MODERN CONTROL ENGINEERING

Douglas A. Cassell
Inconix Corporation

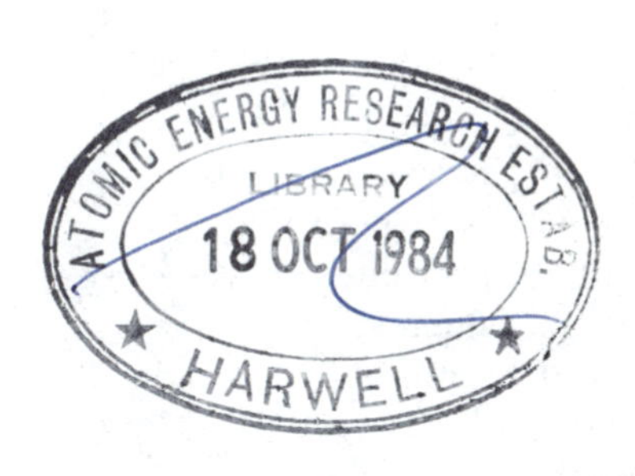

Reston Publishing Company, Inc.
A Prentice-Hall Company
Reston, Virginia

1983

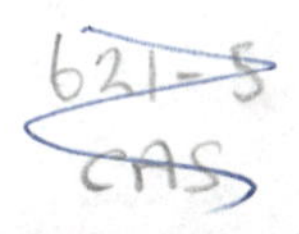

Library of Congress Cataloging in Publication Data

Cassell, Douglas A.
Microcomputers and modern control engineering.

Bibliography: p.
Includes index.
1. Automatic control. 2. Microcomputers. I. Title.
TJ213.C294 1983 629.8'95 82-25040
ISBN 0-8359-4365-8

10 9 8 7 6 5 4 3 2 1

Printed in the United States of America.

To PAT, KIM, and KEVIN

Contents

Preface

Prefaces in books about microcomputers are supposed to begin with a pronouncement that microcomputers are sensational, revolutionary, or miraculous and assert that they will change the world for the better. I feel obliged to say something different. (Though I agree that they are pretty good and, more likely than not, will result in an improvement in our situation.)

Industrial control and instrumentation engineering is a serious matter. Its practitioners are well known for a conservative and serious (even solemn) approach to their work. I am hard put to think of an equally dour group of people.

Conservative caution moderates rapid acceptance of new methods by this group. Frankly, in this business there is a substantial risk in trying unproven methods. Mistakes can have grave consequences. These people have an understandable inclination to view radically new ideas coolly. Until such ideas are proven, they are reluctant to accept them. This is founded on good engineering practice and is as it should be.

But now, I detect a remarkable excitement about new methods. It has grown amid almost universal moaning about productivity statistics and the state of the economy. It has grown among fears of foreign domination of our industrial markets. This is hardly fertile ground. Pessimism is rarely the parent of exciting invention or innovation. This makes the excitement all the more remarkable.

Microcomputers are only part of the excitement. The larger part is our realization of the difference they can make. We have a new tool. But it is what we can do with it that excites the imagination, not the tool itself. There are almost too many ideas for different ways to use it. Variation upon variation rushes into the mind, begging for immediate attention. And in combination, they add up to something even greater and more powerful than the sum of the individual ideas. Anyone who has ever gone out to the hardware store and bought a really "neat" new tool knows what I am talking about.

The excitement is the effect of many creative minds, reflected in innovative products and ideas. The microcomputer is only a vehicle for realizing ideas; it is the painter's new canvas, the sculptor's new marble, the musician's new instrument. It is a mirror in which creative minds are reflected and amplified. That is its contribution. By itself, it is only a clever combination of silicon, chemicals, and metals. What we can do with it is what is important. We "map" our own minds into it when we use it and contribute to the excitement of other creative minds.

The excitement has been spread by the unquestioned success of many microcomputer-based industrial control and instrumentation systems. I hope this book will also help spread this excitement by providing its readers with a liberal education in microcomputers and their application to control engineering.

I imagine its audience to be divided into three parts:

1. Practicing members (of all types: engineers, managers, and technicians) of the industrial community who have limited experience with microcomputers.
2. Practicing members (again of all types) of the electronics and computer science communities who have limited experience with industrial applications.
3. Students who are interested in both microcomputers and industrial applications but have no experience in either area.

In this book, I have tried to avoid dry detail and too much mathematics and to concentrate on principles and practices. It is meant to be read, as opposed to being studied in the usual manner. It does not depend on any particular microprocessor or microcomputer. It was kept independent so that it would be applicable to the study, evaluation, and use of any of the microcomputing products available today, and for some time in the future. If you have an understanding of this material, you can quickly learn the details of any microcomputer system with which you have to deal.

I have worried, from time to time, that the academic training of so many students is compartmentalized and directed toward "mechanical" solutions to problems. Few exams solicit creative solutions. The ability to analyze circuits and prove theorems, though important, does not contribute nearly as much to our advancement as does the stimulation of creative activity.

Creativity and innovation require knowledge of much more than facts and methods. They require a perspective that permits us to see problems from many levels. They require the ability to move from one level to another in the search for solutions. We must understand common threads that run among the levels. We must understand at many levels.

For this reason, many levels of this subject are discussed in this book and related to one another. Topics range from characteristics of solid-state devices to distributed networks of computers; from industrial sensors to the organization and implementation of software systems. And, since we are also a part of the systems we create, you will find some discussion of human, as well as technical, matters in this book.

Like its readership, this book is also tripartite. The first part is about the architecture of industrial control systems. After looking at these systems from a high level, sensors, signal conditioning and measurement methods, control methods, and actuators are examined. In effect, we are working horizontally from input, through decision-making and control, to output. The first part ends with a discussion of distributed systems.

The second part is about microcomputers. First, the general organization of computers is described. This is followed by a discussion of microprocessors and then memories and interfaces are described —in effect, we are working from the inside to the outside. While inside, the opportunity is taken to probe microcomputers as solid-state devices, as integrated circuits, and as integrated subsystems. The second part ends with a discussion of busses, the framework that ties the subsystems together.

The third part is about software. First, programming is discussed in terms of software and systems engineering, components of programs (data structures, algorithms, procedures), and software design techniques. Next, programming languages are examined in a very general way to give an idea of what they are like and what distinguishes one from another. Finally, the discussion closes with operating systems. Operating systems are to software what busses are to hardware—they tie it all together, software with software, software with hardware, and the complete system with the things that it monitors and controls.

I hope that you will enjoy this book and profit by reading it.

Douglas A. Cassell

Sherborn, Massachusetts
August, 1982

Acknowledgements

We are enmeshed in the strands of a web. It connects us all and transmits the reflections of our movements to those around us. Sometimes, our movements cancel or interfere with those of others. A fortunate few of us are entangled in a special way, a way that enhances movement. We exchange knowledge and psychological sustenance along the web's strands. This becomes most apparent when one embarks on a long and difficult, but personal, task. I find myself in this pleasant situation. I discovered it while working on this book. There were lots of clues.

Everyone who has written a technical book has felt the beneficial vibrations imparted by technical discussions with professional associates. This is not uncommon—I felt it amply from my associates at Inconix Corporation. What is not so common is the intense personal interest they took in the book. I always knew they wished it well, but the full extent of their interest and good wishes was firmly engraved on my mind the day the manuscript was sent in. I looked up from my work and found myself surrounded by the entire company, bearing bottles of champagne and having nothing more on their usually sober minds than celebrating the completion of a project that they had all watched with great interest for a year and a half.

They are too numerous to name them all, but they all have my thanks. Some contributed more than others and deserve special gratitude. I have been working with them for a long time (some for twenty years) and would like to single them out: Warren Mayhew, Tom Pye, Barry Kover, and Al Vitale deserve special mention. Others who contributed substantially are Ed Ledoux, Dwight Hardin, Bob Mann, Al Haynes, Tom Brady, Sharon Donovan, Alan Day, Bob Sonnabend, Jerry Templer, and Chuck Stires.

Other clues to the web's existence came from outside my circle of coworkers—from the people I "play" with. The most outstanding examples are the members of the Sherborn Yacht Club. Many times, while crossing tacks in the middle of a yacht race, I was hailed, not with cries of "Starboard Tack!" but with "How's the book going?" Although responding to such a hail is not conducive to the concentration necessary to winning yacht races, it certainly made me know that they were interested in more than beating me to the finish line.

I thank Larry Benincasa of Reston for suggesting the project and for a great deal of helpful advice and support along the way. I thank all the others at Reston with whom I have dealt during this time for being unfailingly attentive, polite, and professional in responding to my doubts, delays, and stubborn opinions.

Most of all, I thank my family. They provided both support and tolerance. Much of the book was written in the early hours of the morning and late hours of the evening. Although *I* had the pleasure of seeing some magnificent sunrises, *they* had to put up with the disturbance of my early risings: Bach on the radio, the buzz of the printer, and the click of the keyboard. Excepting only Brownie, the guinea pig, who tended to squeak loudly when disturbed, their tolerance was outstanding and their support was magnificent. My wife is an old hand with computers. My children are just developing an interest in them. The idea that Dad was writing about computers produced endless streams of extraordinary questions. (Brownie, on the other hand, had only criticism to offer.) More movement in the web's strands, these very close at hand.

I extend my thanks to all members of the web. If they had not been there, I probably would not have done it.

Credits

Quotations from Rudyard Kipling used with permission of The National Trust for Places of Historic Interest Or Natural Beauty.

Chapter 1 quotation from Douglas R. Hofstader, *Gödel, Escher, Bach: An Eternal Golden Braid,* © 1979 by Basic Books, Inc., used with permission of Basic Books, Inc.; used in the British Commonwealth with permission of Harvester Press, Ltd.

Quotations of definitions in Chapters 2 and 3 from *Process Instrumentation Terminology* (ISA-S51.1), used with permission of the Instrument Society of America.

Quotation from Maurice Evan Hare from *The Complete Limerick Book* by Langford Reed (Jarrolds London Ltd.) used with permission of Hutchinson Publishing Group, Ltd.

All other quotations taken from *The Oxford Dictionary of Quotations,* 3rd edition (1979) with permission of Oxford University Press.

PART I

INDUSTRIAL CONTROL SYSTEM ARCHITECTURE

"There are nine and sixty ways of constructing tribal lays,
And–every–single–one–of–them–is–right!"

Rudyard Kipling
In The Neolithic Age

The most important driving force in industry is *information.* Some might argue that economics deserves this role, but I would disagree. It can be argued that money is really "information." There are strong motivations, of course, from other sources as well: survival, social problems, politics, war, pestilence, famine, the desire we all seem to share for a life that is somehow better than the one we presently enjoy or suffer. The relations among these are remarkably complex. But one aspect of our wrestling matches with these issues appears to be invariant: We always end up applying information to the solution of our problems. From information comes *technology.* Technology is inextricably tied to the concept of *tools.* Engineers and scientists develop new technology and use it to make new tools. This is what they do. But their stock in trade is *information.*

Our ability to create tools, through the application of technology, has had the most immediate and recognizable influence on our development. The use of these tools in industry results in our ability to transform our lives, and our political and social systems, into something better than what we had before.

Industrial control systems are the tools we use to manage the manufacture of our other tools and goods. They have been with us since before the Dark Ages and have influenced our present condition and past history in ways that stagger the imagination. Information and its descendant technology are the critical elements in an understanding of control in all of its many forms, particularly as manifested in modern industrial systems. We do not use information and technology simply because they are there. We use them because they are the outcomes of our thinking processes, tangible results of the way our minds work when we solve problems—when we "think."

As we have developed, and as we have thought about "thinking," we have imagined ways in which we might manufacture "thinking machines." This dream has been manifested often in our history and has been the basis of the sciences of artificial intelligence and cybernetics, which have contributed much to the development of control systems we now use and will contribute much more to the control systems we will use in the future.

Some people harbor fear of these sciences. From time to time, the suspicion is voiced that perhaps we should not come to know ourselves too well. We see this in the fear of computers that many express—the fear that automation and the development of "thinking machines" may somehow rob us of our purpose and meaning. This fear was at the heart of the Luddite uprising in the nineteenth century, and it is still present in the attitudes of many toward technology.

But how we think, how we process information in our minds, how we form models and manipulate them, first in our minds and then with the machines we build—these are the things that are uppermost—at the highest "levels"—when we address ourselves to industrial control system design.

In Part I, we survey industrial control technology as it stands now; with emphasis on its "informational" aspects, viewing it from an unusual perspective. This treatment of the subject is rather different from those found in most other texts on these subjects. There are many ways to look at, and to think about, industrial control systems. And if Kipling was correct, perhaps this one may be numbered among the useful.

CHAPTER 1

General Organization of Industrial Control Systems

"Things sometimes look complicated from one angle, but simple from another. He gave the example of an orchard, in which from one direction no order is apparent, but from special angles, beautiful regularity emerges. You've reordered the same information by changing your way of looking at it."

Douglas R. Hofstader
Gödel, Escher, Bach: An Eternal Golden Braid

In this chapter, a generalized view of industrial control system organizations is presented. It is rather different from the views most usually employed. As this view develops, we shall see deeper and deeper detail of the internal organization of industrial control systems and note similarities with the organization of other kinds of systems—among them business, political, social, and economic systems. These similarities (a mathematician willing to stretch the bonds of rigor might call them isomorphisms) can be useful to control system designers as sources of ideas.

TWO GENERATIONS OF ORGANIZATIONAL VIEWS

Many books on industrial control begin their study of system organization by considering examples drawn from particular industries. A typical steel-making or petrochemical process is illustrated. The flow of material through the process is explained, with commentary to point out things that must be measured and controlled to ensure satisfactory operation. While considering several examples, and as particularly illustrative cases come up, the authors note general principles to be used—methods, views, and aspects of control system design that constitute good practice.

This approach is usually found in books by authors who have their backgrounds in the traditional aspects of industrial control. They obtained their experience in the development of the first systems built during the expansion of American industry from around the time of World War II until well into the 1950s. They were the pioneers, the "first generation."

The "second generation" emerged during the initial applications of digital computers to industrial control. These authors engineered the applications of the first commercially available computers to manufacturing automation. Their activity began strongly in the early 1960s and continues into the present, where a new transition is taking place as a result of the development of large scale integration (LSI) in electronics and the invention of the microcomputer.

Second-generation treatments often begin with block diagram illustrations in which the process equipment is shown in one of two ways: (1) a single box in the block diagram labeled "Process," or (2) several boxes, variously labeled "sensors," "switches," "pumps," "motors," "valves," and so on. In the center of the illustration, or near the top, or in some other prominent position, is a large box labeled "Digital Computer" (to distinguish it clearly from the analog variety that was associated with control systems of the previous era, and to emphasize its importance). Lines radiating from the digital computer join it to the process and the elements of which it is composed, suitably labeled with the names of parameters and variables. Figure 1-1 is an example.

Second-generation treatments are still found in modern books on industrial control systems, even those which include discussions of microprocessors. The microprocessor, being new, has an aura of unfamiliarity, even mystery; it is important to dispel this by explaining how it is "interfaced" to the process in ways that are familiar. Second-generation treatments devote much attention to computer system components and interfaces. These are examined in detail and illustrated with photographs of actual examples.

Generally speaking, first- and second-generation treatments use reductionist approaches, concentrating attention on "localized" issues: how to convert signals from their natural analog form to the digital form needed by the computer, how to program the computer in a step-by-step way, and so on. In this way, they provide a comprehensive body of practical information needed by the industrial control system designer. It might be said that they concentrate on nuts and bolts issues which are unarguably practical in nature.

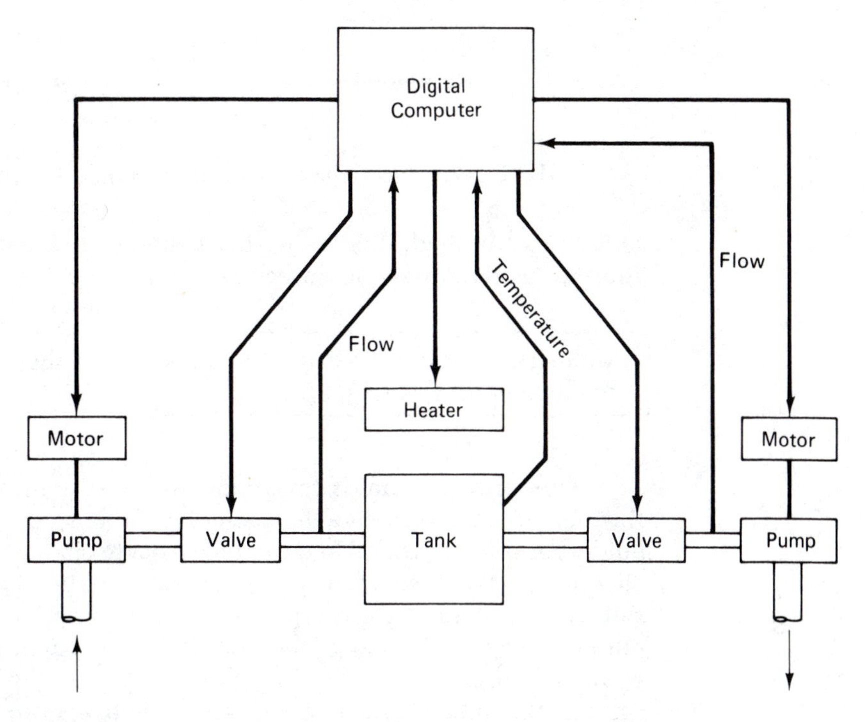

FIGURE 1-1. Example of a second-generation block diagram of a computer-controlled process.

A THIRD-GENERATION VIEW

The discussion of industrial control system organization that follows is sufficiently different from first and second generation treatments that I wish to be permitted to call it an early prototype of a "third-generation" view. I use a photographic metaphor, imagining that we are developing the image of a system that has been recorded with an unusual camera—one that permits us to see the system from a special angle.

Process and Controller Purposes

Imagine, as the print has been in the developing tray for a few moments, that the image that appears is that shown in Fig. 1-2. The system appears as a single rectangular block. Material (input) enters the system from the left and leaves to the right, as output.

We might suppose that the purpose of the system is to transform its input into something that is different in some way. Certainly this must be so—what would be the point of the system if it did not do

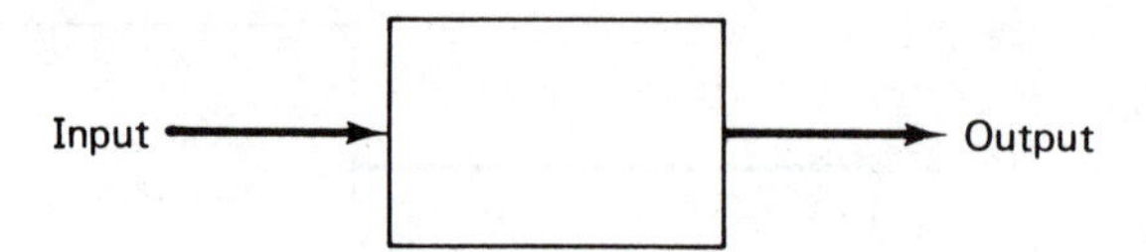

FIGURE 1-2. First image—input is transformed into output.

something? Indeed, this is the first general principle of third-generation industrial system architecture:

> Industrial systems exist to create *change*. If they have nothing to change, they have no purpose.

This principle holds for almost every system we can imagine. Businesses exist to change the economic circumstances of their shareholders and employees. Governments exist to change the lives of their citizens. In the case of most systems, the purpose is improvement. But some systems have the purpose of preventing potential deterioration. They exist to change the rate of change that might occur if they were not present.

In the specific case of industrial control systems, the changes that constitute their purposes are often such things as improving product quality and quantity or the resolution of situations which humans find difficult, uncomfortable, or dangerous.

As the image further develops (Fig. 1-3), we see that the rectangular block is composed of two parts: a "Process" and a "Controller." At this stage, we might imagine that the process is that part of the system that effects the transformation—that does the "work," in a manner of speaking—and that the controller is that part of the system that directs and manages the process in a way that, it is to be hoped, beneficially furthers its performance.

Why should there be a need for a controller? Is it unreasonable to expect the process to go about its business in a workmanlike manner, without interference? In the general systems context, these questions are laced with deep philosophical issues. We shall not pursue

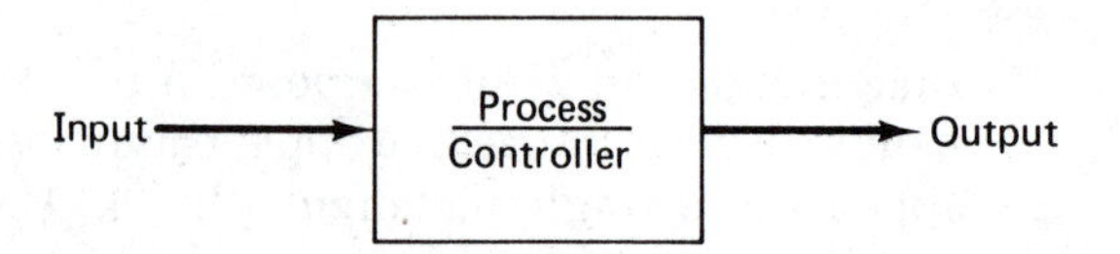

FIGURE 1-3. Second image—the process effects the transformation; the controller directs and manages.

them now. Instead, we simply note, by empirical evidence, that controllers always seem to be necessary and record this observation as the second general principle:

> Left to its own devices, no process operates quite in the way we wish it would.

Applied elsewhere, this principle might be subject to some degree of philosophical argument regarding its absolute truth. In our present context, it embodies the fact that optimization is always with respect to a set of parameters that may have been chosen with some degree of subjectivity. We can always imagine one more parameter, one more criterion for improving the performance of a system, that might be considered. In other words, there is always room for improvement.

Information

In the emerging detail of the image, we see that two streams cross the boundary between the process and the controller (Fig. 1-4). One appears to be information about the process and the other to be instructions passed to the process from the controller. We soon see the reason for this—the controller must "know" something about the state of the process in order to decide what instructions must be given to it, and the process, of course, must receive its instructions in order to act on them.

These two streams are similar in an interesting way. While it is natural to see that one contains *information* (as Fig. 1-4 indicates), we also see that *instructions* can be thought of as *information*, but with something added—the instructions have been distilled from the

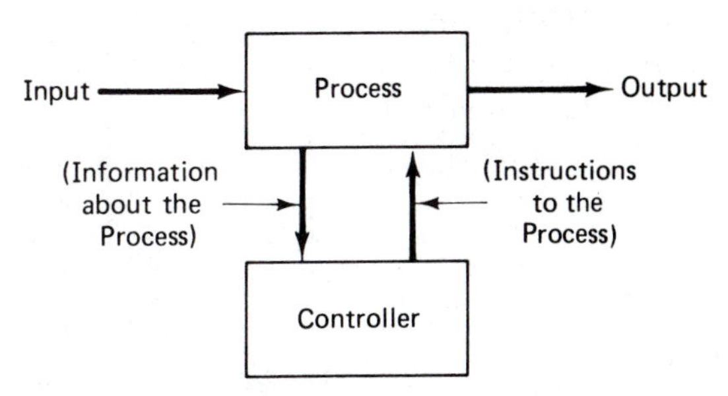

FIGURE 1-4. Third image—information enters the picture and is "processed" by the controller to become instructions to the process.

information about the process. Process information flows through the controller and comes out as instructions. It is as if the controller were a process as well, but a processor of information rather than of material.

Just as an industrial system must have material to change, so a controller must have information with which to work. This leads to the third principle:

> Controllers exist to change information. If they have no information to change, they have no purpose.

We might speculate about the symmetry between this and the first principle. To a degree, it is like the symmetries between time and frequency, functions and their Fourier transforms. This is a symmetry between material and information, processes and controllers. As a function determines its transform, does the process determine the controller?

As we were reflecting on the purposes of processes and controllers and the potential for waywardness inherent in a process, another level of detail emerged in the developing metaphorical image (Fig. 1-5). We now observe that information about the process is composed of two parts—one part being about input to the process and the other about its output.

Is there no information about the process *itself*—does the controller see only what goes in and out and have no knowledge of what happens *inside*? Is it possible to exercise control over something without knowing details of its inner workings? Suppose the purpose of the process was simply to heat water, and the controller measured the

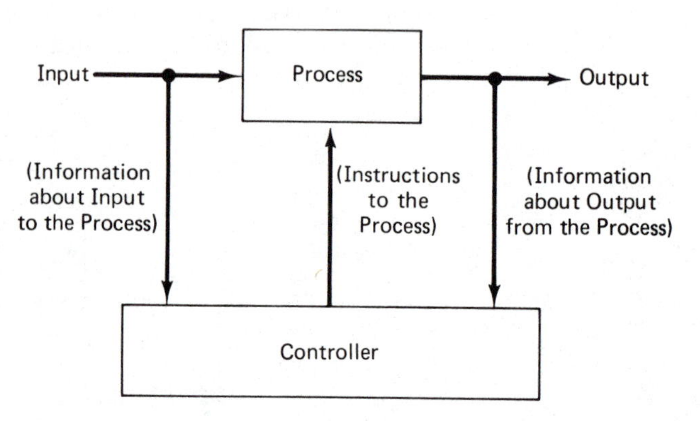

FIGURE 1-5. Fourth image—process information is composed of two parts, one about input and one about output.

temperature of the pot. Wouldn't the controller be receiving information about the *process*?

No. The pot is not the process. What is really going on is that the controller is *inferring* the temperature of the water from data about the state of the pot. In this way, it is obtaining information—in the form of an indirect measurement—about the output of the process (heated water). This may seem a specious distinction, but it is, in fact, very important. The indirectness of the information that controllers must use is an important aspect of control system design. It is entangled with concepts of meaning, noise, observability, and controllability and leads us to the fourth principle:

Controllers must always work with "imperfect" information.

There is always something about the information the controller uses that is less than ideal. When this fact is ignored, the design of controllers becomes a relatively straightforward exercise in mathematics, the solution of control theory equations that are well-defined and tractable. This is a mature field of mathematics and engineering with a history going back to Laplace. It has similarly well-behaved analogs in the field of digital control theory.

But practical control systems must always consider imperfections in the information on which they operate. Although this aspect of control system design is amenable to analytical treatment, detailed analysis is often extremely complex. It must often consider the stochastic character of controller input and the possibility that the controller may also contribute to the imperfections of its information.

Processes within the Process

Meanwhile, further detail has emerged (Fig. 1-6). It seems there is more to the process than was apparent earlier. It appears to have decomposed into several copies of itself, each with connections to the controller similar to the ones we saw at the previous level. We see "processes within the process," each with its own input and output. Each internal process receives separate instructions from the controller and provides separate information about its input and output to the controller.

We see that the original process is composed of smaller processes (let us call them "subprocesses" or "stages"), arranged in series, with the output of each serving as input to a succeeding subprocess. It is

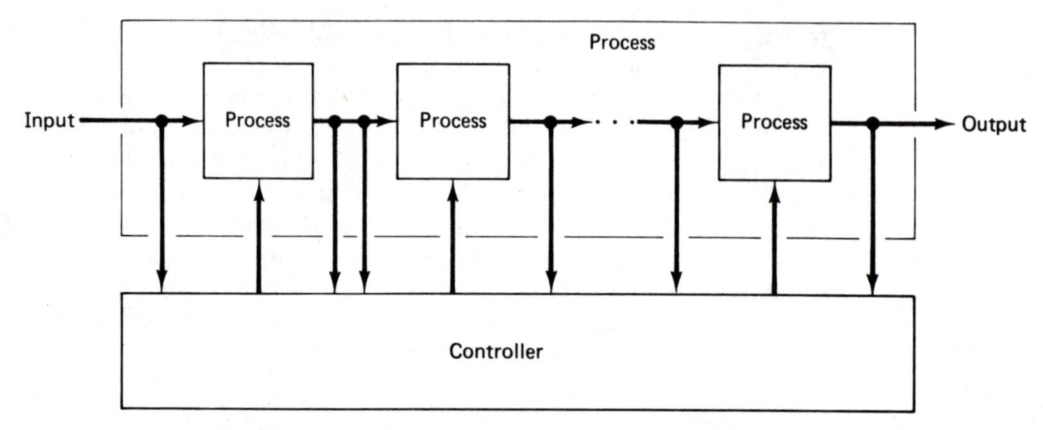

FIGURE 1-6. Fifth image—process decomposition—"processes within the process".

not clear how many subprocesses might compose the complete process. The figure implies a certain generality.

We might suppose, in the example of boiling water, that individual layers of water in the pot might correspond to subprocesses—each acting to conduct the flow of heat from lower layers to higher layers. We might consider that such a model is susceptible to a virtually infinite number of stages of "serial decomposition" and that such decomposition could be applied to virtually every real process.

This leads to the fifth principle:

> Processes are subject to decomposition into a virtually arbitrary number of stages.

As we shall soon see, process decomposition can be "multidimensional." Although the figure illustrates serial decomposition, there are also paths of "parallel decomposition." Further, just as mathematicians sometimes find it useful to multiply a term by 1 in the form of the ratio of a different term with itself, we may utilize "null" subprocesses when it suits our convenience—subprocesses which do nothing but impart a certain structural symmetry to the diagrams of the overall process.

Lateral serial decomposition of some processes could be extended almost indefinitely. In steelmaking, for example, we need not stop at the boundaries of the steel mill. We could also consider stages to the left by which raw materials (coke, ore, energy) are obtained as input and stages to the right where mill products are used to manufacture other goods. These stages might be considered to fall under the umbrella of an immense controller that regulates the pool of resources, economic conditions, availability of mining equipment, political situations in countries that supply resources, competition, and developing technologies that find both new uses for steel and ways to avoid its use.

The extent to which one might wish to decompose a process depends on the control engineering application. Generally, the process is decomposed only to the extent needed for the purposes of practical control design. For example, it is not necessary to extend the decomposition of boiling water to infinitely thin layers of water. Even if we are interested in rather close control of temperature distribution in the pot, a few layers are usually sufficient.

While at this level, we note another interesting aspect of the image: controller input and output could be said to be in a two-to-one ratio. For each subprocess stage, there are two items of controller input (information about the input to the process input and information about the output from the process) and one item of controller output (instructions to the process). This characteristic is seen frequently in real process control systems, even when individual bits of information are counted. Although there are certain variations to this ratio at deeper levels of decomposition, it is generally true that this ratio holds sufficiently well that manufacturers of computers sometimes use it for production planning purposes.

Controllers within the Controller

In the meantime, more detail has developed in the image of the controller itself (Fig. 1-7). We see that the controller is also composed of several "stages," or "subcontrollers." We see "controllers within the controller," each related to the subprocesses seen in the previous image. Clearly, the mission of each subcontroller is the regulation of the subprocess with which it is associated. As we inferred a principle about process decomposition, we also infer a principle about controllers for this image, although one somewhat more restricted:

> Controllers are subject to decomposition to a number of stages which are related to the decomposition of the associated process.

There is a strong correlation between topological decomposition of the process and topological decomposition of the controller. There

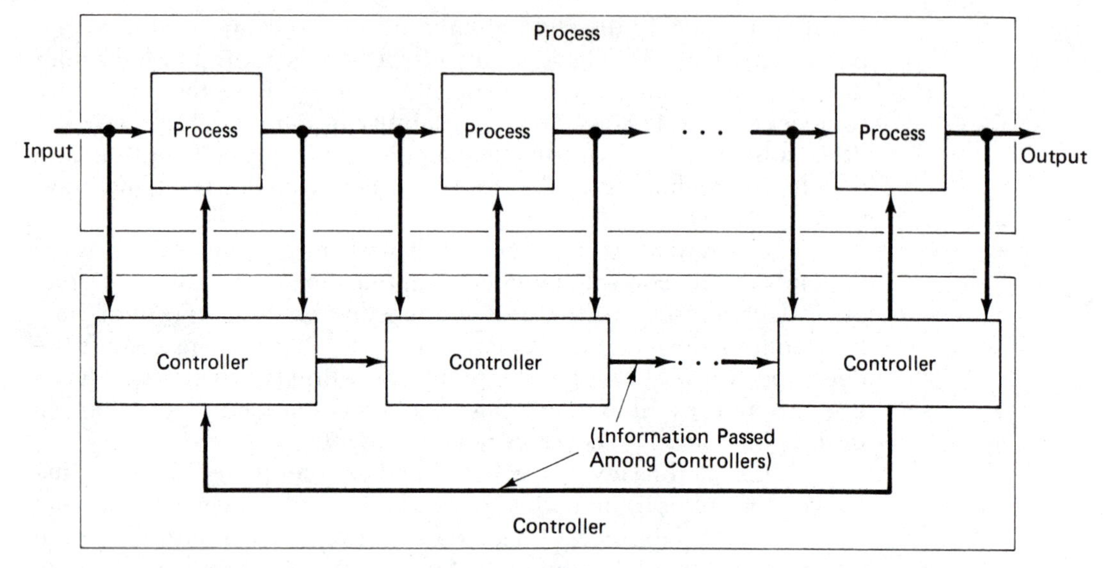

FIGURE 1-7. Sixth image—controller decomposition—"controllers within the controller".

is a subcontroller for each subprocess. This does not mean that subcontrollers have no further available detail of internal organization. Deeper detailing of the internal organization of subcontrollers is performed during their design, in order to implement the control action required by the application. For example, the nature of the desired control action determines the form of the controller's transfer function. Terms in its transfer function form the basis for identifying deeper levels of controller organization.

In theory, such decomposition is only weakly limited in its extent. As in the dcomposition of processes, we carry out controller decomposition only to the extent necessary to the practical requirements of the application.

Coordination, Coupling, and Linkage

We also observe a certain degree of coordination among subcontrollers. This is represented by flows between the subcontrollers. In this case, it is information (not material) that is passed among them. In the majority of control systems, subcontrollers coordinate their control actions on the basis of information they pass among themselves. Control action is often "coupled" between controllers.

However, coupling can also result from the effects of the output of earlier process stages. In some practical examples, there may be little information flow among the subcontrollers. Such a "disjointed" topology would be representative of systems which were coupled only through the effects of the input and output of the subprocesses that they controlled.

Systems which appear disjointed at one level of detail may be coupled at higher levels. For example, it is unlikely that there is any "electronic" linkage in the serially decomposed system represented by the combination of processes used by a steel manufacturer and an automotive manufacturer. Yet at a higher level, perhaps between the purchasing agents of such organizations, there might indeed be such a linkage.

Such linkages are not ordinarily considered in the design of an electronic process control system, but they might be. Regulation of this linkage might be beneficial to the performance of the combined system. This example illustrates the relationships that may exist between control systems and other systems that are not usually considered to be within the domain of industrial control system theory.

Figure 1-7, selected for this illustration because it demonstrates a typical information network topology, shows a circular flow of information among the subcontrollers. Many different topologies are actually used for practical control systems. After discussing the present image, we discuss some variations on these. There will also be further elaboration on system topologies in the chapters on communication systems and multiprocessor systems. In some topologies, separate information paths are established only among subcontrollers that must communicate with one another. In other topologies, a single communication path is shared among all of the subcontrollers.

In this figure, we also see that the ratio of subcontroller input to output has changed to a ratio of three to two. As control system composition is viewed from higher levels, this ratio behaves in a way that is reversed from the one we observed for subprocesses. At the very highest "levels," the number of bits of information that must be passed among the controllers is often enormous, so much so that it becomes impractical to use individual wires for its transmission.

At these levels, information is transmitted serially, one bit at a time. As we proceed to deeper levels, however, it becomes more practical to transmit information in parallel, and we find data being exchanged in units that contain more information. For example, among the individual elements of a microcomputer (its central processor, memory, and input/output devices) we often find information transferred in units containing eight bits each (called "bytes" in the jargon).

The image in Fig. 1-6 was described as a serial connection. For reasons of simplification, we did not note then that other process topologies are also possible, even common. This is an aspect that should be considered now. Coincidentally, we observe (in Fig. 1-8) that this aspect has appeared in the developing image.

Figure 1-8 seems quite complex compared to the image with which we began. When viewed at a detailed level, it is not unusual for the apparent complexity of a system to grow alarmingly. We are fortunate in being able to consider structures from different levels and viewpoints. This ability seems uniquely human and is a decided advantage that we have over machines (at least at the present stage of development of the science of artificial intelligence). Using this ability, we are able to form mental models of the systems we design, at different levels of detail, according to our convenience in thinking about them, and thereby see the orchard that the trees compose.

Performance and Other Topologies

There are many reasons for using nonserial topologies, among them the need to improve performance and reliability. Another is a requirement sometimes imposed by the nature of the work the process is to perform.

It helps to understand how performance improvements might be achieved by considering practical aspects of the elements that typically compose a process. At the level of architectural detail most usually used in thinking about a system, these elements are such items as machine tools, distillation columns, pipes through which material flow is controlled, and so on. Each element has a finite capacity. The machine tool can machine only so many parts per hour. The distillation column has a certain volume. Chemicals must be in it for a certain time and at certain temperatures in order for the chemical processes to reach their conclusion. The pipe has a certain diameter and the material has a certain viscosity, so the amount of material that can pass through in a given time is limited.

One way in which system performance can be improved is through duplication of elements, in a parallel arrangement. This kind of arrangement is shown in Fig. 1-8. Pairs of elements within the decomposed image of the process are copies of one another, each performing the same function as its partner, in order to achieve an increase in the amount of "change" that the process creates per unit time.

Such an arrangement has other benefits. Since nothing is perfectly reliable, each element of a process (and of a controller) has a certain probability that it will fail to operate satisfactorily at some time. One task of the industrial control system designer is to ameliorate the potential effects of unreliability wherever this can be done at reasonable cost.

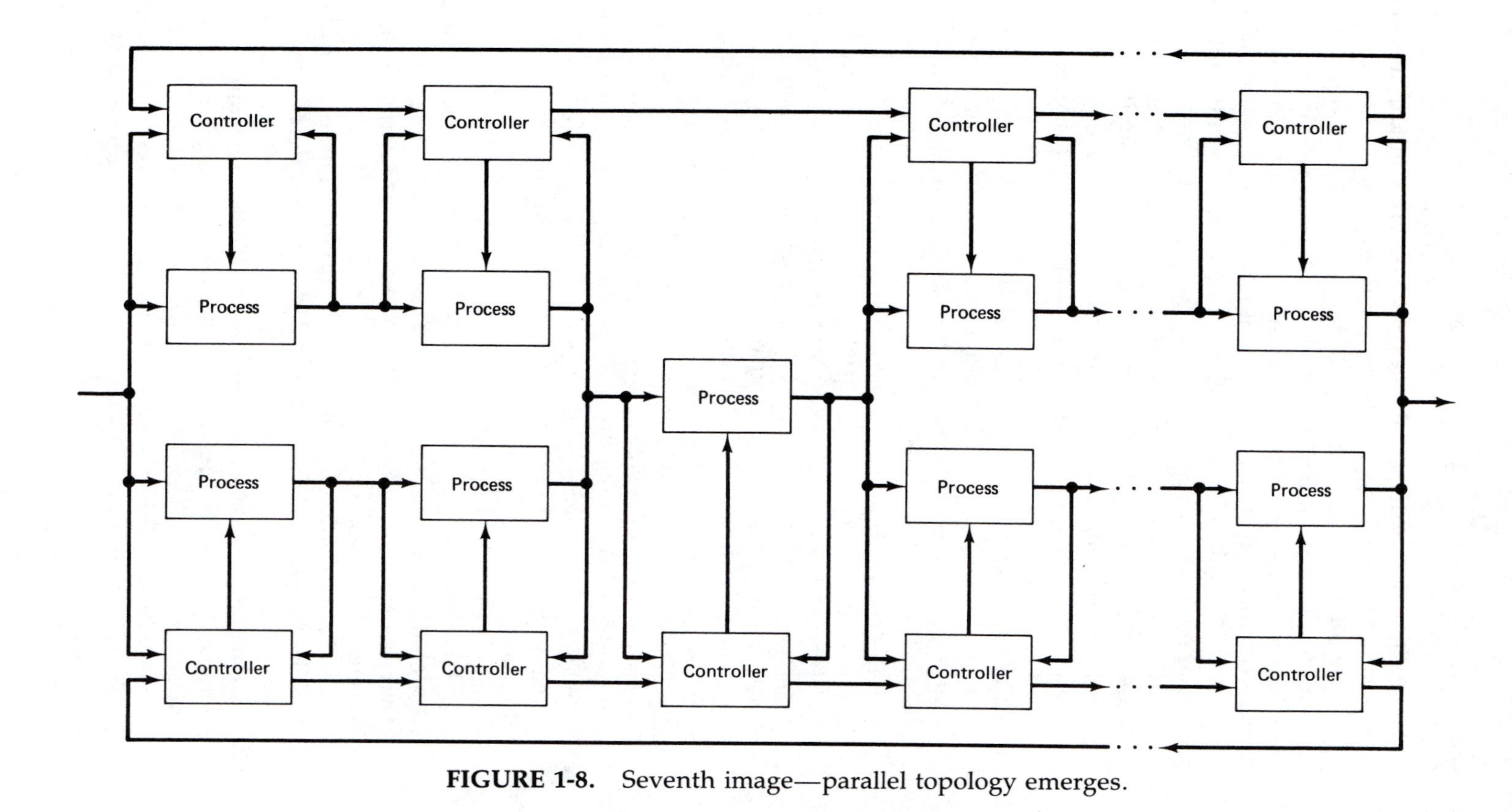

FIGURE 1-8. Seventh image—parallel topology emerges.

There are two general ways of doing this, and they may be used in concert. One is to design and select elements that have a high degree of inherent reliability. Microcomputers and other electronic components are particularly useful in this regard, since they enjoy relatively high degrees of inherent reliability. Another is to design the system so that it contains redundant elements—elements connected in parallel fashion which can take over the functions of their counterparts if they fail to operate satisfactorily.

There are two arrangements in which redundant elements are often employed. In the first, they "stand by," ready to take over in the event of a failure. In the second, they maintain an "active" role during the operation of the system, working alongside their counterparts and performing the same functions. In the event of a failure, they are able to prevent a complete shutdown of the process by continuing to provide an appreciable portion of the input material required by succeeding process stages, even though they are unable to restore the process to the performance level at which it was operating prior to the failure.

Another reason for nonserial topologies is the nature of the work the process performs. For example, consider a process composed of three subprocesses, the first two of which are in a parallel arrangement. The output materials from these two subprocesses are combined by the action of the third subprocess, which is connected to the first two subprocesses in serial fashion. A practical example would be one in which two different materials needed to be prepared (perhaps cooked at different temperatures) and then mixed in a controlled ratio. Here, the nature of the process imposes a constraint, requiring a decomposition that is partially parallel.

Building Blocks

Looking back to Fig. 1-5, we see that the fundamental form there is repeated in Fig. 1-8 many times. The structure in Fig. 1-8 appears to have been assembled from many copies of the simpler form we saw in Fig. 1-5.

This similarity of form—these "building blocks"—are found in the stucture of many systems, and at several different levels within these systems. We have seen this in the decomposition of an industrial system and we will see it later, in the structure of complex computer programs and in the organization of multicomputer systems and communication networks.

Because of this, there are several useful ways of viewing a complete system and decomposing it into its constituent parts. A large industrial manufacturing system can be viewed as consisting of many processes, which may be decomposed individually or in combination into subprocesses connected in serial and parallel arrangements. The collection of computers that might be controlling the system can be viewed as parallel and serial arrangements of individual computers. In modern highly integrated computers, many circuits are found to have internal organizations similar to the structures we have seen in the figures in this chapter. Programs within each computer may be viewed as subprogram building blocks, also arranged in serial and parallel structures. Subprograms and combinations of subprograms may be viewed as composing individual controllers in ways that are conceptually related to traditional forms of closed-loop controllers.

Because these similarities of form are often repeated at several levels, principles stated in this chapter can be applied at different levels of system design.

At the deepest levels of detail, these building blocks display the most variety and become not so much structural as "atomic." There are analogues to this in other fields. The stage in chemistry at which we pass from molecules to atoms is one example. The stage in organization theory at which we pass from groups to individuals is another example. In the theory of industrial control systems, we reach this stage when we come to consider primary elements, transducers, summing and multiplying control elements, and actuating elements.

The methods we use to think about these things are qualitatively different on each side of this boundary. As we approach the deepest levels, we have more analytical and scientific methods at our disposal. At the higher levels, our methods are more abstract, less precise, and less analytical. Nevertheless, these higher-level views are useful in devising strategies for the design of industrial systems.

In some ways, this view of system organization is like the "top-down" approach advocated for the design of modern, complex computer programs. It is distinguished from traditional approaches, which modern computer programmers might call "bottom-up" approaches, in that traditional approaches examine first the components of the system at a low level and then consider how they might be connected to one another.

It might be assumed that such approaches could be expected to "meet" somewhere in the middle, yielding identical designs. In fact, such well-matched meetings are rare. At each level of either approach, whether decomposing from the top down or assembling components from the bottom up, there are diverging paths that might be taken, leading to very different designs.

It is important to understand both approaches. The architect of a building must have an overall plan, but he must also understand the nature and form of the materials that he plans to use—bricks, mortar, wood.

Classifications of Industrial Systems

It is the practice of the industry to divide industrial systems into three broad classifications: *continuous, batch,* and *discrete.*

The first two classifications are applied to manufacturing processes in which input and output materials are commonly measured according to their volume or weight or some equivalent measure. In this manufacturing system class, the process acts on (and produces) liquids, semi-solids, solids measured by volume (e.g., cubic yards of steel, linear feet of wire, gallons of gasoline), and gases.

Continuous processes operate "continuously." Material flows into and out of the process on a more or less continuous basis. This does not mean that the process might not be stopped from time to time for various purposes, such as maintenance or changes to the way in which it operates.

Batch processes operate on "batches" of input material, producing "batches" of output material. In a batch process, a tank or processing vat might be filled with a mixture of input materials and subjected to a certain temperature profile—"cooked," as it were. When the batch is "done," it is removed from the tank, the tank is cleaned up and refilled, and the operation is repeated to produce the next batch.

The term *discrete* is applied to manufacturing processes in which the materials acted on or produced can be more or less conveniently "counted"—that is, they are discrete pieces. In some cases, input materials might well be measured in terms of volume or weight and entered in batches; the manufacture of molded plastic parts is an example. Other examples of discrete manufacturing are the filling of bottles and containers, the manufacture of automobile parts, and operations that involve the working of metal (of which the principal automated examples are to be found in numerically controlled manufacturing operations). The assembly of discrete items into a different manufactured object is a subclassification of discrete manufacturing operations. These assembly operations are frequently automated and are in some cases significant opportunities for the application of robots.

CHAPTER 2

Primary Elements

"When *I* use a word," Humpty Dumpty said, in a rather scornful tone, "it means just what I choose it to mean—neither more nor less."

"The question is," said Alice, "whether you *can* make words mean so many different things."

"The question is," said Humpty Dumpty, "which is to be master—that's all."

Lewis Carroll
Through the Looking Glass

Words and meanings are important. Yet there are "local dialects" within the many communities of people working in industrial control, each with different shades of meaning for certain terms. Different industries, and different manufacturers of equipment used in these industries, use different terms. We often find that they are really talking about the same things.

The *words* themselves are not as important as what they *mean*, but being able to speak the local dialect is important. If we equip ourselves with a good grasp of meanings and fundamental concepts, and if we are prepared to recognize and deal with variations in terminology, it will go a long way in helping us.

TERMINOLOGY

Fortunately, there have been substantial efforts to standardize terms. Among the most important are those of the Instrument Society of America (ISA). The ISA Standard *Process Instrumentation*

Terminology (ISA-S51.1) is a useful tool for anyone in the field of industrial control systems. It is particularly valuable to anyone beginning the study of process control. It provides dictionary-like definitions of most of the terms commonly used, and, just as a dictionary is a useful tool for one who is learning a new language, this standard "dictionary" should be in the library of anyone interested in industrial control systems.

Because it is so useful, several definitions from this Standard are quoted below.

Element—A component of a *device* or system.

Element, primary—The system *element* that quantitatively converts the *measured variable* energy into a form suitable for measurement.

Device—An apparatus for performing a prescribed function.

Sensor—See *transducer.*

Transducer—An *element* or *device* which receives information in the form of one quantity and converts it to information in the form of the same or another quantity.

NOTE: This is a general term and definition and as used here applies to specific classes of *devices* such as *primary element, signal transducer,* and *transmitter.*

Variable, measured—A quantity, property, or condition which is measured.

NOTE 1: It is sometimes referred to as the measurand.

NOTE 2: Common measured variables are temperature, pressure, rate of flow, thickness, speed, etc.

Definitions of *signal transducer* and *transmitter* are deferred to the next chapter, since it is there that they are more pertinent.

Note the difference between the definitions of *transducer* and *primary element.* The operable phrase is "a form suitable for measurement." In typical use, *transducer* is the broader term. There is an implication of "practicality" in the definition of *primary element.* Many things transduce information but not always to forms susceptible to practical applications. Even so, the terms are often used interchangeably.

CLASSIFICATION AND SELECTION

The selection of primary elements and transducers for an industrial control system must consider many things. Primary elements may be classified according to (1) the variables they are intended to measure, (2) the transduction principle, and (3) individual characteristics which make them more or less suitable for a particular application. Listed below are some ways in which primary elements may be organized.

With respect to the type of measured variable

Temperature, force, fluid characteristics (rate of flow, pressure, etc.), moisture, radiation, position and its derivatives, and so on.

With respect to the transduction principle

Resistive, strain-gage, potentiometric, capacitive, inductive, electromagnetic, reluctive, piezoelectric, photoconductive, photovoltaic, and so on.

With respect to individual characteristics

Size, accuracy, repeatability, temperature stability, cost, ease of use, gain, resolution, noise immunity, dynamic range, and so on.

For given combinations of variables to be measured and transducer characteristics, there is usually an optimum choice. But a primary element that is optimum with respect to one consideration may be less than optimum, even undesirable, with respect to other considerations. It is difficult to find an absolutely optimum primary element for a given measured variable. From time to time, considerations having little to do with process theory or physics must be given substantial weight in making a choice. For example, the best primary element may be unobtainable because of a long delivery time or high cost. In this case, a second-best choice must be made.

Although primary elements are not so free in their choice of meaning as Humpty Dumpty seems to be, they are not always so plainspoken as we might wish. In many cases, the output of the primary element is used to *infer* the value of the measurand. Rarely can we measure a variable directly. For example, electromotive force (emf) is the actual output of a thermocouple. From emf measurements, the temperature surrounding the thermocouple is inferred. Force changes the resistance of a strain gage. From this resistance change we infer the magnitude of forces acting on the gage. There may be several links in this chain of inference, each subject to complications of accuracy, resolution, stability, and so on.

It impossible to discuss every type of primary element used in modern instrumentation systems within one book, much less a chapter. The rest of this chapter describes a selection, organized by measured variable (usually our first consideration). Anyone who is already familiar with this subject is likely to have favorite transducers. I ask their forgiveness if their favorites have been omitted.

SENSORS OF TEMPERATURE

Temperature measurement is required in most industrial instrumentation applications. One reason is that the rate of chemical reaction is strongly affected by temperature. Thus, measurement and control of process temperature are important. Temperature measurement is also important in discrete manufacturing; for example, in monitoring tool temperature to prevent overheating and to detect wear.

Temperature is measured by observing its effect on materials that are sensitive to temperature. Common temperature sensors are grouped into the following types, based on the physical principles involved: bimetallic, fluid-pressure, resistive, thermocouple, and radiation. Of these, resistive and thermocouple types are most commonly used in computer-based instrumentation systems.

Bimetallic

Different metals have different coefficients of thermal expansion. In bimetallic temperature sensors, two dissimilar metals are bounded to one another. As the temperature of this assembly changes, each metal expands to a different extent, distorting its original form. Temperature is inferred from the extent of distortion. Bimetallic strips can be formed into coils which move indicating pointers on scales calibrated to match the correspondence between distortion of the strip and ambient temperature. This is the usual design of the common wall thermometer.

This distortion can be used to actuate control elements. The familiar *thermostat* operates on this principle. The bimetallic strip in a thermostat moves a temperature indicator and also opens and closes contacts which control heating and cooling equipment.

Some bimetallic devices are formed so that temperature-induced stress is built up until the assembly deforms suddenly, with a snapping action. These *bimetallic switches* are sometimes used as equipment safety controls to prevent overheating and overcurrent conditions in motors and circuit breakers. A portion of the conducted current heats the switch, causing it to change position when the current becomes too great.

Fluid Pressure

Temperature sensors based on fluid-pressure principles include the common mercury thermometer. These operate by detecting pressure changes caused by thermal expansion of gases or liquids. Another fluid-pressure temperature sensor is the gas-filled *Bourdon tube*, in which gas expansion due to temperature change is sensed through motion of the tube. These are described in more detail in the section on pressure sensors.

Radiation

Sensors based on radiation principles are used to measure very high material temperatures—temperatures too great to permit the presence of a sensor within the material without damage to the sensor. An example is the measurement of molten metal temperature. These devices are commonly called *radiation pyrometers.*

Radiated energy from the material is focused on sensors which are sensitive to this energy (e.g., photocells, thermocouples, thermopiles). Knowledge of relations between material temperatures, radiated energy, and the behavior of the radiation pyrometer sensor enables the instrumentation to infer material temperature from the pyrometer sensor temperature.

Resistive

Temperature sensors based on resistance principles are divided into two classes: (1) those using temperature effects on the resistance of metals (commonly called *resistance temperature detectors,* abbreviated RTD), and (2) those using temperature effects on the resistance of semiconductor materials (of which the thermistor is an example). Devices in the former class generally have positive temperature coefficients, while semiconductor-based devices have negative temperature coefficients.

Resistance temperature detectors are commonly made of nickel or platinum alloys, depending on the temperature range to be measured and the accuracy required. Platinum RTDs offer the greatest accuracy and are usable over the widest range of temperatures (−265°C to +1100°C), but are the most costly. Nickel RTDs may be used over temperature ranges from −100°C to +300°C. Various standards specify RTD temperature coefficients and resistance at particular temperatures. These help to simplify their use since standardized signal conditioning circuits can be designed for different types of RTDs.

Resistance temperature detectors are assembled according to different application requirements. For example, they may be fabricated as encapsulated coils of wire bonded to an electrically insulating material that may be attached to the surface at which the temperature is to be measured. They may also be placed in probes, for insertion into fluids. A variation uses metal films. These are more rugged and less costly, but do not exhibit the same sensitivity.

The *thermistor* is based on semiconductor thermal effects. These are fabricated from a combination of metal oxides and formed into small beads, disks, or cylinders. They are characterized by small size and quick response, making them useful for measuring rapidly changing temperatures. They are very nonlinear, however, which makes it difficult to design signal conditioning circuits for them. They are usually applied to measure temperatures in the range of −100°C to +300°C, which is typical of the generally usable range of semiconductor-based sensors. Semiconductor-based devices are unsuitable for relatively high temperatures, because they suffer from crystal lattice breakdown at temperatures greater than a few hundred degrees Celsius.

Silicon crystals are also commonly used semiconductor-based temperature sensors. These have temperature coefficients which are positive in the upper part of their range and negative in the lower part. Small signal diodes and transistors exhibit similar characteristics. Thermal characteristics of semiconductor devices are used in many situations, not only to measure temperature, but also to control its effects in signal conditioning circuits through combinations of devices which have offsetting or counteracting thermal behavior. They are used in cold junction compensation circuits for thermocouple inputs to electronic and computer-based instrumentation systems.

Thermocouples

The most commonly used device for industrial temperature measurement is the *thermocouple.* These devices use an effect, discovered in the nineteenth century by Charles Seebeck, which causes an electromotive force (emf) to be generated at the boundary between dissimilar metals. The amount of emf is proportional to the temperature difference between the boundary and a point of circuit completion located elsewhere.

Figure 2-1 illustrates the basic arrangement of a thermocouple circuit. The point at which the dissimilar metals are joined is called

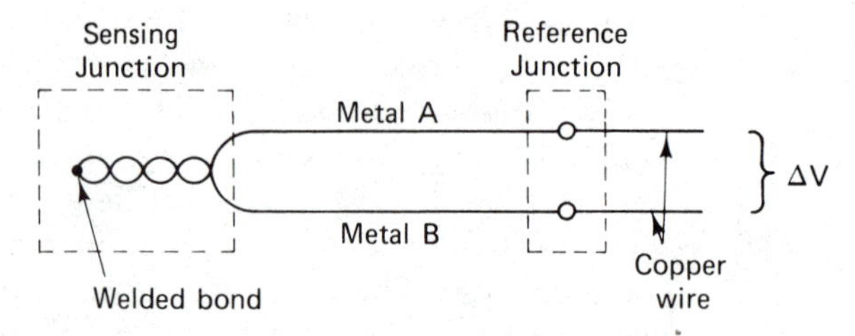

FIGURE 2-1. Basic arrangement of a thermocouple circuit. A temperature-dependent emf is generated at the junction of dissimilar metals.

the *sensing junction.* Wires of each type of metal then lead to a point called the *reference junction*, or the *cold junction*, where the wire usually becomes copper wire of the type commonly used in most electronic circuits.

The Seebeck effect acts at each dissimilar metal junction. Not only is a Seebeck emf generated at the thermocouple wire junction, where temperature is to be measured, but also where these wires join the copper wires of the signal conditioning circuitry. The purpose of *cold junction compensation* is to sort out the emf at the sensing junction from the combined emf sensed by the circuit through its copper wires.

Various combinations of metallic alloys are used in the manufacture of thermocouples, selected for the particular temperature ranges to be measured and accuracies and response times required. Some alloy names are trade names of the manufacturers who first developed them, for example, Chromel-Alumel. Other combinations are iron-constantan, copper-constantan, and platinum-platinum/rhodium. These have been given letter designations (J, K, T, E, R, S, etc.), and standards have been established for their characteristic emfs as functions of temperature. Published standards describe the behavior of each type.

Each type has a different emf versus temperature curve which is generally nonlinear over the thermocouple range. For certain combinations of thermocouple type and specific temperature range, reasonably linear approximations to the curves may be established so that, if carefully chosen, they may be used with relatively simple signal conditioning equipment. For wider temperature ranges, it is necessary to linearize the thermocouple output signal to relate it to actual temperature accurately. Through extensive computer analyses, the National Bureau of Standards has prepared tables of polynomial coefficients for use in linearizing thermocouple signals. These are used in the design of electronic linearization circuits and programs which perform this function.

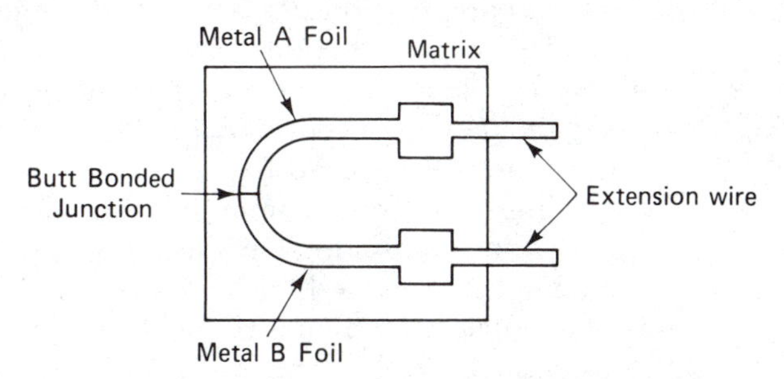

FIGURE 2-2. Construction of a foil thermocouple.

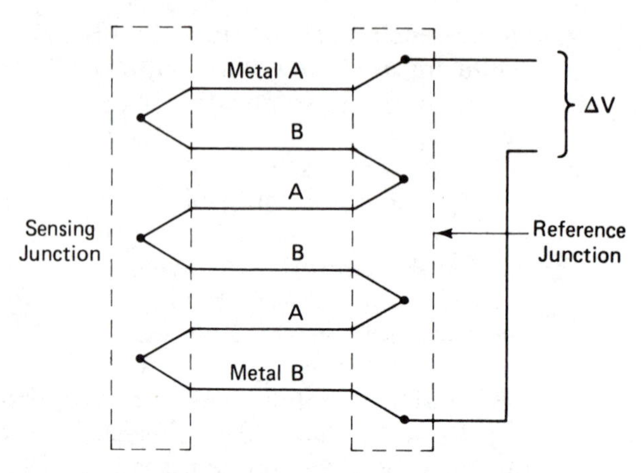

FIGURE 2-3. A thermopile; series thermocouples for amplifying the emf.

Thermocouples are also made in other forms, such as the one depicted in Fig. 2-2. This is a foil in which the alloys are arranged for easy attachment to a surface. Another is the *thermopile,* depicted in Fig. 2-3, a series arrangement of many thermocouples of the same type which amplifies yielded emf.

SENSORS OF FORCE

Primary elements used to measure forces generally employ either resistive or piezoelectric transduction principles. The principal example of the resistive type is the *wire strain gage.*

Wire strain gages use the principle that the resistance of a wire is proportional to its length and inversely proportional to its cross-sectional area. As a wire is stretched, its length increases and its cross-sectional area decreases, producing an increase in resistance. Conversely, compression of the wire (through a decrease in length and increase in cross-sectional area) produces a decrease in resistance. These resistance changes are sensed by signal conditioning circuits designed for this purpose. (Bridge circuits, described in the next chapter, are frequently used.)

There are two common arrangements of wire strain gages: bonded and unbonded. In the *bonded wire strain gage,* the wire is arranged in a back-and-forth pattern (Fig. 2-4) and bonded to a flexible backing material which is then attached to the material in which strain is to be measured.

An application of force sensing is in weighing systems, where a force-sensing primary element called a *load cell* is often used. Load cells are often constructed in the form of a steel rod or column, set within a housing designed to prevent lateral movement or buckling. A bonded wire strain gage is fastened to the side of the rod with epoxy cement. The container to be weighed is mechanically linked to the load cell, often by using the load cell to support a known fraction of the container weight. The force on the load cell induces compression in the rod which, in turn, compresses the strain gage wire and decreases its resistance. From this, the instrumentation system infers the container weight.

Load cells may be "inverted" mechanically to measure tension forces. Here, the strain gage is cemented to a column from which the load is suspended. Tension induced in the column by the load is reflected in a stretching of the strain gage wire and a corresponding increase in its resistance. Principal applications of bonded wire strain gages are in the measurement of moderately large forces.

Unbonded wire strain gages are used principally to measure relatively small forces. Figure 2-5 shows one possible arrangement of an unbonded wire strain gage. Resistance wire is arranged around pairs of posts, one of which is embedded in a fixed frame and the other in a movable spring-loaded armature. As force moves the armature, strain on one wire pair is reduced while that on the other is increased. From this change, the amount of force is inferred. Unbonded wire strain gages are often arranged so the resistances can serve in a bridge circuit.

Semiconductor materials replace wire in some strain gages. Of these, both resistive and piezoelectric types are to be found, the transduction principle depending on the type of semiconductor material.

Resistive semiconductor strain gages are made from silicon, a fundamental material employed in the construction of integrated elec-

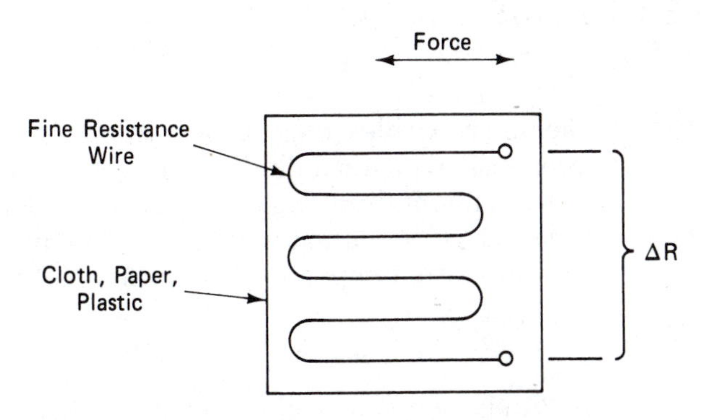

FIGURE 2-4. Construction of a bonded wire strain gage. Lateral force increases the resistance of the wire.

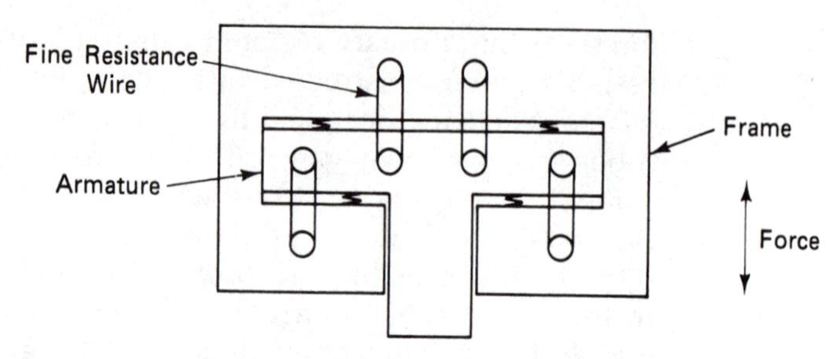

FIGURE 2-5. Construction of an unbonded wire strain gage. Armature motion increases resistance in one pair of wires and decreases resistance in the other pair.

tronic devices. In these devices, a silicon crystal wafer is bonded between a pair of electrode plates. Force on the plates compresses the crystal, causing a resistance change from which load force is inferred. *Piezoelectric semiconductor strain gages* are constructed in a similar fashion, but use materials (such as quartz crystal) which exhibit piezoelectric effects. These produce a voltage when subjected to mechanical compression. Semiconductor strain gages have very high sensitivities and are, therefore, employed principally for measuring relatively small forces, as in weighing systems for laboratory applications.

Semiconductor properties which make them useful for measuring temperature may interfere with their application to force measurement. For this reason, semiconductor strain gages are sometimes made with two differently formulated materials which have temperature coefficients of opposite polarity, so that their output signals may be used to compensate one another as their temperature changes.

SENSORS OF FLUID AND GAS VARIABLES

Several variables must be measured by controllers responsible for processes which handle fluids or gases. One is the total amount (mass or volume) of fluid or gas that has passed a given point in the system. Another is the rate at which it is flowing. Other important variables are pressure, temperature, and the level at which fluid stands.

Fluid Flow

The total amount of fluid or gas which has passed a given point is generally termed *total flow*. The rate at which it passes a given point is termed *flow rate*. Flow measurements are generally in terms of the mass of the fluid (mass flow) or its volume (volumetric flow). Common flow rate sensors operate by obstructing the fluid flow in some way. Such obstructions create local differential pressures which are related to the flow rate and from which it can be inferred. From flow rate, and from knowledge of the pipe diameter and fluid density, mass flow can be inferred.

There are several methods of obstructing flow to create differential pressures. One is the *Pitot tube*, illustrated in Fig. 2-6. This assembly is formed from two concentric tubes, the centermost of which is in line with the direction of flow. The outer tube is closed in the direction of flow, but has a small opening perpendicular to it. As fluid passes the tube, different pressures are induced in each tube. Other types include the *Venturi tube* and the *nozzle*, illustrated in Figs. 2-7 and 2-8. These force the fluid to flow through a constricted area, resulting in pressure differences on either side of the constriction.

When flowing fluid is required to turn a corner, pressure differences are also induced. Since practical installations often require pipes to change direction, advantage can sometimes be taken by measuring pressure differentials induced by these direction changes. Flow rate sensors that use this principle are called *centrifugal sensors.* By locating pressure sensors on the inner and outer sides of a bend in the pipe, the pressure difference induced by the fluid direction change can be measured and flow rate can be inferred. If no convenient bend in the pipe is available, one can be installed by curling the pipe about itself in a full circle and installing pressure sensors on its inner and outer circumferences.

Flow rate also can be measured by using kinetic energy in the flowing fluid to turn a vane or paddle wheel. *Turbine flowmeters* are

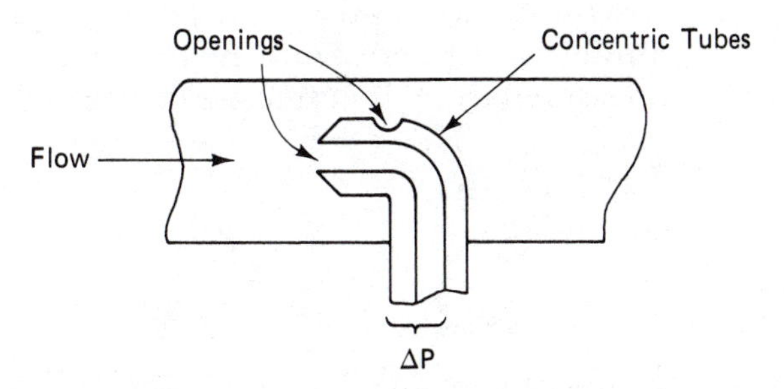

FIGURE 2-6. Flow rate creates proportional differential pressure in the concentric tubes of this Pitot tube.

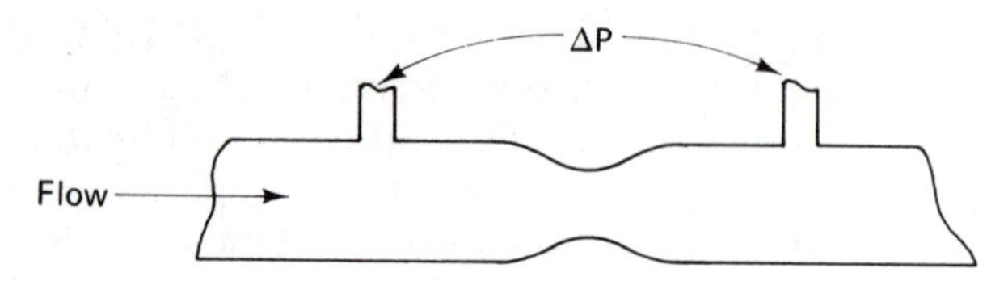

FIGURE 2-7. In this Venturi tube, the constriction creates a difference in pressure that is proportional to flow rate.

an example. A vaned rotor is mounted coaxially in the pipe. Fluid flow around the rotor turns it at a speed proportional to the flow rate, which is then inferred from the rotor speed. If mechanical links to the rotor are impracticable, electromagnetic sensing can be used by equipping the tips of the rotor vanes with permanent magnets. As the tips pass a magnetic pickup coil installed in the pipe wall, pulses are induced in the coil. These pulses may be counted by the instrumentation system and, from this information, flow rate may be inferred.

Flow measurement can be mathematically complex. The behavior of flow rate sensors is affected by viscosity, turbulence, and specific gravity of the fluid. All of these must be considered in these measurements.

Pressure

Most methods of pressure measurement detect effects which fluid and gas pressures have on the vessels which contain them.

Bellows are common examples of pressure sensors. A simple bellows pressure sensor is shown in Fig. 2-9. Pressure causes the bellows to expand when pressure inside the bellows exceeds outside pressure and to contract when outside pressure is greater than inside pressure. This expansion and contraction is proportional to the difference between these pressures. This device measures what is known as *gage pressure,* expressed as "pounds per square inch, gage" (abbreviated "psig") in English units. Bellows expansion and contraction may be measured by motion sensing devices (discussed later in this chapter),

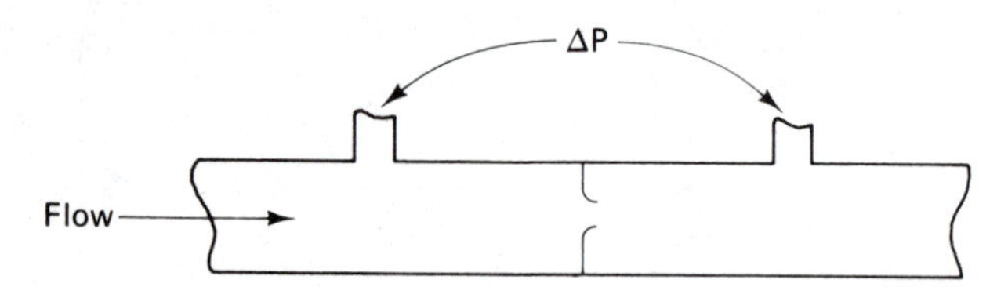

FIGURE 2-8. The nozzle flow sensor creates a differential pressure that is similar to that of the Venturi tube.

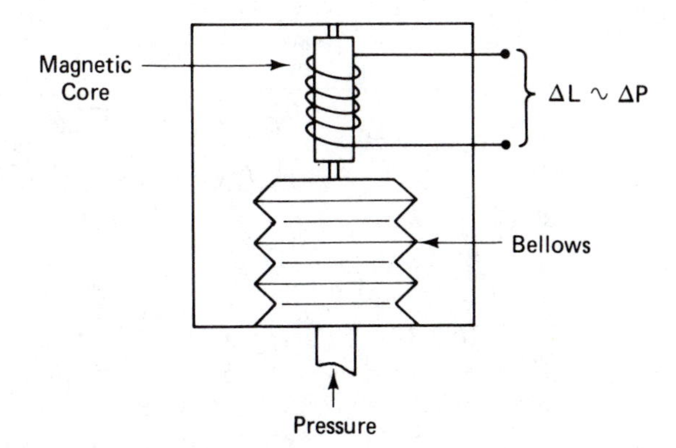

FIGURE 2-9. A bellows pressure sensor. Bellows motion is proportional to the pressure differential between the inside and the outside of the chamber. Bellows motion is measured inductively in this example.

and the amount of gage pressure may be inferred from the motion sensor output. An inductive sensor is illustrated in the figure.

A pair of mechanically connected bellows permits measurement of *absolute pressure,* expressed as "pounds per square inch, absolute" (abbreviated "psia") in English units. This is accomplished by arranging the bellows so that pressure within one bellows (exposed to outside ambient pressure) acts to oppose expansion of the other bellows, which is exposed to the fluid or gas in which the absolute pressure is to be measured. The bellows pair is enclosed in an evacuated housing so that the net effect is a subtraction of ambient pressure effect from the total displacement of the bellows pair. A mechanical link to the point at which the bellows meet is then used to measure its position, from which absolute pressure is inferred.

Bourdon tubes, diaphragms, and capsules are other pressure sensing devices which use similar principles—in a way, each can be thought of as a "bellows" of a different form. The *Bourdon tube* is a tube of spring metal, closed at one end and formed into a curve, or a helical or flat spiral. The open end of the tube is exposed to the pressure to be measured, and the other end is mechanically linked to a position sensor. Pressure inside the tube deforms the tube, and pressure is inferred from a measurement of this deformation.

Diaphragm pressure sensors are cylindrical devices, one face of which is a spring metal diaphragm. Pressure to be measured is admitted to the interior of the cylinder through a fitting on the opposite face. This pressure extends the diaphragm, which is linked to a position sensor. The movable part of this device (the spring metal diaphragm) is often corrugated, to increase its surface area. Since pressure

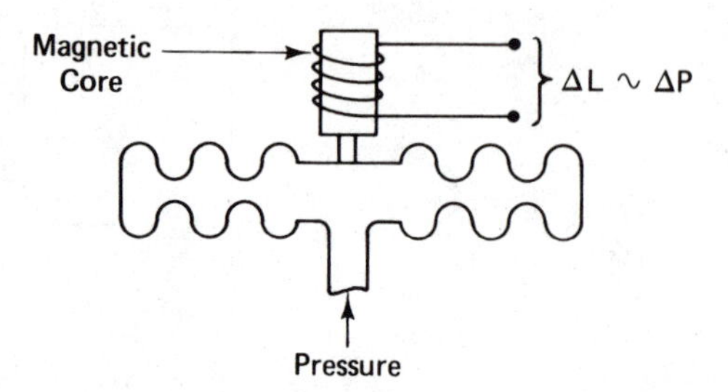

FIGURE 2-10. A pressure capsule. Corrugation of its spring metal form effectively increases the available "signal" to the motion sensor.

is in units of pounds per square inch, the larger the effective surface of the diaphragm, the greater the force acting upon it and the greater the amount of mechanical force available as its output "signal." This is an example of providing additional "gain" from the transducer. Generally, the more "gain" available from a transducer, the better it is for measurement. We discuss this further in the next chapter.

Pressure capsules also employ this concept of added gain at the source of the measured variable. As seen in Fig. 2-10, which illustrates a pressure capsule in cross section, virtually the entire device is corrugated spring metal.

Strain gages are used in pressure measurement, in devices called *pressure cells.* A shaped tube, closed at one end, is exposed to the pressure to be measured through the other end. The tube is shaped to accept a bonded wire strain gage cemented to the tube. The strain gage is used to measure distortion of the tube, which is proportional to the pressure standing within it.

The *manometer*, illustrated in Fig. 2-11, operates on the principle that pressure differences can lift a column of suitable fluid contained within a tube of proper shape. In this device, the fluid (often mercury) is displaced by the difference in pressures at each end of the tube. The fluid must, of course, be essentially incompressible and the device must be arranged so that the fluid will not spill.

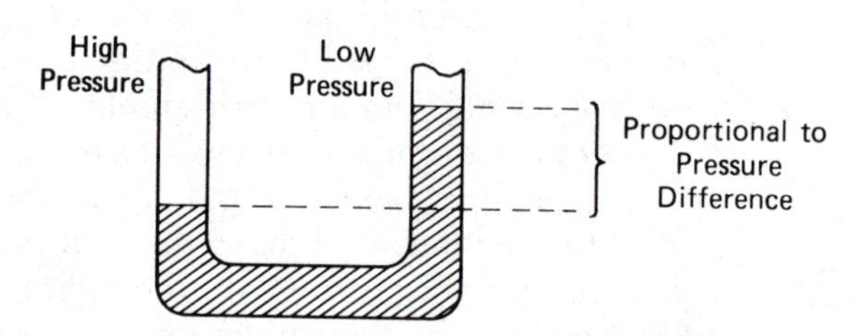

FIGURE 2-11. In a manometer, pressure difference lifts a fluid column.

The oldest forms of manometers used a scale, scribed in a visible position on the side of the tube. There are various clever ways of designing manometers to use hydrodynamic principles to obtain greater gain from the device. For process instrumentation systems where the manometer measurement is used to drive other indicators, or as input to a recording or automatic control device, a float is employed. The float rides on one end of the fluid column and carries an object whose position can be measured electrically or magnetically. For example, the object might be the core element of a linear variable differential transformer (described later in this chapter). Such an arrangement is capable of detecting very small variations in pressure very accurately when a stable standard pressure is provided on the opposite side of the manometer.

Liquid Levels

Another important variable is the level at which fluid stands in tanks and vats. This is measured by level detectors, of which there are two principal types: (1) those which provide a more or less continuous quantitative measure of level and (2) those which provide a discrete signal when the fluid reaches a specific level. The amount and level of fluid in a tank may be determined from the geometry of the tank, the specific gravity of the fluid, and the pressure that the fluid exerts on a sensor at the bottom of the tank.

Liquid levels can also be measured through their effects on the holding tank as a circuit element. If the fluid is conductive, its level will affect the resistance between a pair of electrodes inserted into the tank. Measurement of changes in this resistance provides data from which the level can be inferred. If the fluid is nonconducting, its effect on the capacitance between an electrode and a metallic tank can be used to infer its level.

Note that for each case, we need *a priori* information. For a pressure-based level sensor, we must know the tank geometry and the specific gravity of the fluid. For electrically based measurements, we must know the resistivity of the fluid or its dielectric constant.

If we need only to know when the liquid is at a certain level, such as when the tank is full, a discrete signal (a signal that is either "on" or "off") is all that is required. This can be obtained by arranging a float to actuate a switch as the level rises. If you have an automatic sump pump in the basement of your home, it may operate in this way. When water reaches a particular level, the float arm activates a switch which turns on the pump.

Floats may be used to provide continuous fluid level measurements by linking them to position sensors. Float level measurements require less *a priori* information, but their output signals may be disturbed by surface waves on the fluid, which may cause the switch to close and open several times before the waves settle or cause a position sensor output to oscillate. Thus, its electrical or mechanical signal conditioning should include some degree of hysteresis or damping.

SENSORS OF MOISTURE AND HUMIDITY

Process controllers must often control moisture in the materials used in the process. The manufacture of textiles and paper generally requires moisture content to be measured and controlled. Drying processes often must be controlled so that they do not take place too quickly or too slowly. The ability of air to absorb moisture is a function of temperature, pressure, and the amount of waper vapor it already contains. On the basis of measurements of moisture in drying air, a controller can inject water vapor or control temperature in order to control drying time indirectly.

Devices for measuring humidity (the amount of water vapor present in a gas) fall under the general name of *hygrometers*. There are several ways to quantify humidity. *Absolute humidity* expresses the mass of water vapor in a unit volume of gas. *Specific humidity* is expressed as the ratio of water vapor mass to sample mass. *Relative humidity* is the ratio of actual vapor to the amount that would be present when the gas is saturated and can absorb no more. The temperature at which the gas is saturated is called the *dew point*. It is at this temperature that vapor begins to condense.

Humidity is generally measured by sensing the effect of water on some object. Its effect on the resistance of a material is among the most commonly used methods. In this case, a pair of electrodes is bonded to an insulating substrate and surrounded by a conductive film, as illustrated in Fig. 2-12. The film is usually formulated from compounds containing hygroscopic salts (chemicals which readily absorb water vapor). As these salts absorb vapor, the resistance of the film changes and this change is measured.

In some applications, the material whose moisture content is to be measured (such as paper) is positioned between electrodes. In a paper mill, paper might pass between two conductive rollers which serve as the electrodes. As the paper moisture varies, resistance between the rollers changes, and the moisture content of the paper is inferred from these data.

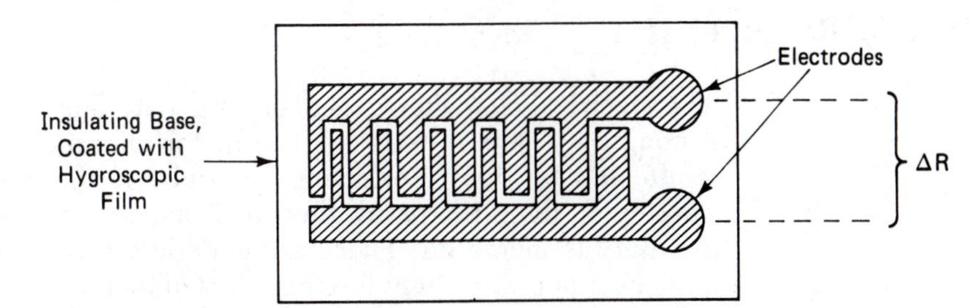

FIGURE 2-12. Construction of a resistive humidity sensor. The resistance of the hygroscopic film changes as it absorbs water vapor.

Human hair is sensitive to moisture, as you may have noticed on a humid summer day. In devices based on this principle, strands of hair are opposed by a spring mechanism. As humidity increases, the hair is stretched by the spring. As humidity decreases, the hair contracts and draws the spring. This relative motion is then measured, and the relative humidity of the surrounding air is inferred from it.

Humidity measuring instruments also use the fact that cooling rates are affected by the amount of water vapor in a gas. These are called *psychrometers* and operate by measuring the temperature at two locations (one maintained in a dry condition and the other in a moist condition) as the gas is passed across them. The temperature at the dry location is called the *dry bulb temperature*; that at the wet location is called the *wet bulb temperature*. If the gas is dry, evaporation around the wet bulb cools it more quickly than the dry bulb. Relative humidity may be inferred from this rate of cooling. Quantitative readings of relative humidity depend on atmospheric pressure, temperature of the dry bulb, and differences between bulb temperatures.

Dew point is measured by detecting condensation and recording the temperature at which it first appears. Condensation is forced by cooling the dew point sensor. When the sensor cools down to the dew point, water vapor condenses on it. For some dew point sensors, condensed water vapor is sensed through a change in the resistance of a conducting grid deposited on the sensor surface. In other types, reflective characteristics of the surface are used. For example, the surface might be mirrored and used to reflect a beam of light. At the dew point, condensed water vapor fogs the mirror and drastically reduces the reflected light by scattering the beam. A sharp drop in the output of a photosensitive device upon which the light beam was focused indicates that the dew point has been reached. The temperature reading at that time is then recorded as the dew point.

SENSORS OF LIGHT

The electrical characteristics of certain materials are affected by light. Among these are certain semiconductor materials, whose resistance is affected by light. For some materials, as the energy of light impinging on them increases, their resistance is lowered and their conductivity is increased. These are generally termed *photoconductive* devices. For others, light has the effect of increasing resistance. These are generally termed *photoresistive* devices.

Different materials vary in their degree of response to different light frequencies and must therefore be selected according to the frequency content of the light being measured. In the devices discussed above, the frequency of the light was not important (we simply wished to know whether the beam of light was present or not); but in some instruments the relative amplitudes at specific frequencies contain the information the instrumentation system seeks.

Fourier spectrometers are examples of such instruments. In these devices, the principles of the Michelson interferometer are used to deduce the chemical constituents of a gas or liquid by computing the spectrum of the light passed through the sample. Absorption lines in the spectrum enable a chemist, or a suitably programmed computer system, to determine what chemicals are present in the sample and their relative amounts.

Diode and transistor semiconductor devices are also sensitive to light and may, therefore, be employed in its detection. These devices are similar in principle to the devices used in the fabrication of electronic logic, but are packaged to permit light to impinge on their sensitive areas. Diodes so packaged are called *photodiodes* and transistors so packaged are called *phototransistors*. The characteristics of these devices are affected by their ambient temperature. For this reason, they are sometimes unsuitable for quantitative measurement of light except in laboratory conditions. For simply detecting the interruption of a light beam, however, they are generally satisfactory and frequently used.

Light-sensing elements which generate power when exposed to light are called *photovoltaic* devices. Solar cells used in heating systems are examples of this kind of device, although made on a scale much greater than that used in process instrumentation devices, since their purpose is to generate large amounts of power. These devices are fabricated from layers of selected semiconductor materials. Doped silicon materials are used for silicon photovoltaic cells, one layer being made rather thin so that light can penetrate to the junction between the differently doped silicon layers. This light excites electrons and causes a potential across the junction which generates a sufficiently large current for measurement with suitable signal conditioning circuits. Other types of photovoltaic cells are fabricated from selenium and cadmium oxide and produce similar measureable effects.

MEASUREMENT OF POSITION AND ITS DERIVATIVES

Position comes in two basic "flavors"—linear and angular. Linear position is distance along the axes of a rectangular coordinate system from some origin. Angular position is defined by the relationship between two coordinate systems, one fixed in space and the other fixed relative to the geometry of the object whose angular position is being measured.

Measurement of position is a prominent feature of most discrete manufacturing systems, especially those which include numerically controlled machine tools. Measurement of position is required less often in continuous process control systems, being generally used to measure the degree to which valves are opened or closed or to monitor motor speed.

Absolute Position

Most position transducers provide a basis for inferring absolute position by indicating that the observed object has moved over some short distance. That is, the transducer is used to measure small amounts of *relative* motion, as opposed to absolute position. The instrumentation system then determines the amount of total motion by summing these smaller amounts of motion. This is necessary because the dynamic range of most position transducer signals is small compared to the potential dynamic range of the observed object's absolute position.

By measuring motion over an interval of time, the instrumentation system can reach conclusions about speed; by keeping track of changes in speed over a given interval of time, it can reach conclusions about acceleration. There is usually little need for position derivatives of higher order in industrial process control applications. Speed is often the highest order derivative required, although acceleration is sometimes needed as well.

To measure absolute position over moderate dynamic ranges, the instrumentation system must be able to determine the initial position of the object. This may be assumed as an initial condition or provided through an independent input. For example, a machine tool operator may "inform" the controller that stock to be machined is at the initial (home) position by pressing a button. Alternatively, the controller may detect that the stock is at the home position by observing the output of a limit switch.

Switches, in the form of single-pole/single-throw limit switches, are often used to detect the presence of objects at particular points. The point is defined by the location of the switch, which is closed or opened when the object comes into contact with it. Information about the location of the switch is assumed in the design of the control system.

In most situations, electromechanical switches are used to detect infrequently occurring limit positions, such as overtravel on the bed of a machine tool. It is good practice to minimize the number of electromechanical switches, since they are subject to wear mechanisms that can degrade the performance of the system over extended periods of use. In places where the switching function must be utilized often, it is best to use solid-state switches. These may be optically based detectors, in which the object interrupts a light beam, or switches which employ magnetic principles (generally called "proximity detectors").

Encoders are used to measure position over a range of values. Some are electromechanical and suffer from the same wear problems as electromechanical switches. The most reliable encoders use optical or electromagnetic transduction principles. Generally, encoders use any of the transducers of motion described in the following sections and carry out the initialization and summing functions needed to provide direct position readouts.

Motion

There are several classes of linear and angular motion transducers, characterized by the transduction principle they employ. Examples are the following:

Potentiometers—A mechanical link is established between the object whose motion is to be measured and the wiper on a potentiometer (a variable resistor). As the object moves, the resistance of the potentiometer changes, as illustrated in Fig. 2-13. Potentiometric position sensors are conceptually simple, but suffer from several deficiencies. They must be excited by a voltage or current in order for the resistance change to be measured; they are mechanical and therefore are subject to wear when used to detect frequent or rapid motion, and they have limited resolution and dynamic range.

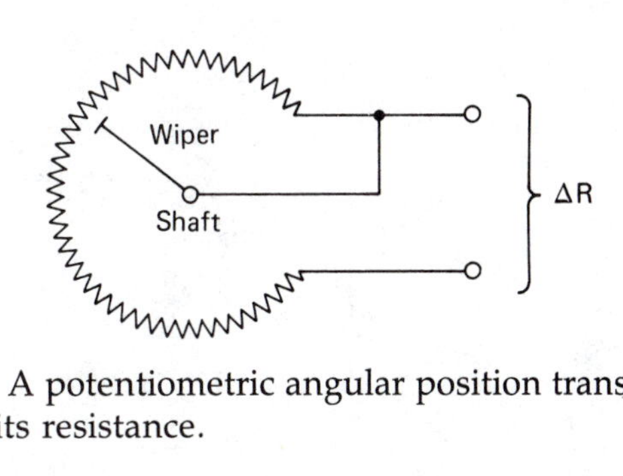

FIGURE 2-13. A potentiometric angular position transducer. Shaft rotation changes its resistance.

Inductors—A conductive core material is mechanically linked to the object. It moves the core within a coil, resulting in a change in inductance, as shown in Fig. 2-14. These are more reliable than potentiometers, because they do not involve mechanical wiping action. Their principal disadvantage is the complexity of their signal conditioning circuitry. It is more difficult to measure inductance than resistance.

Capacitors—A set of plates, one linked to the object, is used to induce a capacitance change as the object moves. The advantages and disadvantages of capacitive position sensors are similar to those of inductive sensors.

Variable differential transformers—These are found in two main types, according to the type of position being measured (linear or angular). The first is the *linear variable differential transformer* (abbreviated LVDT) and the second is the *rotary variable differential transformer* (abbreviated RVDT). With these devices, a mechanical link is established between the object and a core which is a movable part of a transformer, as shown in Fig. 2-15. The transformer primary is excited with an alternating current signal. As the core position changes, the signal from the transformer secondary changes.

Signal conditioning circuitry may detect changes in amplitude and phase relationships of the ac output signal or base meas-

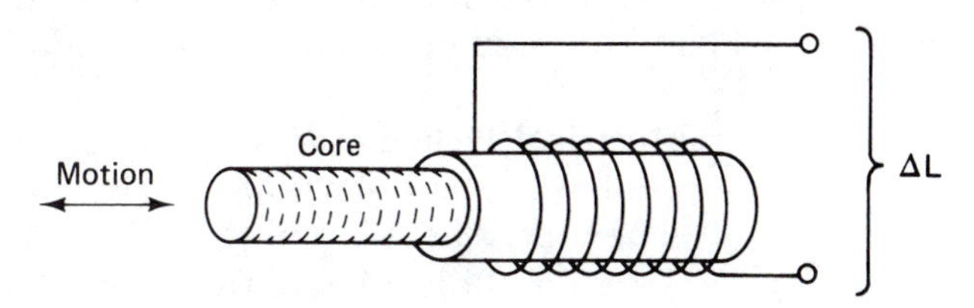

FIGURE 2-14. An inductive linear position transducer. Motion of the core material changes its inductance.

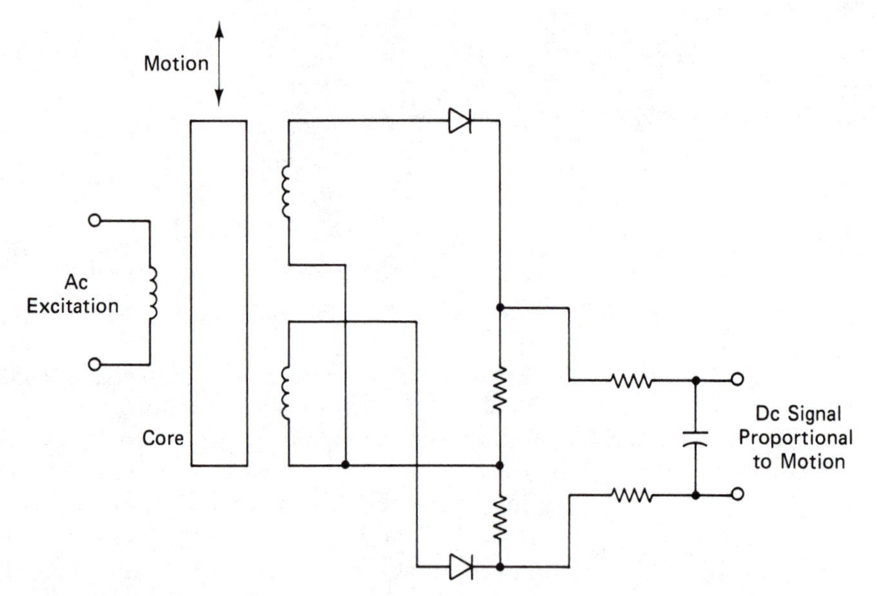

FIGURE 2-15. A linear variable differential transformer. Motion of the core material changes the amplitude of the output signal.

urements on a direct current signal derived from the ac signal. From these outputs, the amount and direction of motion are inferred.

The LVDTs and RVDTs are highly reliable and capable of very accurate measurements. Dynamic range is limited by their mechanical design, as is also the case with potentiometric, inductive, and capacitive motion sensors.

Speed of Rotation

Sensors for the first derivative of angular position (speed of rotation) come under the general name of *tachometers*. The angular position and motion sensors described above can be used to measure rotational speed through a suitable combination of mechanical linkage and signal conditioning, but some sensors produce more direct indications.

Electrical tachometers—These sensors couple the rotating object to the shaft of a suitably sized direct or alternating current generator. The amplitude or frequency of the generator output is then an indication of the speed of rotation.

Optical tachometers—These employ principles similar to those used in optical position encoders. The rotating object is coupled to a device designed to interrupt a light beam as the object turns. Interruption of the light beam is sensed by a photosensitive device, the output of which is a series of pulses with frequency proportional to the speed of rotation.

The light beam interrupter may be fabricated in several ways, depending on the desired performance of the sensor. In the simplest, but lowest, performance devices, it might be a perforated disc. In the best devices, it is fabricated from transparent material on which opaque markings are placed using precision screening techniques. These are capable of very high accuracy and resolution. By using two sets of markings, one offset from the other, and two light beams, each with its own optical sensor, the device can also provide information about the direction of rotation.

Magnetic tachometers—These are similar to optical tachometers, using magnetic sensing in place of optical sensing. The rotating object is linked to a magnetized disc, machined to a shape designed to induce pulses in a nearby coil/permanent magnet circuit. The frequency of the pulses induced as the object rotates is proportional to its speed of rotation.

The best choice of angular motion sensor is often determined by the nature of the process instrumentation. For instrumentation that is analog in nature, sensors that provide analog signals are preferred. Such instruments readily accept analog signals as input, but require additional circuitry to convert pulse trains into the continuous analog forms they naturally prefer. Conversely, instruments which are principally digital (such as microcomputers) readily accept pulse trains or digital inputs from direct-reading encoders, but require additional circuitry to convert continuous analog signals into the digital forms with which they deal most naturally.

Acceleration and Vibration

The second derivative of position (acceleration) is not often required in process control, but is sometimes needed to measure and test the quality of manufactured products. Acceleration is the principal variable used in testing the ability of a product to survive shock and vibration. Vibration testing applies sinusoidally varying accelerations to the device under test. Shock testing subjects the device to falls from a particular height or to calibrated blows.

Acceleration sensors are attached to the device under test or to the fixture that carries it. Some use a calibrated mass, arranged to allow motion along the axis in which acceleration is to be measured, attached to the sensor body by a spring or reed. As the sensor body moves, the mass moves relative to it in a way which is proportional to the acceleration. The mass may be arranged to act as the core of an LVDT, so that sensor body vibration produces signal variations of the type found in motion-measurement LVDT applications.

Other acceleration sensors use piezoelectric transduction. With these, a calibrated mass is bonded to a suitable crystal arranged between electrodes. As the sensor moves, the mass exerts force on the crystal, causing it to generate a voltage which varies with acceleration. The amount and direction of acceleration is inferred from the magnitude and polarity of this voltage.

SUMMARY

In the manufacture of transducers and primary elements, special note should be taken that their characteristics must be closely controlled in order to have devices which provide useful output signals. The composition of a Bourdon tube and the forming of its shape must be closely controlled. The wire strain gage is subject to stringent requirements in the composition and arrangement of its wire. These considerations apply to very nearly every type of primary element. Although it is often easy to understand the physical principles by which transducers operate, to manufacture them, while maintaining reasonable costs and high levels of quality, is quite another matter entirely.

There are also many subtle matters which must be considered when using these devices. Designers must consider nonlinearity, drift, accuracy, and stability in order to produce instruments that perform well. The cost, quality, and performance of primary elements has been, and continues to be, one of the most important factors in the design of industrial process control and instrumentation systems. We make special note of this because the use of microcomputers in industrial systems will have important effects on the requirements placed on primary elements in future systems. This is because the microcomputer can sometimes compensate for undesirable characteristics in primary elements through clever programming.

Before leaving this chapter, we must remind ourselves once more that in almost every case we cannot directly measure the variable we wish to observe. Instead, we must observe its effects on some object and, in some cases, carry indirection and inference through several levels until we arrive at the desired quantitative expression of the variable. There are many opportunities along the way for complication, error, and loss of accuracy and resolution. As a result, the design of signal conditioning equipment is far from being simple and straightforward. It requires a unique combination of knowledge of materials, physics, electronics, and "system sense."

CHAPTER 3

Signal Conditioning and Measurement

"Est modus in rebus, sunt certi denique fines,
Quos ultra citraque nequit consistere rectum."

(Things have their due measure; there are ultimately fixed limits, beyond which, or short of which, something must be wrong.)

Horace

"Round numbers are always false."

Samuel Johnson

Imagine that we are faced with the situation illustrated in Fig. 3-1. We have attached transducers to the process that we wish to monitor, and we have a computer system to collect and act on information from them. What goes in the space between? In part, the answer depends on what is already there. There may be electromagnetic "noise," which can distort information from the transducers. There may be time-varying ground potential differences. Long distances may be involved.

The methods by which transducer signals are converted or rearranged, in order to put them into forms that can be transmitted to computers and instrumentation and handled by them, are generally known as *signal conditioning*. It can be thought of as "outfitting" the signal for a rather dangerous trip, from its origin at the transducer to its destination at the computer, and providing it with roads to travel on.

To transport the signal to the computer, we must consider what it is like when it leaves the transducer, what threats and hazards

FIGURE 3-1. The measurement situation—what goes in the space between the transducers and the computer system?

may be present along the way, and what the signal must be like when it gets to the computer. We also may have to consider how the signal might be "repaired," if it should arrive in "damaged" condition.

Measurement is the means by which we grasp the meaning of the signal. As noted in the previous chapter, we must often infer the value (the "meaning") of the measured variable from the signal that is received at the computer.

Preservation of *meaning* is the most important factor in signal conditioning and measurement, yet it is rarely discussed prominently, perhaps because precomputer systems involved people so directly in control decisions. The meaning of "meaning" was clear and seemed to require no further discussion. In those systems, information *had* to be arranged for display to people. It was necessary to convert signals into *engineering units,* units of measurement conveniently perceived by people.

Modern systems still must consider this when people are involved in control decision-making. So long as any human is involved in operating the system, process variables must be available for display in engineering units. But properly programmed computers have more freedom in the units they use. They are able to handle numbers that are related to measured variables in ways which, although efficient for computers, are difficult for people. This gives instrumentation designers more latitude in signal conditioning and measurement designs.

When one is setting out on a trip, it is useful to have a map. For the trip we propose, Fig. 3-2 is the beginning of such a map. It shows the outline of the territory to be traveled. The map is crude, much like the rough maps used by Lewis and Clark. Only broad outlines are depicted. The map shows the measured variable as an input to the transducer, which converts it to an electrical signal. The signal is

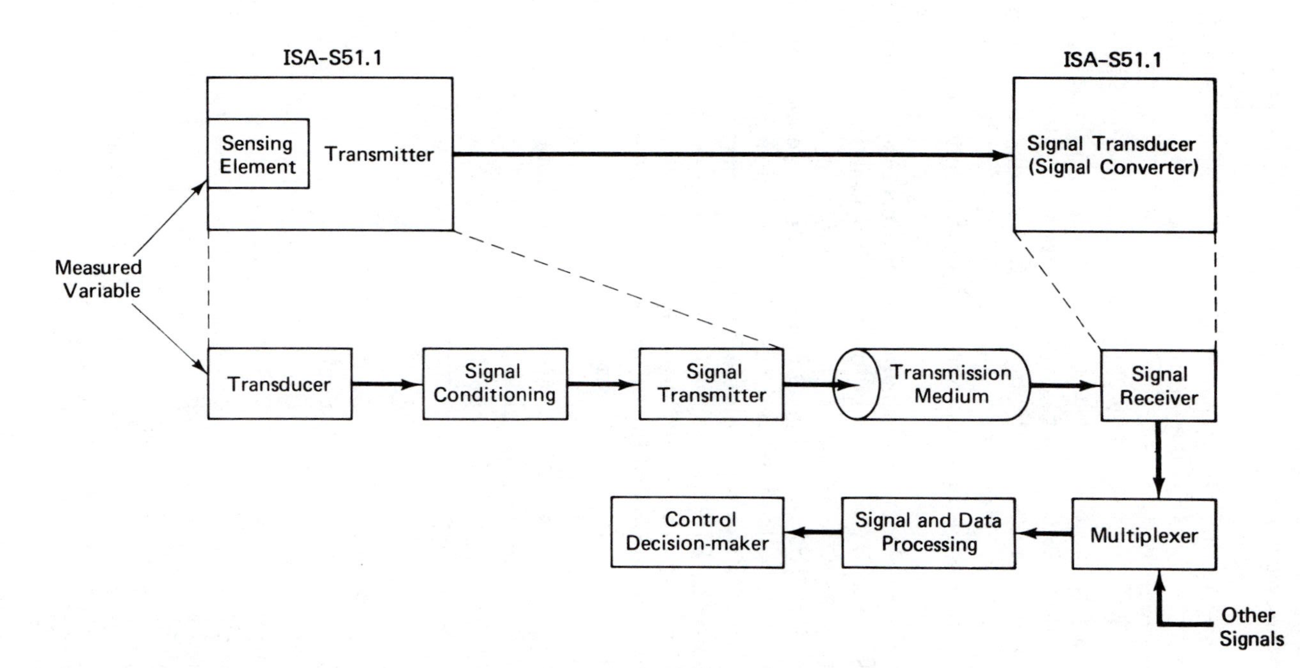

FIGURE 3-2. A map of the territory between the transducer and the control decision-maker.

then transformed by signal conditioning equipment so that it may be transmitted by the signal transmitter over some suitable transmission medium (usually wire).

The signal is received by the signal receiver, which converts it into a form suitable for use at the receiving end of the transmission path. In some cases, so as to share signal-handling equipment at the receiving end, equipment is provided so that other signals can be switched in. This is the purpose of the multiplexer. The signal data processing function may include filtering and conversion from analog to digital form.

The resulting information is then used by the control decision maker to carry out the required control functions which yield commands to be distributed to actuating elements in the system.

Signal conditioning and measurement terminology differs among various dialects of the process control and instrumentation language, as it does for primary elements and transducers. The ISA Standard, ISA-S51.1 (1976) *Process Instrumentation Terminology,* is useful here also. Figure 3-2 also illustrates an alternate form of the transmission portion, using terminology of the ISA Standard. The following are some definitions from this standard that are pertinent here.

Element, sensing—The *element* directly responsive to the value of the *measured variable.*
NOTE: It may include the case protecting the sensitive portion.

Parameter—A quantity or property treated as a constant but which may sometimes vary or be adjusted.

Signal—Physical variable, one or more *parameters* of which carry information about another variable (which the signal represents).

Signal transducer (signal converter)—A *transducer* which converts one standardized transmission *signal* to another.

Transmitter—A *transducer* which responds to a *measured variable* by means of a *sensing element,* and converts it to a standardized transmission *signal* which is a function only of the *measured variable.*

TRANSDUCERS AND SIGNAL QUALITY

There are many things we need to know about the transducers and the signals they emit. We must also be conscious of how we value the "quality" of the work done by the signal conditioning and measurement apparatus. The terms used to assess original transducer signal quality, signal quality during the trip to the computer, and signal quality when it arrives at the computer are important, but have varying meanings.

Just as a person may be said to be of good moral fiber, honest, reliable, and so on, similar things may be said of a transducer, but with different words. For example, we might say that a transducer is "stable," or that it is "linear," or that it has "high gain." These are all desirable qualities in transducers, just as honesty and reliability are desirable qualities in people. Why they should be so has to do with their preparedness for the metaphorical trip we are considering.

Accuracy

An accurate transducer does not "lie"—at least not very much. As there are degrees of truthfulness among people, so are there degrees of truthfulness among transducers. A transducer and measurement system is inaccurate if, for a given process condition, it sometimes says one thing and sometimes another, and if exactly what it will say cannot be predicted practically. The degree to which the things it says differ from one another and from the true value is a measure of its inaccuracy. Most sources of reading error are combined into a single number, offered as the "accuracy" of the device.

This interpretation places more emphasis on the stochastic nature of measurement accuracy than does the usual dictionary definition. This is important in the context of computer-based measurement systems because they are more able than people to cope with *consistent* differences between the true value of the measured variable and the reading the instrumentation yields. Traditionally, if an instrument always produced a reading that differed from the true value by, say, +1 percent, it would have been said to have been inaccurate to that degree. In modern systems, such consistent differences can be removed by autocalibration or suitable programming. The importance of *consistent* reading "error" becomes less.

Resolution

Accuracy is often confused with a related term: *resolution.* Resolution has to do with the degree to which the transducer is able to distinguish two narrowly separated conditions of the measured variable. The resolution of a transducer is said to be poor if the measured variable can change substantially without a corresponding change in the transducer output. In a sense, the transducer is not lying—it just is not telling the whole truth.

You can see that the difference between accuracy and resolution can be characterized as subtle. It is sometimes said of a thermocouple that it cannot be used to measure a temperature "down to a tenth of a degree" because it "is not that accurate." What is really meant, not always consciously, is that a difference of a tenth of a degree in the temperature surrounding the thermocouple is not great enough to produce a sufficiently large change in its output signal to be measured by practical equipment. That *some* change takes place is an unalterable fact of physics—most temperature-sensitive transducers are not discrete devices. It is just that the change is too small for us to measure with our equipment.

This subtle difference has carried over into common, but not official, terms used in casual conversation among some engineers when they are talking about the operation of converting analog signals into digital form suitable for input to a computer. In these conversations, they frequently speak of the "accuracy" of an analog-to-digital converter when they really mean its resolution. They are heard to say that "a 12-bit converter is not accurate enough" when they really mean the converter resolution is not fine enough to distinguish changes of the magnitude that they want to measure. The resolution of the converter is the characteristic that, since it is a 12-bit converter, it divides the full range of its input signal into 2^{12}, or 4096, parts.

Stability

If what the transducer says varies in ways which have nothing to do with the value of the measured variable, then the transducer is not *stable.*

As noted in the previous chapter, the output signals of some transducers used to measure pressure may change in ways that depend on their temperature. If we know this temperature, we can adjust its output signal ("compensate" it) so that it conforms more exactly to the actual value of the variable.

Sometimes, lack of stability in one device can be used to compensate for lack of stability in another device. This technique is sometimes applied by taking advantage of the fact that some devices have temperature coefficients of opposite polarity. If both devices produce corresponding signals, if they share a constant of proportionality with the condition being measured, but their constant of proportionality increases with temperature for one (a positive temperature coefficient) and decreases for the other (a negative temperature coefficient), then the output signal of one can be used to offset stability error in the other.

Repeatability

If a transducer always says the same thing for a given value of the measured variable, then it has perfect *repeatability.* Lack of repeatability may be predictable or unpredictable.

For some transducers, the history of their input affects their output in a predictable way. As the input signal of the transducer increases to a certain point, it may produce one reading. Suppose now that its input signal increases beyond that point, then decreases until it returns to that point. If the transducer output is not the same for the latter condition as it was for the former condition, the transducer is said to have *hysteresis.* For some applications, this is undesirable. Even where it is predictable, it complicates the task of inferring the actual value of the measured quantity. But when the amount of hysteresis is related to the desired resolution in a certain way, it can be useful, because it tends to suppress small amounts of noise.

Range

The term *range* describes several transducer characteristics. There are two principal variations on this term. One addresses the range over which the transducer has the ability to survive and still produce meaningful readings. At high temperatures, for example, we must consider whether the transducer might suffer physical damage. There are degrees of such damage; actual destruction is the most extreme, and serious deterioration of signal quality is less extreme, but important.

If the application is willing to bear the cost, a "kamikaze" transducer might be used, sent in to collect as much information as it can before its surroundings do it in. The Mariner landing missions to Venus were examples of this method. These are extreme methods, rarely used in day-to-day process instrumentation. Usually, we are interested in the range of conditions in which we can place a transducer and still expect reasonable output signals from it. The term for this is *useful range.*

Dynamic range is the range of signals received as output from the transducer. The dynamic range of a type T thermocouple, for example, might be −5602 to +20869 microvolts over a temperature range of −200°C to +400°C. Knowledge of the transducer dynamic range is used, along with knowledge of the range over which measurements are to be taken, to assess the work to be done by signal conditioning equipment and the degree to which various threats along the signal transmission path may alter or obscure the measurement. The term *dynamic range* is, in fact, applied to virtually every element along the way, from the transducer to the final digitized value presented to the computer or the instrumentation system for analysis. *Span* is often used as a synonym for transducer dynamic range.

Linearity

If the relationship between the physical phenomenon being measured and the output of the transducer can be described, up to the resolution and accuracy desired, by a linear equation, then the transducer is said to be *linear*. A linear equation has the following form:

$$y = Ax + B \tag{3.1}$$

where x is the output signal from the transducer, y is a representation of the measured variable, and A and B are constants (or, sometimes, parameters). Constants A and B are sometimes formed so as to yield the quantity y in appropriate engineering units.

Except in the case of the more straightforward position transducers, almost every practical transducer exhibits some nonlinearity. That is, there is no single relation like Eq. (3.1) which relates the physical phenomenon to the transducer output over the useful range of the device.

However, there are two possibilities that often simplify nonlinear transducer applications. The first is that, up to the required resolution and accuracy, a sufficiently linear approximation may exist over the *portion* of the useful range in which measurements are to be made. If it does not, sometimes the range can be divided into several segments over which different, but nevertheless linear, relations can be used to convert readings to measurements. These are known as *piecewise linear* approximations.

In many cases, tractable nonlinear approximation relations may exist between the transducer signal and the measurement, and these may be used to convert readings into measurements. An example is the use of polynomials to convert thermocouple voltage outputs to temperature readings. The National Bureau of Standards has prepared tables of coefficients of second, third, fourth, fifth, and higher-order polynomials. These are designed for various thermocouple types, with corresponding ranges of application and degrees of accuracy. These coefficients have been determined by computer analyses of thermocouple behavior in which the polynomials are fitted mathematically to the temperature versus voltage curves.

The second possibility is related to whether there really is a need for conversion to engineering units. Sometimes a nonlinear transducer output can be used directly in the control equations. The computer does not need to do all of its "thinking" in engineering units. It requires only that there be a consistent relationship between the measured variable and the reading and that this relationship be incorporated in its programming. That the relationship may be complex is a problem only to the programmer during the preparation and testing of the program. Its complexity is rarely an impediment to good control system performance. In fact, the control system may be capable of higher performance if it is not required to burden itself with unnecessary engineering unit conversions.

Gain and Sensitivity

Gain is the extent to which the transducer "amplifies" the physical phenomenon that it is measuring. Gain is entangled with the ideas of resolution and accuracy. To the extent that Eq. (3.1) is descriptive of transducer output signal behavior, the gain of the transducer is represented by the term A in Eq. (3.1).

A transducer with high gain has at least two advantages. The first is that it is sensitive to smaller changes in the measured variable. For this reason, the terms gain and *sensitivity* are sometimes used interchangeably, although they are not the same in every case. For example, since hysteresis can produce quantization effects, a transducer might be relatively insensitive to *certain* changes in the measured variable, yet still produce an output signal of relatively high amplitude. Thus, although the relation between sensitivity and gain is often such that they can be used interchangeably in casual discussion of transducer characteristics, we should remember that they are actually different and should be careful that distinction is made where it is important.

The second advantage of a transducer with high gain is that its signal has a higher degree of immunity to noise threats. Because signals from transducers with high gain can be sorted out more easily from interfering noise, practical measurements are easier when they are used.

STRATEGY, TACTICS, THREATS, AND COSTS

The following is a brief summary of the strategy, tactics, and "threat" and cost considerations related to the work of outfitting the signal for its trip.

Form (Analog vs Digital)

We know that a computer is the final destination in the signal's metaphorical trip. We also know that computers operate most conveniently on digital information. When presented with analog information, they must convert it to digital form before beginning serious work on it. We may choose to convert the transducer signal to digital form before we transmit it to the computer, or we may transmit it in analog form and postpone the conversion to digital form until the signal has arrived at the computer. This conversion is the purpose of analog-to-digital converters, which are discussed in Chapter 10.

Noise

Noise is any unwanted component of the received signal. Although noise is often thought of as "random," most noise is actually caused by the unintended effects of other pieces of process control equipment and wiring. Interference with transmitted signals is most usually due to a combination of common mode noise and normal mode noise.

Common mode noise is impressed on both the signal wire and its return wire "in common" (equally). It may be due to differences in the ground potentials at the transmitter and at the receiver, or it may be coupled into the signal circuit by fields that interact with the signal wires and ground loops that may exist between the receiver and the transmitter.

Normal mode noise is impressed onto the signal wire and its return differentially, also by fields and conduction, and is therefore summed with the intended transmitted signal.

Algebraically, these may be represented as follows (Fig. 3-3):

$$\begin{aligned} \text{Voltage on signal wire} &= V_c + V_n + V_s \\ \text{Voltage on return wire} &= V_c \end{aligned} \tag{3.2}$$

where V_c is the common mode voltage, V_n is the normal mode voltage, and V_s is the intended transmitted signal.

Amplification and Filtering

The use of differential and isolation amplifiers as receiver circuit elements helps to protect against common mode noise. The use of filter circuits in the receiver helps to protect against normal mode noise.

Transducer signals are likely to be weak compared to some of the threats that exist along the way. Noise signals can be orders of

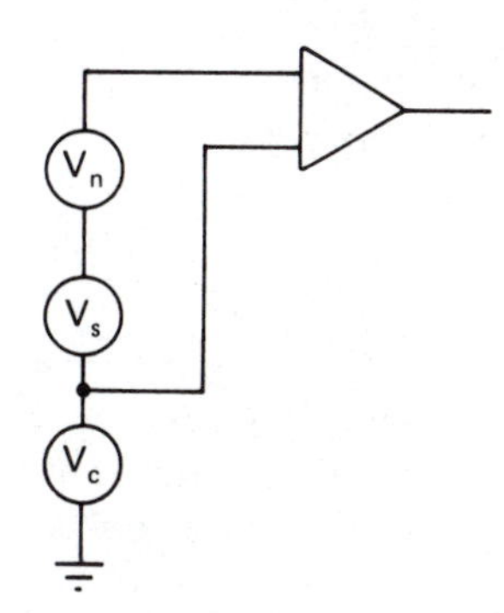

FIGURE 3-3. The relationship between common mode and normal mode noise and the intended transmitted signal.

magnitude greater than the transducer signal. Most practical industrial instrumentation situations are characterized by significantly high levels of noise threat. For this reason, although high gain transducers are helpful and should be used wherever they can, more must be done to fortify the signal before it is transmitted. It would, therefore, seem to make sense to "fortify" the transducer signal somehow. This is the purpose of amplifiers, and we consider their use later in this chapter.

Isolation

In some cases it is possible to provide protection for the signal by arranging for various degrees of isolation from interfering noise and potentials. Among the available tactics are good wiring, shielding, and grounding practice and various forms of transformer (galvanic), optical, and switched (flying capacitor) isolation.

Disguise

It is sometimes possible to "disguise" the signal in a way that may hide it from some of the threats. This is the purpose of encoding methods used in the preparation of the signal for transmission (and also described in this chapter).

Strength In Numbers

We can also help to get the message through to the computer by sending many copies of it. Techniques of signal averaging are based on this idea. By summing many copies of a small signal, or by integrating the signal, we can come up with an effectively improved ratio of signal to noise at the receiver.

Cost

Instrumentation system cost is rarely treated at any length in textbooks, perhaps because most textbooks are written by people to whom cost is of much less interest than other, more tractable, aspects. Analyzing control system costs over many possible choices of primary elements and signal conditioning arrangements is a monstrously complex task. For most practical situations, it requires the application of more experiential judgment than mathematics and is very difficult to describe and teach.

System cost is a function not only of transducer and signal conditioning equipment prices, but also of the effort that must go into design, maintenance, and installation. Installation costs are affected by complex and poorly understood factors. Among these are the overheads of the organization which installs the system (its owner, seller, or a third party), labor union relations (in some installations, certain types of work must be done by union personnel), and costs of basic raw materials. Market prices, of copper for wires and gold for connectors and contacts, strongly influence the installed cost of process instrumentation and control systems.

Control of the costs of a process instrumentation system often requires consideration of more than the simple costs of the components used. In some cases, the use of "intelligent" devices presents better opportunities for the control of costs than are available from the mere consideration of the costs of individual components.

COMPUTER BENEFITS

The presence of a computer in the instrumentation can lead to significant relaxation of requirements that might be placed on transducers. Although the characteristics we have outlined are still desirable, their absence in a transducer becomes less of an obstacle to its practical application. This comes through the ability of the microcomputer to perform additional digital processing on the signal which enhances measurement quality without requiring corresponding increases in transducer quality.

A microcomputer can be programmed to extract proper measurements, even from highly nonlinear transducers. This programming can be done once and applied to each input and to each example of the microcomputer-based instrumentation system. Without the microcomputer, a separate electronic circuit would be required for each example of the system, if not for every input channel to the device. Thus, systemwide costs may be shared and the "per channel" costs may be reduced drastically.

One reason for using high gain transducers is to overcome the effects of noise threats. To set up a system which works effectively with a low amplitude signal, one should average many copies of the signal, to improve its signal-to-noise ratio, and equip each input channel with a filter designed to reject the types of noise which are most threatening to the measurement. This represents a substantial cost in conventional instrumentation devices, because they must be duplicated in every example of the device.

With a microcomputer-based system, the cost of these functions is the *memory* needed to contain programs which do averaging and filtering functions digitally and the *time* required for their execution. Memory costs money, but very little compared to its value. Execution time limits the work that the microcomputer can do per unit time but not always to such an extent that we find it seriously burdened. Microcomputer capabilities are sufficient for many present applications in this area and show promise of growing to satisfy additional future needs.

What does the microcomputer ask of the transducer in return? Certainly some penalty must be associated with an improvement in cost and performance.

The answer is that the use of a microcomputer places greater emphasis on transducer *repeatability*. We trade off gain and linearity requirements against repeatability when we take fullest advantage of the capabilities of a microcomputer-based instrumentation device.

SIGNAL CONDITIONING AND MEASUREMENT DEVICES

We now review some common electrical and electronic devices used in signal conditioning and measurement for process instrumentation and control systems. These are discussed approximately in the order in which they are usually encountered, as the signal proceeds from the transducer to the instrumentation and control device.

Bridges

As we have seen, common primary element outputs may be small changes in resistance, capacitance, or inductance. Bridge circuits are commonly used to measure small changes, because their output signal is a function of the difference, or change, that has taken place since the bridge circuit was set up and balanced.

A simple resistance bridge circuit is illustrated in Fig. 3-4. The output voltage V of the bridge is related to its input voltage E in the following way:

$$V = \left(\frac{R_1}{R_1 + R} - \frac{R_2}{R_2 + R_3}\right) E \tag{3.3}$$

where R is the varying resistance of the transducer. The bridge is balanced by varying the resistors on the opposing leg of the bridge until the output voltage becomes zero. In most applications, all of the resistance values are chosen to be initially equal. In this condition, the bridge is said to be "nulled."

As the resistance of the transducer changes, V varies. The variation in V is, however, nonlinear. This can be seen by imagining the transducer resistance R_t to be related to the null resistance R by the following:

$$R_t = R(1 + \alpha) \tag{3.4}$$

where α is the change in its resistance. Then the bridge output voltage becomes

$$V = \left(\frac{\alpha}{4 + 2\alpha}\right) E \tag{3.5}$$

If each leg of the bridge is a function of the measured variable, arranged so that the resistance change of opposing pairs of legs are

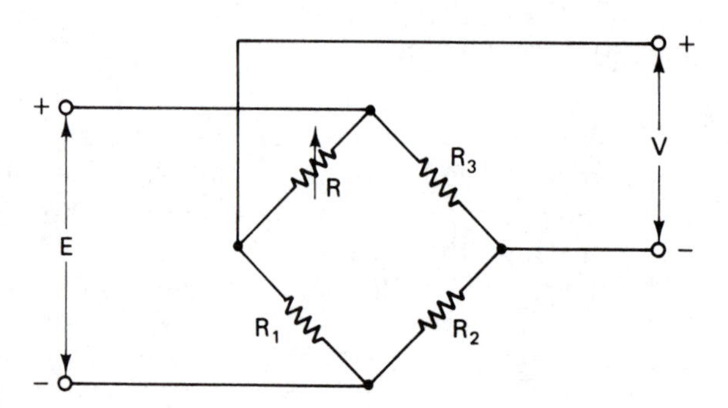

FIGURE 3-4. A simple resistance bridge circuit used to measure small changes in transducer resistance R.

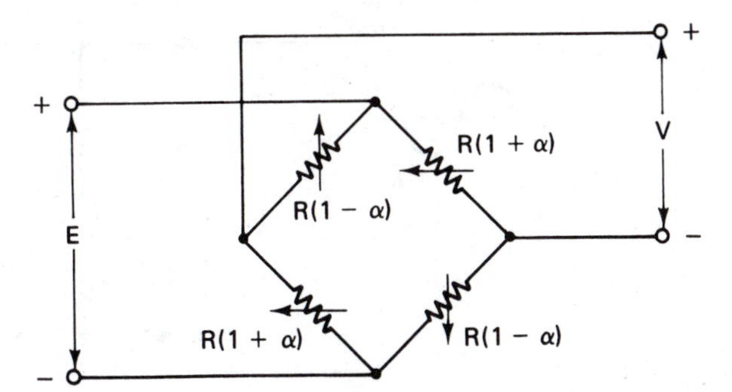

FIGURE 3-5. A bridge circuit in which each leg is changed by the measured variable. This is more linear than a simple bridge and has more gain.

equal and opposite, then a linear output can be obtained. This also has the advantage of providing higher gain. Such an arrangement is shown in Fig. 3-5. This might be used with the unbonded strain gage shown in Fig. 2-5, with each resistance wire serving as one leg of the bridge.

Note that the circuit requires the excitation voltage E. This voltage must be delivered to the bridge wherever it is located and results in the need for additional wiring to the primary element, or a suitable power supply in its vicinity.

Bridge techniques are well-developed branches of instrumentation engineering. There are many refinements of these circuits which are beyond the scope of this book, among them compensation for temperature variations and voltage drop in excitation wiring and handling of bridge output signals in ways that do not disturb the balance of the bridge itself.

AC/DC Conversion

Some transducer output signals (from LVDTs, for example) are alternating currents. Direct current signals are the most naturally acceptable input to computer instrumentation devices. It is, therefore, often necessary to convert alternating current signals to direct current signals. Such conversion circuits are generally made from arrangements of diodes, as shown in Fig. 3-6, which depicts the familiar half-wave, full-wave, and bridge rectifier circuits with their corresponding input and output signals. Some applications call for measurement of peak or root-mean-square (rms) values of these waveforms, requiring additional circuitry to produce these functions of the transducer output.

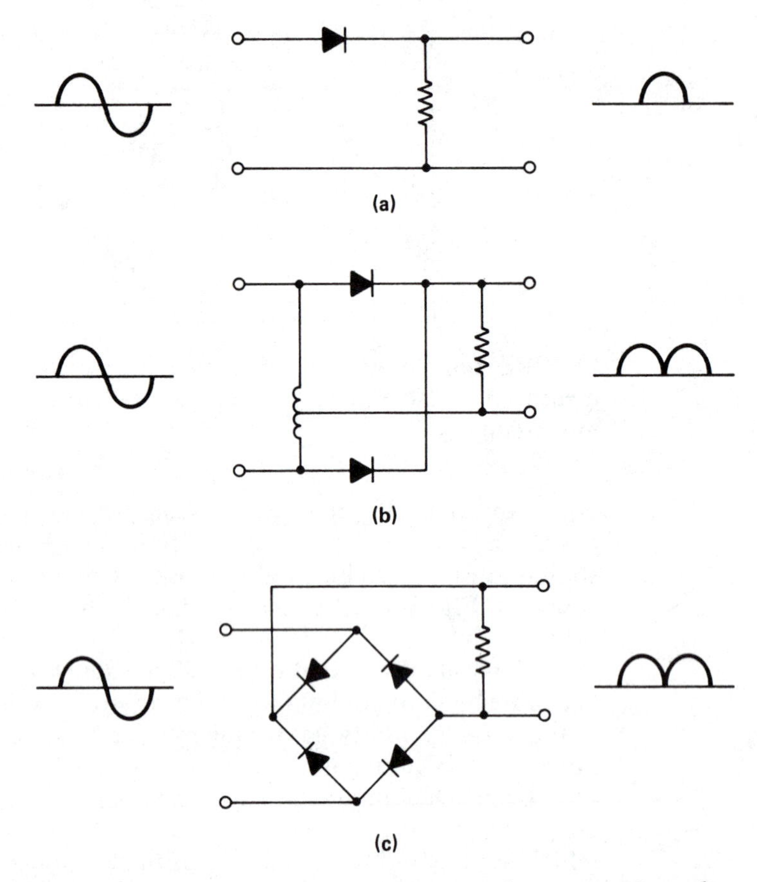

FIGURE 3-6. Diode circuit arrangements for converting alternating current to direct current signals: (a) half-wave rectifier, (b) full-wave rectifier, (c) bridge rectifier.

Amplifiers

General Aspects. The system designer may choose to amplify a signal in order to fortify it for transmission and thereby improve the signal-to-noise ratio of the received signal. This improves the probability of accurately detecting and measuring the signal in the presence of noise which may have been added to it in the transmission system and its environment. Amplifiers may be used at either end of the transmission path. Amplifiers at the receiving end, however, do not improve the ratio of signal to noise unless they discriminate against noise in some way.

If the devices *do* discriminate against noise, they are also considered to be *filters*. Filters are not used on the transmitting end of an instrumentation system unless there is some known local noise source and it is desired to inject a reasonably "clean" signal into the transmission path. The best arrangements (other than those which involve special methods of signal protection, such as encoding) use amplification at both ends plus filtering at the receiving end, although this may not always be practical economically.

Such arrangements are conceptually simple, but they require careful circuit design and thorough understanding of the signal environment. Criteria for design decisions on the transmitting end are based on requirements for received signal-to-noise ratios and knowledge of noise power threats. They are also influenced by the amplitude and bandwidth of signals from the primary element. Moderately strong low bandwidth signals are most easily handled, since less amplification is required and amplifier response characteristics have less effect on them.

If the amplitude of the transducer output is very small, more amplification may be needed. Since the response characteristics of an amplifier cannot be decoupled entirely from its gain characteristics, more care is required in its design as the amount of amplification and dynamic range increases. As transducer signal bandwidth increases, amplifier bandwidth must be increased if it is to yield an accurately amplified reproduction of the signal. But the wider amplifier bandwidth will also pass more noise, so tradeoffs are almost always necessary.

Complete analysis of an instrumentation system must consider transmission path characteristics as well as the transmitting and receiving elements at each end. This is because the transmission system also has frequency-dependent behavior. It will have its *own* influence on the transmitted signal, and its own susceptibility to noise and differences in electrical conditions at each end of the transmission path. Every electronic circuit has a finite bandwidth. Even a simple resistor has characteristics of capacitance and inductance which cause it to have frequency dependent behavior. Care must also be taken that amplifiers do not contribute additional interference through amplification of noise as well as signal.

Single-ended amplification (depicted in Fig. 3-7) is usually practically limited to relatively short transmission distances determined principally by the electrical nature of the environment and differences between electrical conditions at each end of the transmission path. Much can be done to compensate for differences in electrical conditions at each end (of which ground potential differences are the

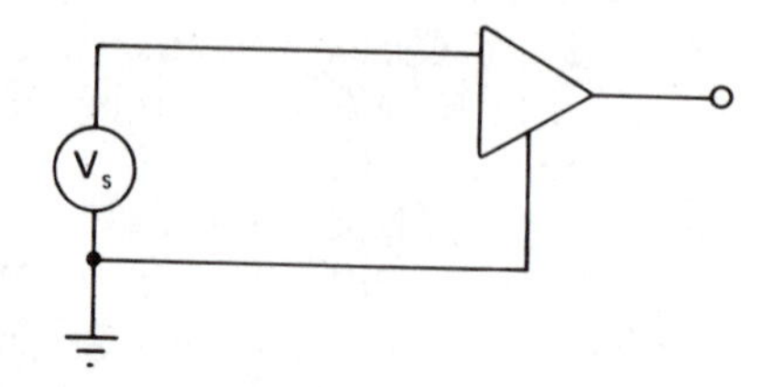

FIGURE 3-7. Single-ended amplification.

principal concern) by receiving the signal in *differential* form, as illustrated in Fig. 3-8. Here, the receiving circuit is designed to amplify only the difference in potential across its input terminals.

Transistor Amplifiers. Some solid-state signal amplifiers use transistor circuits similar to the simplified circuit shown in Fig. 3-9. Several practical considerations in the design of effective versions of such circuits are the response time of the transistor, its loading effects on the transducer output, and its ability to drive the transmission line. These considerations often call for additional components to adapt the circuit to the particular application requirements.

Operational Amplifiers. More modern circuits use operational amplifiers. These may be constructed from individual transistors and other components, but usual practice is to use integrated versions packaged by semiconductor manufacturers. Such devices are commonly called *op amps* in textbooks and the trade press. A wide variety of op amps are commercially available for industrial applications.

Operational amplifiers utilize feedback principles to carry out stable and predictable operations on their input signals. Figure 3-10 shows (a) the basic op amp circuit, (b) its simplified circuit symbol, and (c) idealized open loop input/output voltage characteristics.

One of the principal op amp parameters is its *open loop gain*:

$$A = \frac{\Delta V_O}{\Delta V_{in}} \tag{3.6}$$

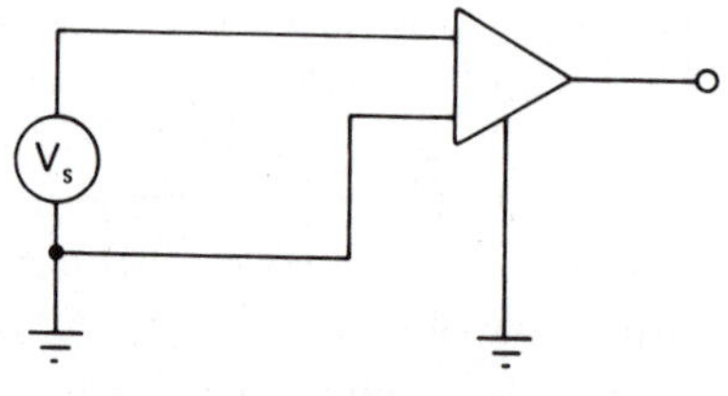

FIGURE 3-8. Differential amplification.

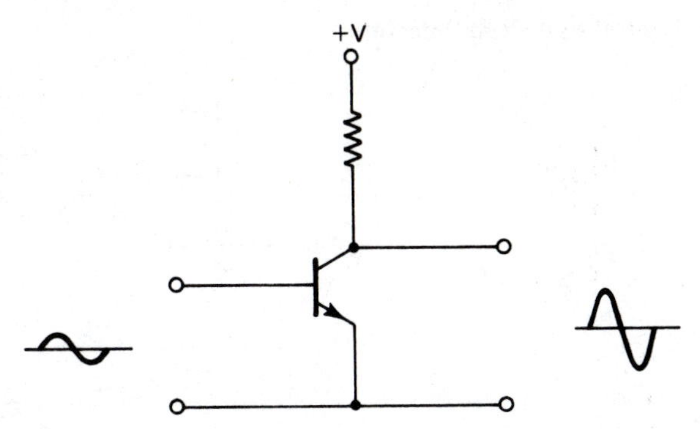

FIGURE 3-9. A simplified transistor amplifier circuit.

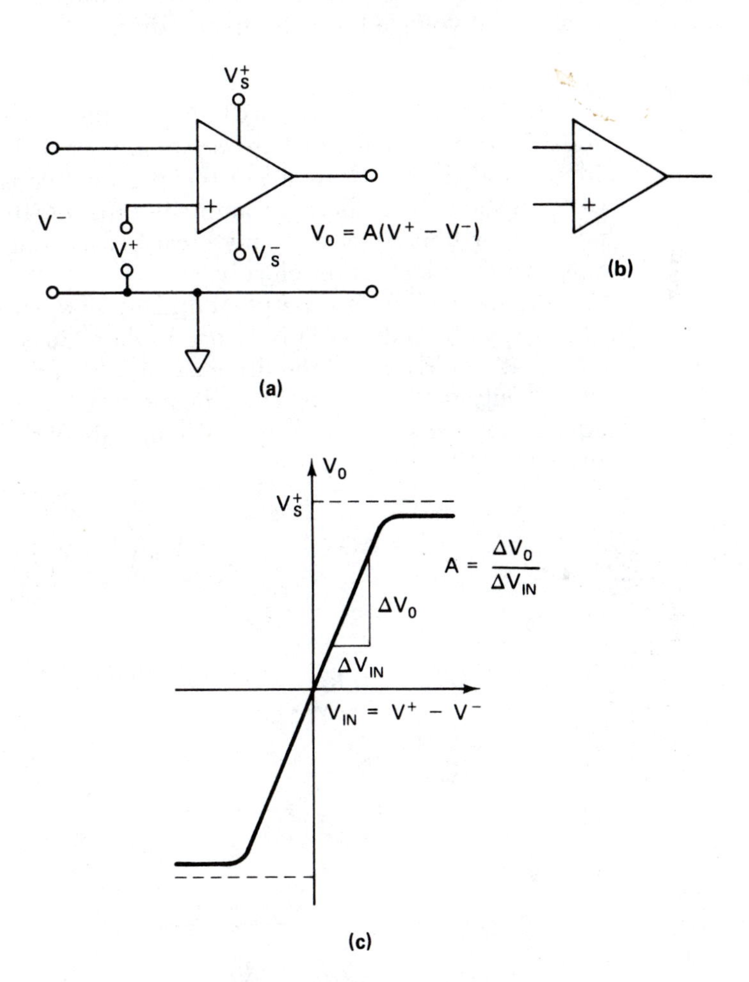

FIGURE 3-10. Operational amplifier: (a) basic op amp circuit, (b) simplified circuit symbol, (c) idealized op amp open loop input/output voltage characteristics.

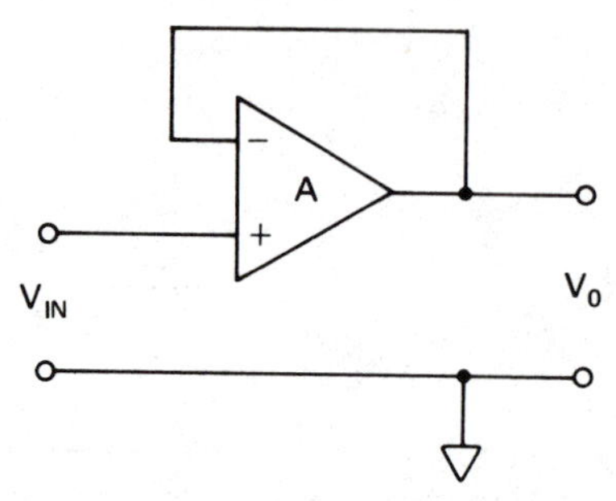

FIGURE 3-11. Op amp unity follower configuration. Output voltage equals input voltage but with greater drive.

Figures 3-11 through 3-13 show three basic configurations of "ideal" op amp circuits that are commonly used. These are simplified circuits and do not show all of the practical considerations of real-world designs. The ideal op amp (like the ideal resistor, capacitor, inductor, or transistor) is a "mythical beast," but practical op amps approach the ideal rather closely.

Figure 3-11 shows the ideal op amp in what is called the *unity follower* configuration. This is used where it is necessary to drive loads that would exceed the drive capability of the primary element's natural output, but where amplification of the voltage level is not required or desired. Since the open loop gain of an op amp is typically very large (10^5–10^6),

$$V_O = A(V_{in} - V_O) = V_{in}/\left(1 + \frac{1}{A}\right) \cong V_{in} \qquad (3.7)$$

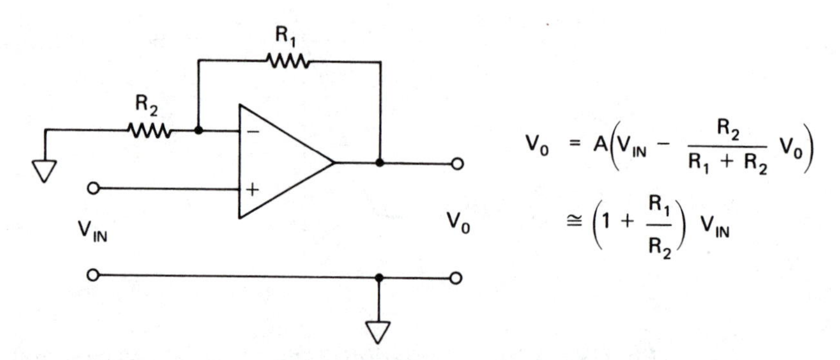

FIGURE 3-12. Noninverting operational amplifier configuration. Gain is determined by resistor ratio.

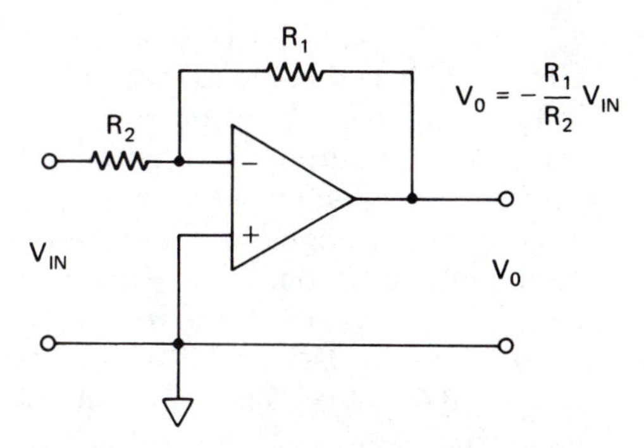

FIGURE 3-13. Inverting operational amplifier configuration. Polarity of the amplified input signal is reversed.

Figures 3-12 and 3-13 show *noninverting* and *inverting* operational amplifier configurations, respectively. Each amplifies its input signal by the factor indicated in the figure. This factor is determined by the ratio of the resistance values in the circuit. These are *single-ended* amplifiers.

The *differential* amplifier shown in Fig. 3-14 is more often used in industrial instrumentation because of its better immunity to common-mode interference. In this configuration, potentials appearing in common on both input terminals are rejected in the ideal case. In practice, there are limits to the degree to which a differential amplifier can reject common-mode signals, but very effective designs can be had at moderate cost and are to be preferred where direct transmission of an analog signal is called for. As with single-ended configurations, the amplifier bandwidths are at issue.

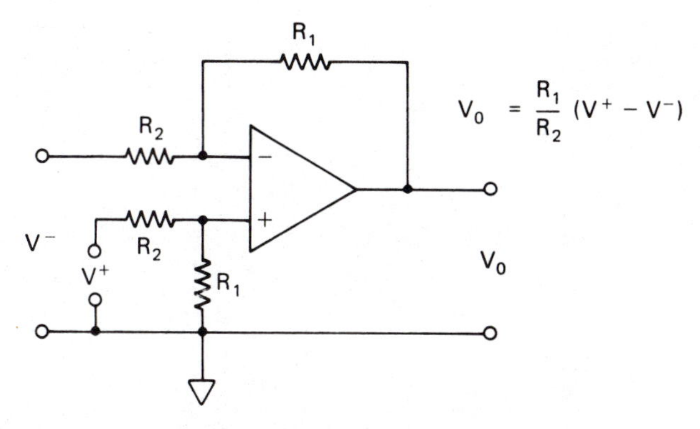

FIGURE 3-14. Differential amplifier. This circuit amplifies the difference between the signals at its input terminals.

The op amp may also be used to construct an integrating active circuit, such as the one shown in Fig. 3-15. This circuit acts as a combination amplifier-filter and is sometimes used to produce an averaged signal for the purpose of improving the signal-to-noise ratio.

In the selection of op amps, the most desirable characteristics are high input impedance, low output impedance, high ratios of common-mode rejection, and the precision of the gain applied to the input signal. In order for the output signal to be related precisely to the primary element signal, it is important that the op amp exhibit very low input offset voltage and offset drift and very low input bias current and drift.

Operational amplifiers are available in several classes. The most commonly used op amps are in the "general-purpose" category, which are lowest in cost and are suitable for most undemanding applications. Op amps with field-effect transistor (FET) input stages exhibit very low bias current parameters. High accuracy, low drift differential op amps complete the collection of the classes that are most relevant to the majority of process instrumentation applications. "Chopper stabilized" op amps are employed where extremely low drift is mandatory. Extremely fast, wide bandwidth op amps are another class, but their application to process control signal conditioning is less frequent because of the low bandwidth of most process variables. They are often used in fast, precision analog-to-digital converters and communications filter circuits.

When protection must be provided against very high common-mode voltages, *isolation amplifiers* are used. In these devices, the input and output stages of the amplifier are isolated from one another, usually by an arrangement of small transformers. Figure 3-16 is a block

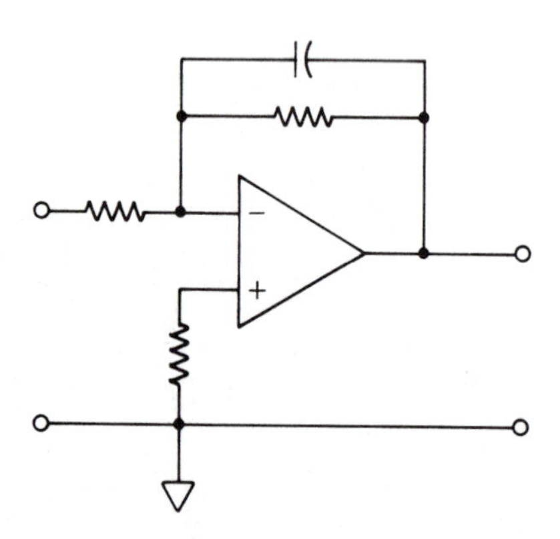

FIGURE 3-15. An integrating amplifier, used to filter and average noise effects out of the desired signal.

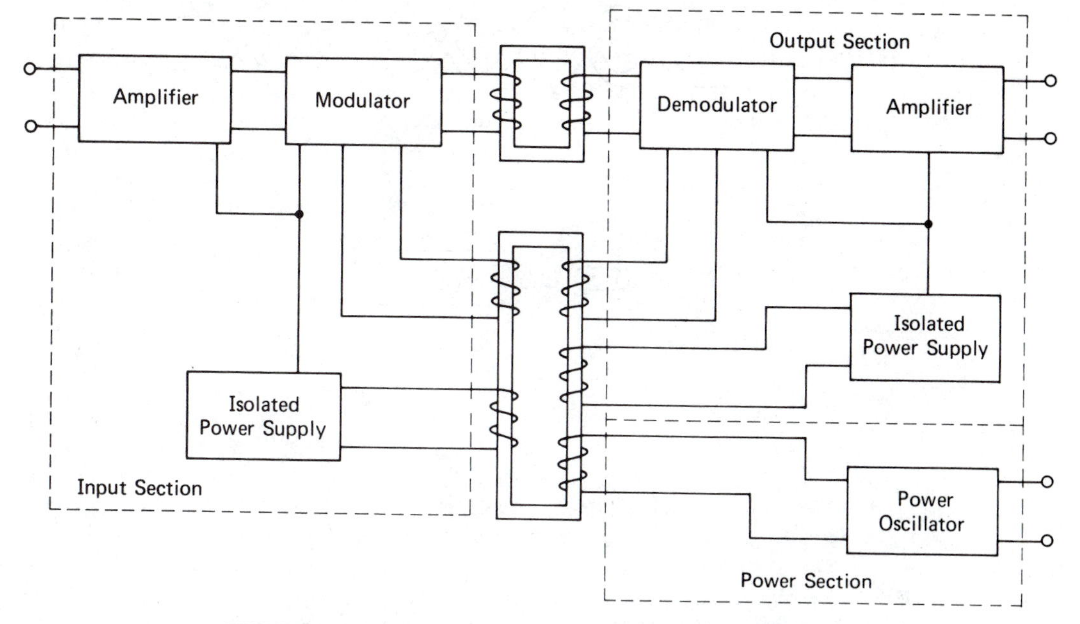

FIGURE 3-16. An isolation amplifier. Transformers provide isolation between the input and output sections. The input section is provided with transformer-coupled isolated power.

diagram of an isolation amplifier that illustrates its principal components. Such devices can provide protection against common-mode potentials in excess of 2000 volts.

Wiring

The principal means by which noise can interfere with a signal are differences in ground potentials at either end of the transmission line and inductive and capacitive coupling between the noise source and the transmission line. Differential amplifiers are a common remedy for the first type of interference. Proper wiring and grounding provides various degrees of protection against coupled noise sources. Three common types of wiring are illustrated in Fig. 3-17.

One low-cost type which is commonly used twists the signal and return wires together. This is called *twisted pair* [Fig. 3-17(a)] and is moderately effective for protection from small noise threats over moderate distances. Another type is *coaxial cable,* in which a central signal wire is surrounded by a braided shield from which it is separated by an insulating material [Fig. 3-17(b)]. The combination of twisted pair and a surrounding coaxial shield, called *twinax* cable [Fig. 3-17(c)],

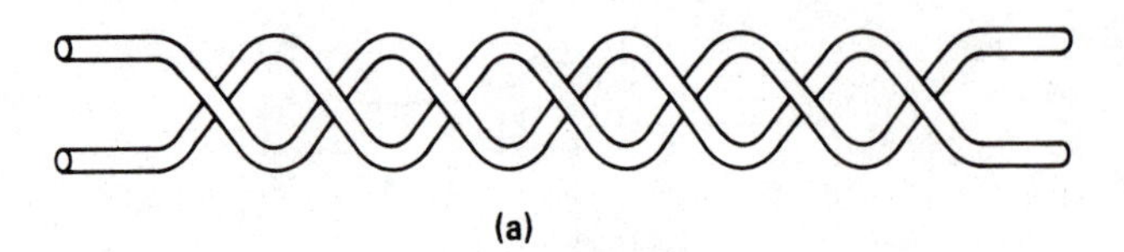

(a)

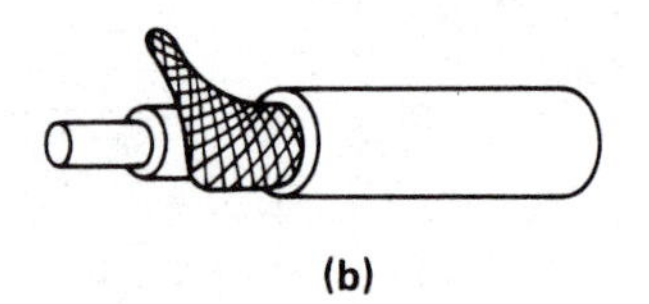

(b)

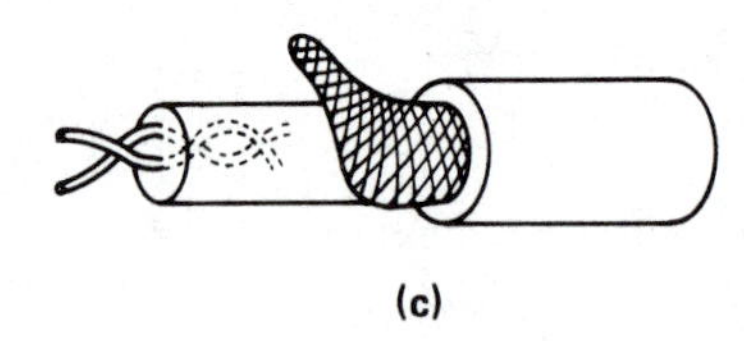

(c)

FIGURE 3-17. Three common wire types: (a) twisted pair, (b) coaxial cable, (c) twinax—shielded twisted pair.

is the best choice for the transmission of analog voltage signals of moderate bandwidth, but it is the most expensive.

For shielded cable to be most effective, it should be arranged so that the shield carries no current. This is accomplished by grounding the shield at only one end of the transmission line. Generally, this is done at the primary element end, where noise threats are most likely to be present, since primary elements are often installed near motors, generators, and other types of equipment which produce substantial amounts of electrical noise in the form of strong electromagnetic fields.

Encoding

Current Loops. With current loop transmission methods, the span of the primary element output signal is used to control the amplitude of a direct current over a corresponding span. The most common current span is one that ranges from 4 mA to 20 mA, generally called a *4-20 mA current loop.* In these, the lower current value is taken to correspond to the minimum output of the primary element (the 0 percent output), and the upper current value is taken to correspond to its maximum output (100 percent, or full scale, output).

The first receiving circuit element on the other end of the transmission line is a resistor. The voltage drop across this resistor can be measured by any of several types of input amplifier/filter circuits to yield the received primary element reading. This voltage is, of course, offset from zero volts when the primary element signal is at its minimum output, because 4 mA will produce a voltage which depends on the resistor value. This characteristic makes it possible for the instrumentation system to detect certain types of breakdowns at the primary element or in the transmission line.

For example, if the primary element fails or the transmission line breaks, the voltage across the resistor will go to zero or to an off-scale voltage. A suitably designed receiver circuit (or computer) then has sufficient information to report that something has gone amiss and to take proper action, such as displaying a warning, holding control outputs at present values, or turning off equipment in an orderly fashion.

Because noise coupled into the transmission lines must transfer more energy to induce a significant change in such a current-based signal, and because the circuits at the transmitting end are designed to control the current, current loop transmission methods are more immune to interference than are voltage transmission methods. However, current loops have more restricted bandwidths than direct voltage transmission methods. The task of modulating a significant amount of current at a moderately high frequency is difficult with practical circuits of reasonable cost. For most typical process variables, which do not have wide bandwidths, current loops are quite practical.

With current loop devices, it is often possible to arrange the circuits so that local devices may utilize moderate amounts of power that are actually delivered to them over the signal transmission lines themselves from a centrally located power supply.

Voltage-to-frequency. Voltage-to-frequency signal transmission methods use the primary element output to modulate the frequency of a transmitted signal. The frequency of the received signal is then converted back to a voltage that corresponds to the original primary element output voltage. A basic center frequency is chosen for modulation which is unrelated harmonically to expected noise threat frequencies and is well matched to the characteristics of the transmission line, so that transmission line effects are themselves minimized. Although the amplitude of the received signal may be attenuated by transmission line effects or may be distorted by coupled interfering noise, the inherent information is carried in the *frequency* of the signal, not in its amplitude.

These methods require local power supplies at the transmitting end and involve circuits at both ends which are more complex than those used for current loop methods. For these reasons, they are more costly than current loop methods. However, they do provide significantly better noise immunity and greater practically achievable bandwidths.

Other Telemetry. Other variations on transmission methods include *pulse counting* and *pulse width modulation* methods. With the former, a number of pulses are transmitted over an agreed-upon period of time, the number being related to the primary element output. Pulse width modulation methods adjust the width of a periodically transmitted pulse in a way which varies with the primary element output.

Isolation

Signal conditioning lines are sometimes isolated to provide protection from common mode potentials and threatening voltages (such as accidental contacts with high voltage wiring). The principal forms of isolation are transformer, optical, and switched.

Transformer isolation (also called *galvanic* isolation) is arranged by coupling signals through a small transformer, as shown in Fig. 3-18. The input signal is modulated ("chopped") by the output of an oscillator, coupled through the transformer, then reconstructed by a demodulator to produce a reasonable facsimile of the original input signal. Such arrangements can provide input to output isolation in excess of 2000 volts. This principle is also used in the isolation amplifier shown in Fig. 3-16.

Optical isolation is arranged by using the input signal to drive a light-emitting diode (LED). A phototransistor exposed to the light from the LED then provides the output signal, as shown in Fig. 3-19.

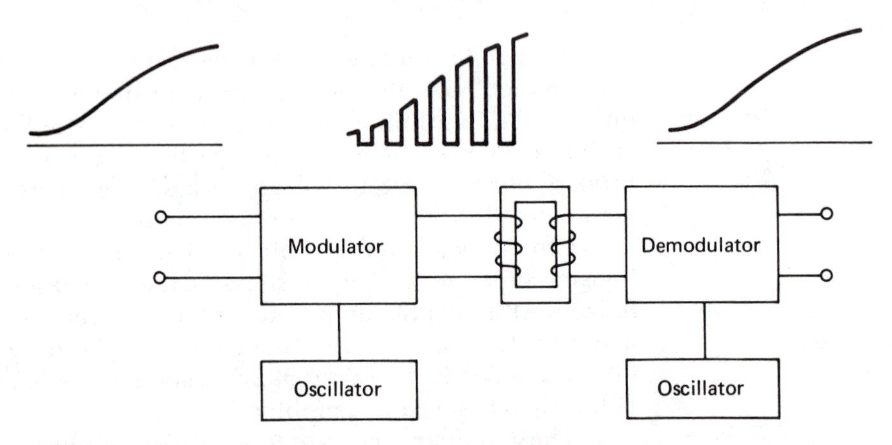

FIGURE 3-18. Transformer (or galvanic) isolation. The modulated signal is coupled through the transformer and then demodulated.

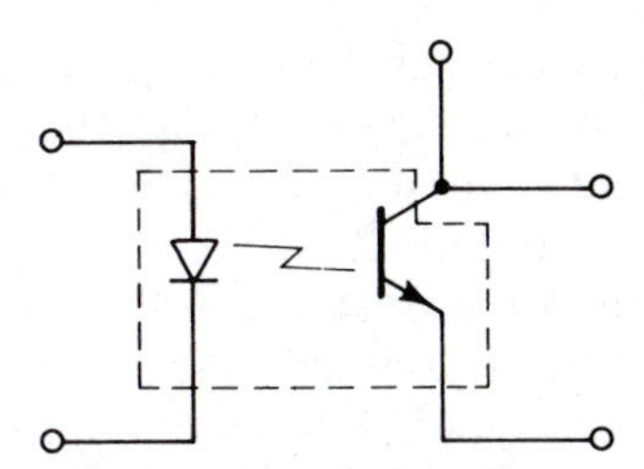

FIGURE 3-19. Optical isolation. The input signal causes a light-emitting diode to produce light that is detected by a phototransistor.

Galvanic and optical isolation devices are commercially available as integrated packages. Optical isolation is usually restricted to use with digital signals, since it is difficult to reconstruct analog signals properly with such devices.

Switched isolation (also called *flying capacitor* isolation) is illustrated in Fig. 3-20. This operates by switching a capacitor between the input signal and the input amplifier. In position A in the figure, the capacitor is charged by the input signal. When the signal is to be read, both terminals of the capacitor are switched (using a relay) so that the capacitor appears across the amplifier input terminals (position B). Isolation is obtained from the fact that the amplifier is never physically connected to the input signal wiring. Switched isolation methods can also be used to multiplex several signals to a single amplifier.

Filters

The behavior of the output signal of virtually every physically realizable circuit depends upon the frequency content of its input signal. This is the basis for *filters*. By designing a filter circuit so that it attenuates unwanted frequency components, it may be used to remove substantial portions of interfering signals.

As a block diagram component, filters may be represented as

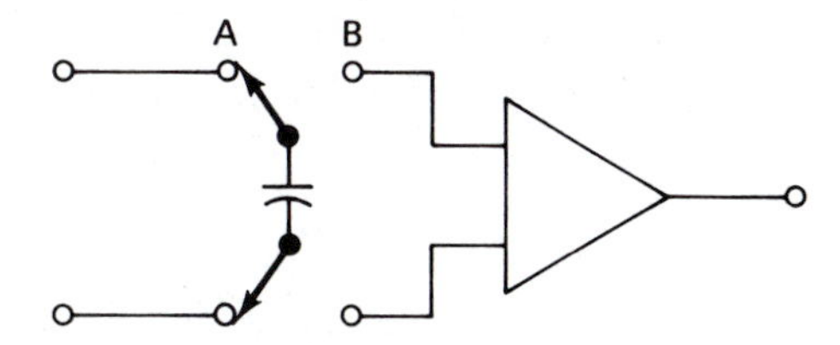

FIGURE 3-20. Switched (or flying capacitor) isolation. The capacitor is charged by the input signal and then switched to the input amplifier.

four-terminal devices, as shown in Fig. 3-21. The relation between the input signal $x(t)$ and the output signal $y(t)$ may be described by using several mathematical expressions, the choice of which is generally based upon the particular type of analysis being done. In many cases, the *transfer function* of the filter is used to describe its behavior. The transfer function is the ratio of the Fourier transforms of the output and input signals when continuous, or analog, filters are being discussed. Ratios of the Laplace transforms of output and input signals are also used to form transfer functions. For digital filters, the z transform is the applicable mathematical transform.

Letting $X(\omega)$ represent the Fourier transform of the input signal $x(t)$ and $Y(\omega)$ represent the Fourier transform of the output signal $y(t)$, where ω is radian frequency, we find that they are related according to the following equation:

$$Y(\omega) = H(\omega) \cdot X(\omega) \tag{3.8}$$

so that the transfer function $H(\omega)$ is

$$H(\omega) = \frac{Y(\omega)}{X(\omega)} \tag{3.9}$$

Fourier, Laplace, and z transforms are closely related functions. They share an important characteristic: For a very broad and useful class of devices (of which filters are important members), transfer functions formed from them are independent of the specific input signal applied to the device. That is, if $x_1(t)$ and $x_2(t)$ are different input signals, with corresponding filter outputs $y_1(t)$ and $y_2(t)$, then

$$H(\omega) = \frac{Y_1(\omega)}{X_1(\omega)} = \frac{Y_2(\omega)}{X_2(\omega)} \tag{3.10}$$

The transfer function describes the gain that the filter applies to the individual frequency components of its input signals, so that a graph of the transfer function shows frequencies affected by the filter and the ways it affects them.

Filters are divided into four basic categories, according to char-

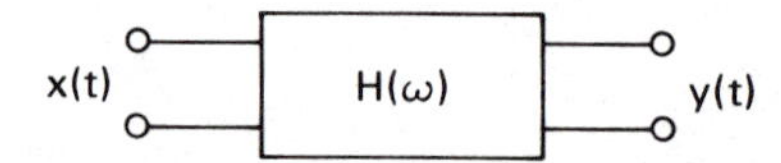

FIGURE 3-21. Block diagram of a filter as a four-terminal device with a transfer function that depends on input signal frequency components.

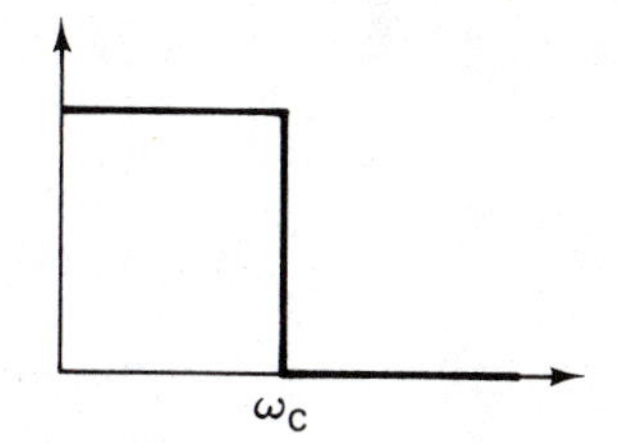

FIGURE 3-22. Idealized low-pass filter transfer function.

acterizations of their effects as functions of frequency: (1) low pass, (2) high pass, (3) band pass, and (4) band reject. Figures 3-22 through 3-25, respectively, are idealized plots of the transfer functions for each type.

The *low-pass filter* allows input signal frequency components which are lower than ω_c to pass with minimal attenuation, while attenuating components of higher frequencies. The *high-pass filter* performs a complementary function, permitting frequencies greater than ω_c to pass and attenuating lower frequency components. The *band pass filter* attenuates signal frequencies on either side of a frequency band of width $\Delta\omega$ centered at frequency ω_c, while passing frequencies within the band with minimal attenuation. The *band reject* (or *band stop) filter* is the complement of the band pass filter, attenuating only those frequencies that lie within a narrow band. The range of frequencies passed by a filter is called its *passband.*

In practical circuit realizations of filters, the sharp frequency cutoffs illustrated in the figures cannot be obtained. No filter totally rejects every frequency component. In practical filters, the transfer function plots fall off more gradually so that, although frequencies outside the passband of the filter are attenuated substantially, they are not attenuated entirely.

Figure 3-26 is a transfer function plot that is more nearly typical of a practical low-pass filter. For most practical cases, logarithmic or semi-logarithmic plots are used because they are more effective at

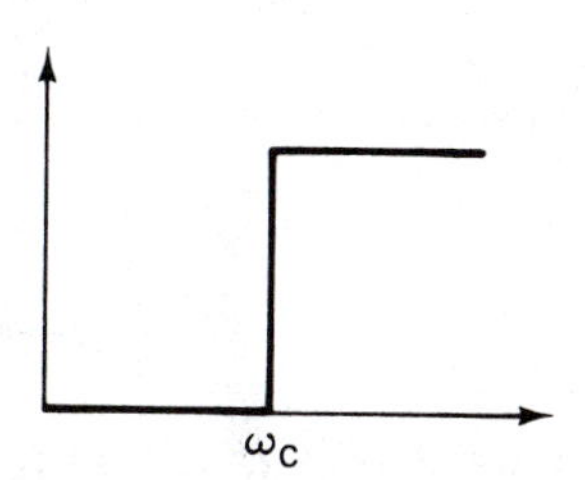

FIGURE 3-23. Idealized high-pass filter transfer function.

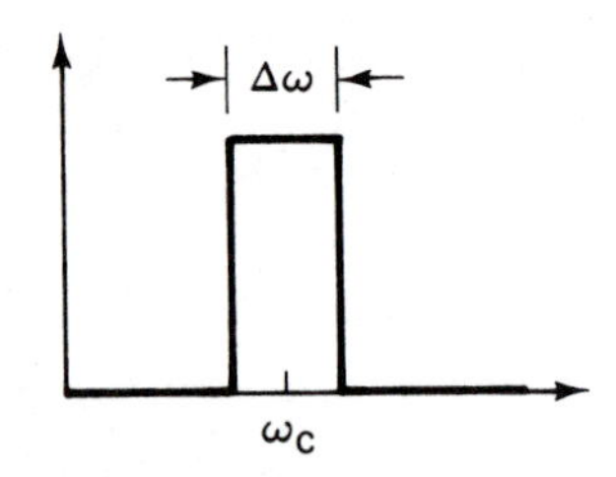

FIGURE 3-24. Idealized band pass filter transfer function.

dealing with the range of frequencies over which this *rolloff* occurs. For many practical filters, the graph is logarithmically linear over much of its extent so that such plots also aid in graphical analysis of filters. Much of the terminology of filters is based on visual characteristics of their graphs. The point at which the graph curves sharply, for example, is sometimes called the *knee* of the filter.

The visual effect of this rolloff on the graph of the band reject filter (an example of which is shown in Fig. 3-27) leads to this kind of filter's being called a *notch* filter. Notch filters are used to reject particular frequencies which would otherwise interfere with measurements. For example, in an application intended to measure a rapidly varying pressure, a notch filter might be employed to reject the 60-Hz noise signals commonly found in process system environments.

The point on the graph that corresponds to an attenuation by a gain of approximately 0.707 is alternately called the *half-power point* or the *3-dB cutoff* point. At this point, the gain of the filter is equal to the reciprocal of the square root of 2 (the value 0.707), the output signal voltage is 3 dB below the value of the input signal voltage, and the power of the output signal transmitted by the filter is one-half the power of the input signal. The frequency at which this occurs is called the *cutoff frequency*.

Filters are also classified by the type of components of which they are made. *Passive filters* are built from passive components (re-

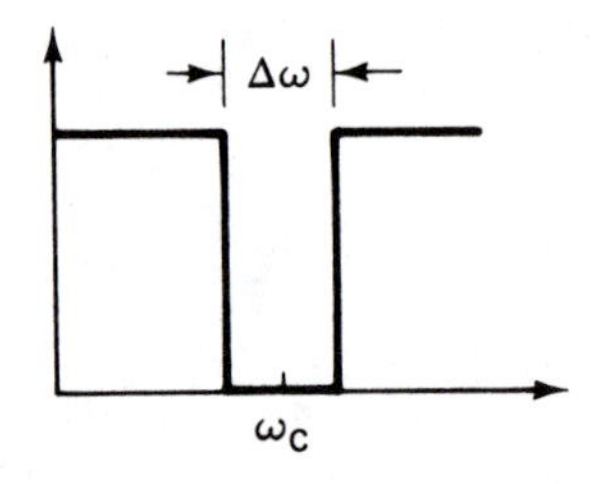

FIGURE 3-25. Idealized band reject (or band stop) filter transfer function.

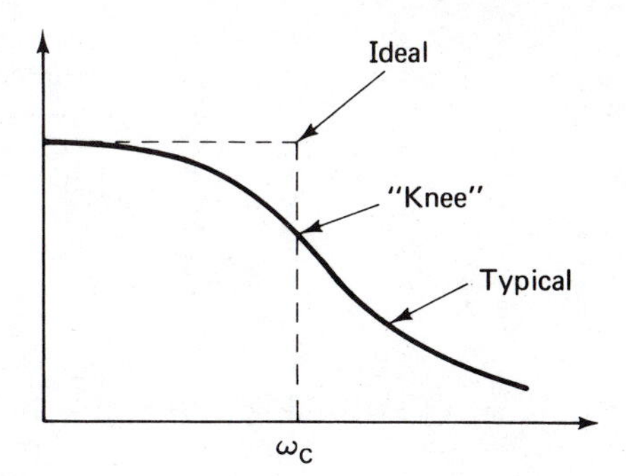

FIGURE 3-26. Typical low-pass filter transfer function compared to the ideal (dashed line).

sistors, capacitors, and inductors). *Active filters* are built from active components. In modern practice, the active components used are op amps.

Passive filters with simple and unchallenging specifications are relatively easy to build. The principal practical types used from time to time in process control and instrumentation applications are low-pass passive filters consisting of resistors and capacitors, as illustrated in Fig. 3-28. Such filters may be bulky, because of the strong relations which exist between the electrical properties of capacitors and their physical size. Except in the simplest cases, active filters are generally preferred for their versatility and performance. The increasing levels of integration with which the components of active filters, as well as complete predesigned active filters, are found make their use in instrumentation system design much more convenient and economically desirable.

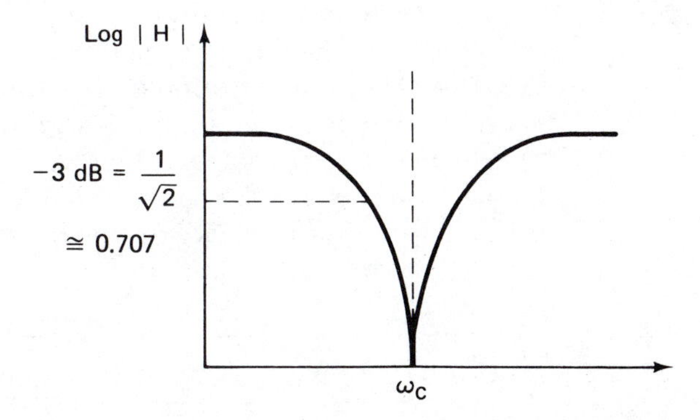

FIGURE 3-27. Rolloff in a band reject filter in which the logarithm of the transfer function is plotted and the half-power point is noted.

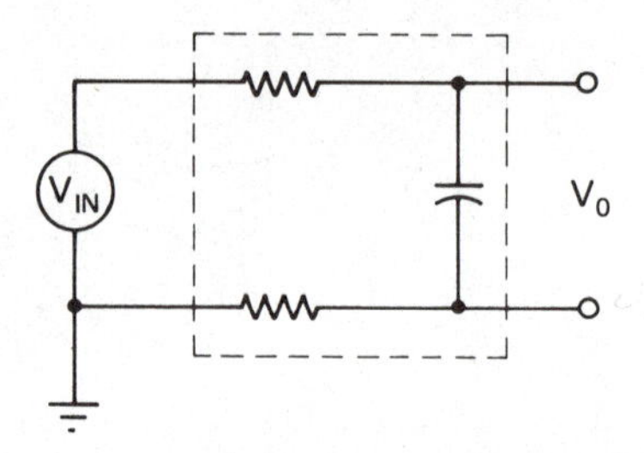

FIGURE 3-28. A simple low-pass passive filter circuit.

The shape of the graph of a filter may be tailored to particular forms through such techniques as *cascading* several filters in two or more stages. This is illustrated in Fig. 3-29. By arranging several low-pass filters in stages, the net transfer function of the combined filter can be shaped to produce such desirable characteristics as a sharper rolloff. That is,

$$\frac{Y(\omega)}{X(\omega)} = H(\omega) = H_1(\omega)\cdot H_2(\omega)\cdot \ldots \cdot H_N(\omega) \tag{3.11}$$

where $H_i(\omega)$ is the transfer function of the ith stage of the filter.

Filters, after classification according to their transfer function type and the components of which they are made, may also be classified according to certain fine details of their transfer functions. With some exceptions, the names of these classifications are the names of the engineers, scientists, and mathematicians who either invented them or the functions used in their mathematical descriptions. The principal examples are quadratic filters, Butterworth filters, Bessel filters, and Chebyshev filters.

Quadratic is a name used for any filter in which the denominator of the transfer function is a quadratic polynomial. Such filters have two poles. The number of poles of a filter is called its *order*. Thus, a quadratic filter is a second-order filter. A second-order passive filter requires two distinct stages—that is, two passive filters arranged so that the output of the first is the input to the second. Quadratic filters may be fabricated from active components in several ways. Figure 3-

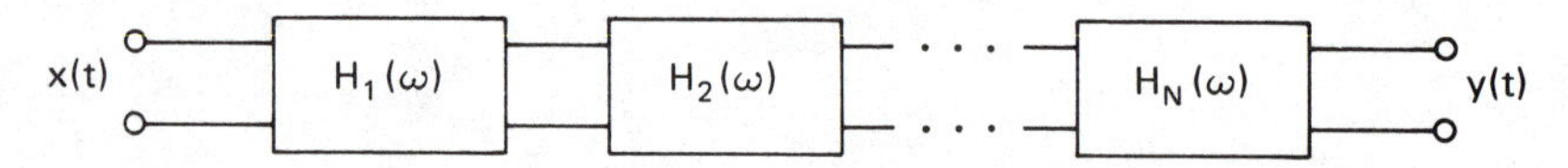

FIGURE 3-29. Cascading filters in order to shape the overall transfer function.

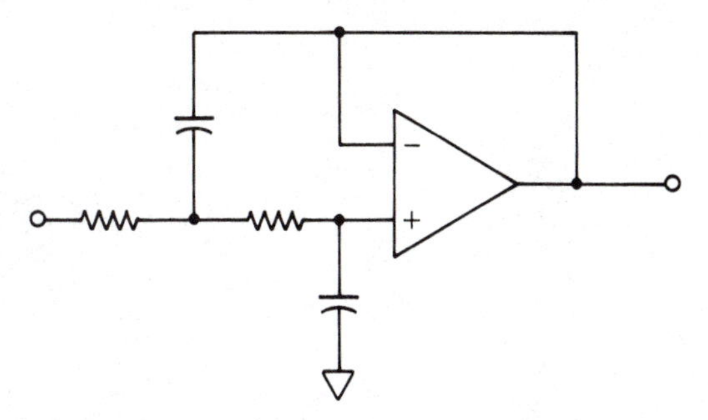

FIGURE 3-30. An active quadratic filter circuit using an operational amplifier.

30 illustrates one common arrangement. Quadratic filters may be thought of as "building blocks" from which other types of filters may be formed.

Butterworth filters are commonly employed in process instrumentation applications requiring low-pass filters because of their characteristic of having a relatively flat amplitude versus frequency response near the direct current region ($\omega = 0$), coupled with a steeper rolloff than is obtainable with a single stage arrangement. The graph of this response is shown in Fig. 3-31 for several different orders of the Butterworth filter. Note that the cutoff frequency is the same and the amount of rolloff is greater for each successively higher degree.

The *phase response* of a filter is often an important consideration, especially if the filter is a component of a control circuit where delays

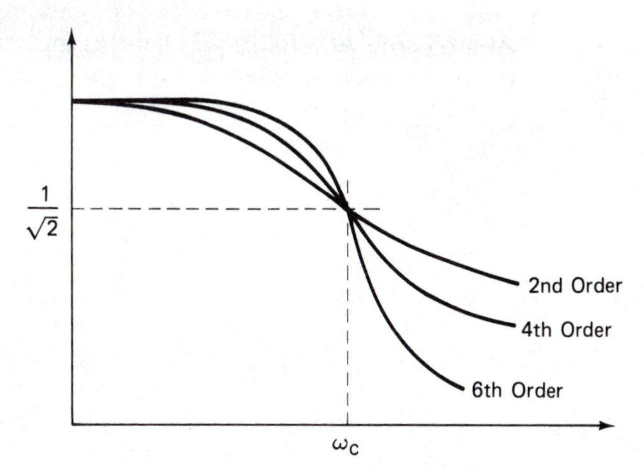

FIGURE 3-31. Graph of the response for different orders of a Butterworth filter. Note that the cut-off frequency is the same for each order.

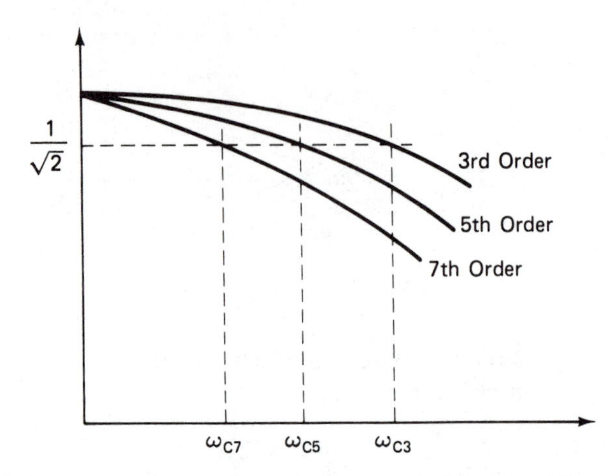

FIGURE 3-32. Graph of the response for different orders of a Bessel filter. Note the change in the cut-off frequency.

in phase can lead to instabilities in control behavior. For instrumentation systems this is less of a concern when the system is collecting data from inputs characterized by low frequency and bandwidth. The measurement of temperature is an example of such an application. The phase response of a Butterworth filter is moderately nonlinear, but not so much so that it is ruled out for applications such as these.

For applications where minimal phase effect is needed, the *Bessel filter* is generally the preferred choice. Of the most popular filter types, the phase behavior of the Bessel filter is most nearly linear, especially in the passband region, and approaches a logarithmically linear falloff asymptotically as the order of the filter is increased. However, as the order of a low-pass Bessel filter is increased, its

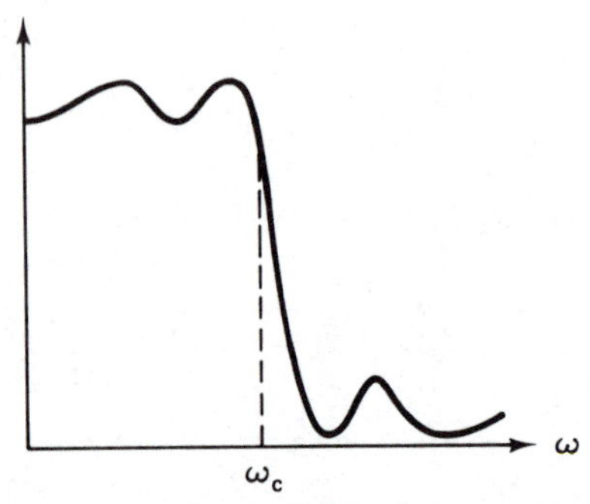

FIGURE 3-33. Chebyshev filter response (exaggerated). Note the rapid rolloff and average flatness of the passband and stopband.

cutoff frequency is reduced, as shown in Fig. 3-32. This is substantially different from the behavior of the Butterworth filter, where the 3-dB point remains stationary as the order of the filter is increased. Bessel filter phase response makes it the most desirable type for use in pulsed circuits because its response to step functions exhibits minimal overshoot. For this reason, it is often employed in instrumentation systems which encode information into pulses.

The *Chebyshev filter* is desirable for its characteristic of having a very sharp transition from its passband to its stopband. Its amplitude response is shown in Fig. 3-33 in exaggerated form. While it is maximally flat in both its stopband and passband, and has very rapid rolloff, it exhibits a substantial amount of ripple. This ripple may be deleterious to the accurate measurement of signals whose frequency varies within the passband. However, it is very useful where required accuracy is not extreme, but where interfering noise signals may be close to the measured signal frequencies, because these interfering signals may be sharply attenuated, thanks to the filter's sharp rolloff.

Compensation

When one is making measurements with certain transducers, it is sometimes necessary to compensate for unwanted, but unavoidable, effects of the actual practical circuits used. Thermocouple-based temperature measurements are the principal example of such a situation.

The process of converting a measured thermocouple voltage signal to a temperature measurement at the intended place becomes complicated, because the junctions between thermocouple and instrumentation system wires form additional thermocouple-like junctions. The sensed voltage (ΔV), as illustrated in Fig. 3-34, is an approxi-

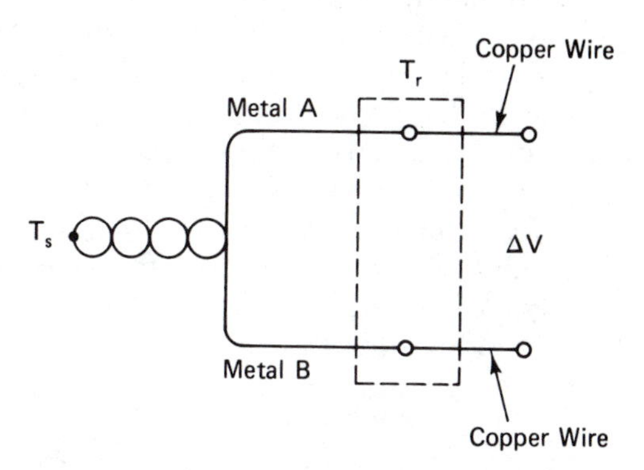

FIGURE 3-34. Relation of sensor temperature and reference temperature in a thermocouple circuit.

mately locally linear function of the difference between the temperature to be measured at the sensor (T_s) and the temperature at the junction of the thermocouple wiring and the instrument's copper wiring (T_r). Taken at face value, this voltage (using the curves for the type of thermocouple being used) would imply a temperature T_m which is different from the actual temperature at the sensor. That is,

$$\Delta V \cong k(T_s - T_r) \Rightarrow T_m \tag{3.12}$$

so that

$$T_s = T_m + T_r \tag{3.13}$$

In order to perform the compensation of T_m to obtain T_s, T_r, commonly called the *reference junction* temperature, must be known.

Modern practice with electronic systems uses a different sensor to measure T_r. Since the range of reference junction temperatures is generally limited in actual installations, T_r can usually be measured with a semiconductor temperature sensor. This measurement of T_r, adjusted for differences between the coefficients of the thermocouple and the semiconductor temperature sensor, may be added to the measurement electrically, in order to provide a compensated reading, or it may be read into the computer, where the compensation may be done by programming.

SUMMARY

We have discussed only a few of many methods of conditioning signals and making measurements. This is a subject of immense variety and complexity, and it is one of the most challenging parts of modern control system engineering. It is also a highly dynamic field, with new methods, circuits, and devices being developed continually. Its dynamic nature is strongly influenced by the use of computers in measurement and control systems, with microcomputers having particularly interesting effects through their ability to distribute "intelligence" throughout the system.

CHAPTER 4

Control and Decision-Making

"Gouverner, c'est choisir."
(To govern is to make choices.)

Duc de Lévis
1764–1830

This chapter provides overviews of the three principal types of control: (1) continuous process control, (2) discrete control, and (3) numerical control. The theory and mathematics of each type is very rich. This chapter gives a hint of this. Complete treatment of any one of them is beyond the reach of most textbooks, much less that of a single chapter. Rigor and depth is eschewed in favor of trying to convey intuitive understanding of their general behavior and principles.

CONTINUOUS PROCESS CONTROL

Continuous process control is exemplified by the closed loop control of essentially continuous process variables (e.g., temperature, pressure, flow rate). Analysis is through the use of Laplace transforms of the differential equations which describe the behavior of these systems. The principal issue in their design is the degree to which close control can be maintained without incurring risk of system instability.

Traditional continuous process control systems are fabricated from electronic amplifiers and discrete components or from mechanical, pneumatic, and hydraulic devices. Analogies may be drawn between the behavior of continuous processes and their controllers and other physical systems, such as circuits of resistors, capacitors, and inductors and mass-spring-dashpot mechanical devices. Such analogies are often useful in explaining their behavior.

Traditional continuous process controllers have been used in conjunction with computer systems for many years. In some examples, the computer is responsible for adjusting the parameters of the continuous (or "analog") controller. Such arrangements are generally called *supervisory control* systems. In more recent examples, computers replace the analog controller and incorporate the control algorithms in their programs directly. Such arrangements are generally called *direct digital control* (DDC) systems.

Process and Control System Models

Continuous control system design and analysis begin with the development of block diagram models. These models help us to visualize the interactions between the component parts of a system. Block diagrams show the parts of the system, the variables of interest, and the ways in which these are combined.

Most systems of interest can be modeled by using components whose behavior can be described using ordinary linear differential equations with constant coefficients having the following nth order form:

$$a_n \frac{d^n y(t)}{dt^n} + a_{n-1} \frac{d^{n-1} y(t)}{dt^{n-1}} + \ldots + \frac{dy(t)}{dt} + a_0 y(t) = bx(t) \quad (4.1)$$

A system having this behavior is said to be *linear*. Many practical systems can be described adequately by using only first- and second-order equations. Laplace transforms are used to make the solution of these equations easier. The Laplace transform of a function of time, $f(t)$, is defined as

$$F(s) = \mathcal{L}(f(t)) = \int_0^\infty f(t)\, e^{-st}\, dt \quad (4.2)$$

where s is a complex variable.

The following theorems of Laplace transformation are necessary in the analysis of these systems:

$$\mathcal{L}(f_1(t) + f_2(t)) = \mathcal{L}(f_1(t)) + \mathcal{L}(f_2(t)) \quad (4.3)$$

$$\mathcal{L}(cf(t)) = c\, \mathcal{L}(f(t)) \qquad c = \text{constant} \quad (4.4)$$

$$\mathcal{L}(f_1(t) * f_2(t)) = \mathcal{L}\left(\int_{-\infty}^{\infty} f_1(\tau) f_2(t - \tau) d\tau\right)$$

$$= \mathcal{L}(f_1(t))\, \mathcal{L}(f_2(t)) \quad (4.5)$$

The Laplace transform of Eq. (4.1) may be written

$$G^{-1}(s)Y(s) = bX(s) \tag{4.6}$$

The factor $G(s)$ is known as the *transfer function* of the subsystem represented by Eq. (4.1). $G(s)$ is a polynomial in s with coefficients a_n through a_0. Typically, $a_0 = b = 1$. The use of transfer functions permits the analysis of the system using relatively simple algebraic methods.

Figure 4-1(a) shows how this subsystem would be presented in a block diagram. The input and output variables are capitalized versions of the time-dependent variables, and the transfer function is written in the block which represents the subsystem. Figure 4-1(b) shows how sums and differences of variables are combined in block diagrams. Points at which these are combined by summing or subtraction are called *summing points*, with the actual operation indicated by "+" or "−" symbols at the arrowheads.

The principles outlined in this section apply only to *linear systems;* that is, systems describable by Eq. (4.1). Many real-world systems are nonlinear or have coefficients which vary with time. In such cases, these principles can often be used to design models which approximate their behavior in regions around their nominal operating points.

In the usual analysis of systems, we are most interested in deviations in the behavior of variables about some steady-state condition. For this reason, the variables used in the equations are usually taken to be deviation variables and initial conditions are translated to a "zero" condition.

In the following sections, the behavior of particular types of simple components are described. These are commonly used to form models of more complex processes and controllers.

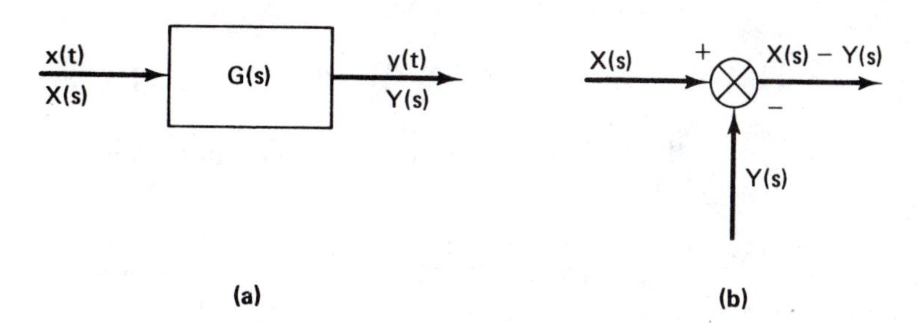

FIGURE 4-1. Continuous control system block diagram symbols: (a) subsystem block symbol with transfer function G(s), (b) summing point symbol.

FIGURE 4-2. System component that multiplies its input by a constant gain K.

Simple Gain

A component consisting of a simple constant *gain* having value K is shown in Fig. 4-2. This might be used to approximate the behavior of a fast-acting measurement element or to represent a change of scale.

First-order Element

First-order elements are describable by equations of the form

$$\tau \frac{dy}{dt} + y = Kx \tag{4.7}$$

These have transfer functions of the form

$$G(s) = \frac{K}{\tau s + 1} \tag{4.8}$$

and are shown as illustrated in Fig. 4-3. Figure 4-4 shows the response of a first-order element to a step change (Δx) in its input. This response is described by the following equation:

$$y(t) = K\,\Delta x(1 - e^{-t/\tau}) \tag{4.9}$$

The term τ is called the *time constant* of the element and is characteristic of such elements. Because

$$1 - e^{-1} = 0.632 \quad \text{and} \quad 1 - e^{-5} = 0.993$$

the time required for a 63.2 percent change in output, in response to

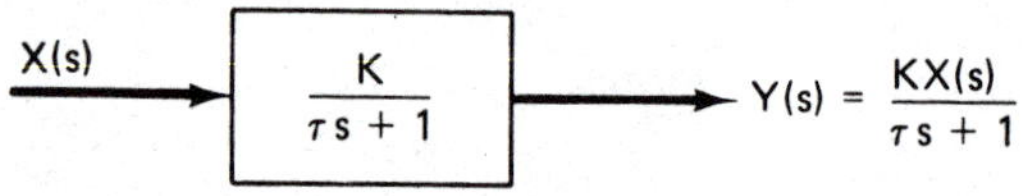

X(s) → $\frac{K}{\tau s + 1}$ → $Y(s) = \frac{KX(s)}{\tau s + 1}$

FIGURE 4-3. First-order element (also called first-order lag, lag, or process reaction delay).

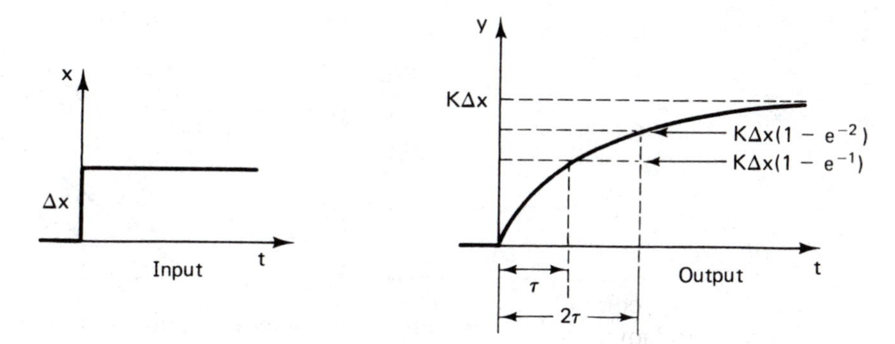

FIGURE 4-4. First-order element response to a step change in its input.

an input step change, is often taken as a reasonable approximation to the time constant of the element; the response is often taken to be essentially complete after five or more time constants have elapsed, since the output is then within 1 percent of its final value.

Examples of real system components having behavior approximated by a first-order element are resistor-capacitor circuits, most temperature sensors, and heated, stirred tanks. As we have seen before, terminology varies among workers in process control. This is also true in the descriptions of continuous process elements. First-order elements are sometimes called *first-order lags, lags,* and *process reaction delays.*

Second-order Elements

Second-order elements are those describable by equations of the form

$$\tau^2 \frac{d^2y}{dt^2} + 2\zeta\tau \frac{dy}{dt} + y = Kx \qquad (4.10)$$

and are shown as illustrated in Fig. 4-5. Figure 4-6 shows the response of a second-order element to a step change in its input. The principal characteristics of the response are determined by the nature of the parameter ζ, which is called the *damping factor.* When $\zeta = 1$, the element is said to be *critically damped.* If $\zeta < 1$, the element is said

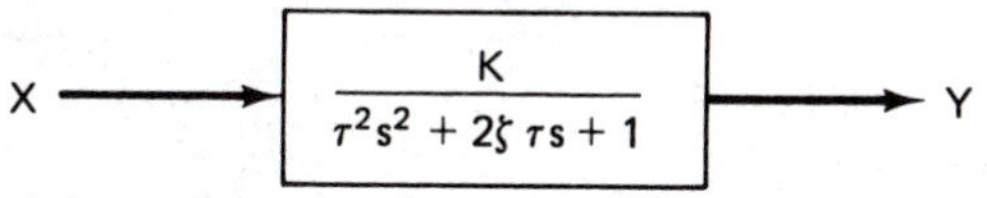

$$X \longrightarrow \boxed{\frac{K}{\tau^2 s^2 + 2\zeta \tau s + 1}} \longrightarrow Y$$

FIGURE 4-5. Second-order element (also called transfer lag).

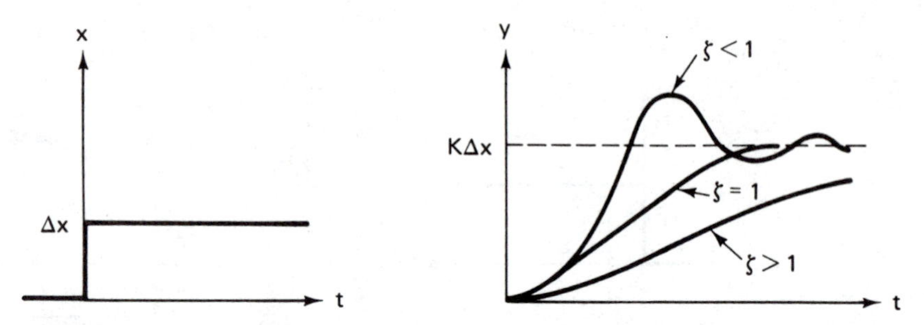

FIGURE 4-6. Second-order element response to a step change in its input for underdamped ($\zeta < 1$), critically damped ($\zeta = 1$), and overdamped ($\zeta > 1$) situations.

to be *underdamped*, and it displays the oscillatory behavior shown in the figure. When $\zeta > 1$, the element is said to be *overdamped*, and it approaches its final value more slowly. The reciprocal of the parameter τ, called the *natural frequency* of the element, represents the frequency of oscillation of an underdamped element as it oscillates about, and closes on, its final value.

Second-order elements have a slow initial response to changes in their input, followed by a significantly more rapid response, and appear to apply a "lag" to their input as the input is "transferred" through the element. For this reason, second-order elements are frequently called *transfer lags*.

First-order Elements in Series

Some transfer lags may be approximated by first-order elements connected in series, as shown in Fig. 4-7. The differential equation describing the combination of these two elements is

$$\tau_1\tau_2 \frac{d^2y}{dt^2} + (\tau_1 + \tau_2)\frac{dy}{dt} + y = K_1K_2x \qquad (4.11)$$

constituting a second-order element with time constant and damping

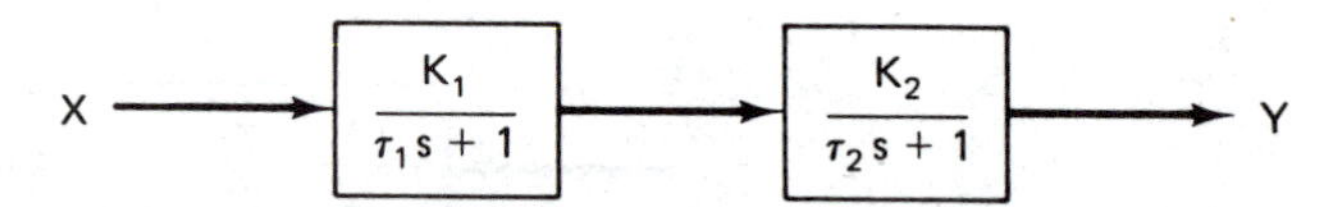

FIGURE 4-7. First-order elements connected in series form a second-order element that is never underdamped.

factor given by

$$\tau = \sqrt{\tau_1 \tau_2} \quad \text{and} \quad \zeta = \frac{1}{2}\frac{\tau_1 + \tau_2}{\sqrt{\tau_1 \tau_2}} \tag{4.12}$$

and transfer function

$$G(s) = \frac{K_1 K_2}{\tau_1 \tau_2 s^2 + (\tau_1 + \tau_2)s + 1} \tag{4.13}$$

Notice that $\zeta \geq 1$, so that such an arrangement is never underdamped and is critically damped when the time constants of the two elements are equal.

Transportation Lag

An element may display lag behavior beyond that implied by Eq. (4.10). This lag is variously called *transportation lag, dead time,* and *transport delay.* It is exemplified by measurement delays caused by distance between the measuring element and the source of the process variable being measured, as might be the case when one is inferring boiler temperature by measuring the temperature of steam in an insulated pipe some distance after it has left the boiler. Transportation lag is described by the following time equation and transfer function:

$$y(t) = x(t - \Delta t) \quad \text{and} \quad G(s) = e^{-\Delta ts} \tag{4.14}$$

where Δt represents the delay time in response to its input, as shown in Fig. 4-8.

Approximating Real Processes

Few processes can be modeled adequately by using only analytical methods. It is often necessary to take measurements of the behavior of the process. These measurements are preferably made on actual process equipment. Where such measurements are not practicable, laboratory or "pilot" models are developed and the resulting measurements are scaled up to conform to the expected behavior of the full-scale plant. These are then verified once the full-scale plant is available. This sequence can be tedious and time-consuming, but it is necessary to ensure high quality control of complex processes.

Generally, the experimental process is first operated in an *open-loop* mode, without controls, but with recording equipment to

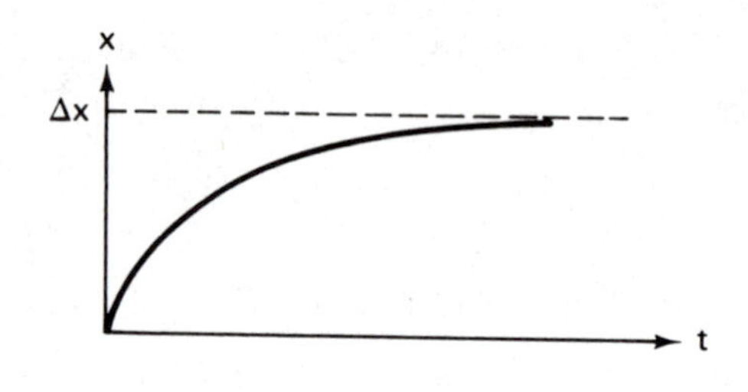

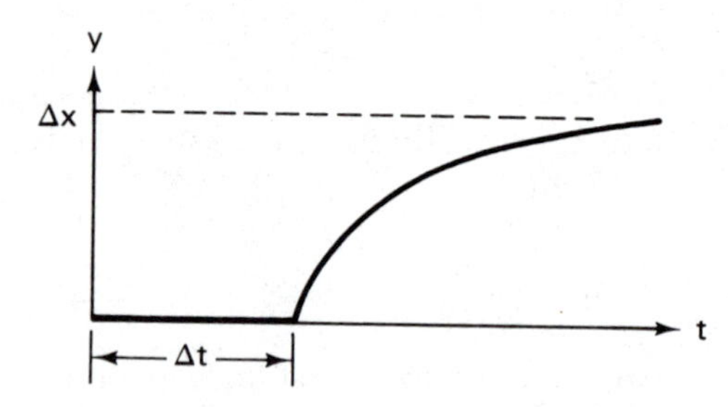

FIGURE 4-8. Transportation lag effect delays the response of an element.

measure the process variables of interest. Safety controls should be provided, however, to avoid accidents. These might take the form of on-off controls that shut down the process if safety limits are exceeded.

One of the usual first steps is to measure the open-loop response of the process to a step change in its input. Figure 4-9 illustrates a

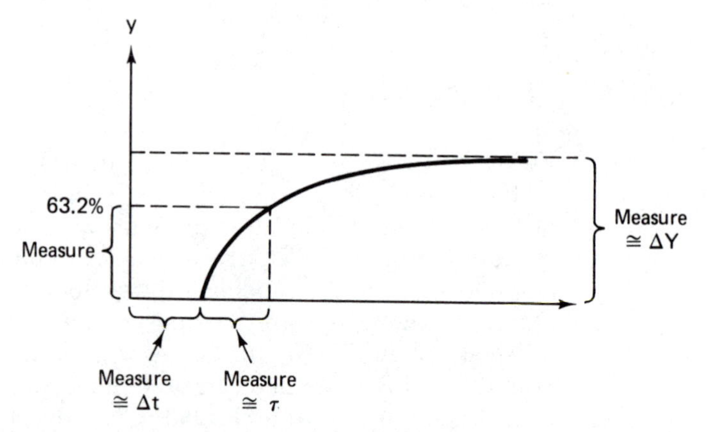

FIGURE 4-9. Measurements made on the open-loop response of a first-order system.

typical response which might be observed in such an experiment and the measurements which would be taken from the recorded output.

Transportation lag is measured by observing the elapsed time between application of the step input and the first indication of process response to the input. The process time constant is measured by observing the elapsed time between the first indication of process response and the time at which the observed variable attains 63.2 percent of its final value. Process response recording must continue long enough to ensure that the process has approached its final value sufficiently closely. Overall process gain is approximated by the ratio of the final output value to the step input amplitude.

The order of the process transfer function can sometimes be determined by observation of the shape of the response. The response shown in the figure is characteristic of a first-order system. Thus, this process might be approximated by the following transfer function:

$$G(s) = \frac{e^{-\Delta t s}}{(\tau s + 1)} \tag{4.15}$$

Determination of the transfer function of second-order systems may require curve-fitting in order to determine parameters to sufficient accuracy. In some cases, second- and higher-order systems may be approximated by several first-order models arranged in series fashion, as illustrated in Fig. 4-10, each having time constant τ/n and gain K. The transfer function of such an arrangement is given by

$$G(s) = \frac{K^n}{\left(\frac{\tau}{n}s + 1\right)^n} \tag{4.16}$$

As the number of series first-order stages is increased, the behavior of the process tends to approach that of a simple gain.

If the process is significantly nonlinear, models are sometimes designed to apply only to regions in the vicinity of the intended operating point, where it may be possible to approximate its behavior

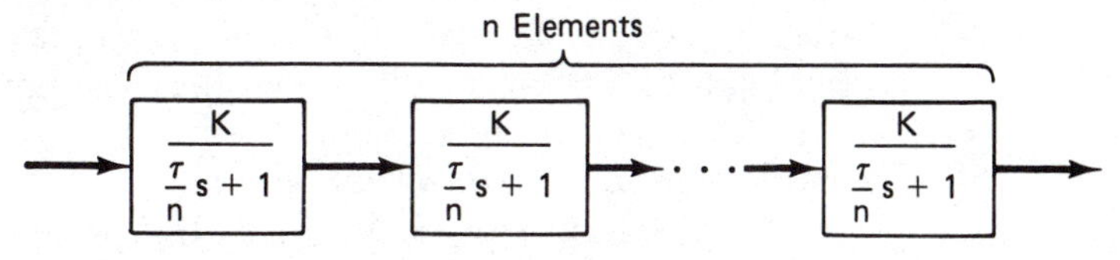

FIGURE 4-10. Series first-order elements may be used to approximate certain higher-order elements.

with a linear model. Although this may be necessary, it complicates the control design task, since the process operating characteristics assumed in the controller design must be adjusted whenever the operating region is changed. The control of processes having these characteristics is often possible with supervisory control systems in which the supervisory computer recognizes departure from the nominal operating region and adjusts controller tuning parameters so that it uses values which conform to the process behavior in the new region.

Modeling the Complete System

The formulation of a complete system model requires the incorporation of load, actuation, and measurement elements and the controller itself. In order to distinguish different types of system elements and their transfer functions, the following symbols will be used:

A = actuation elements
C = control elements
G = "general" elements (not specifically of any one type)
L = load elements
M = measurement elements
P = process elements

Where more than one element of a type is involved, subscripts are used to distinguish them.

Figure 4-11 shows the relationship between process, load, actuation, and measurement elements. The loading variable is shown summing directly into the process "output"—the *controlled variable*. In some systems, this summing point may appear prior to the process, between the actuator and the process, rather than after the process. Such a model would be appropriate if the dynamics of load and process were similar.

The measurement element is shown as having a transfer function since few measurement elements provide direct measured variable readings. It is often the case that measurements are subject to first-order lag effects. Temperature measurements are examples of such an effect. Some actuation elements also display such effects. The response time required to open a control valve to the desired setting would be an example of such an effect.

Figure 4-12 now incorporates the process controller and its *set point* into the block diagram, to form a *feedback control* arrangement. The set point is the desired measured variable value at which the

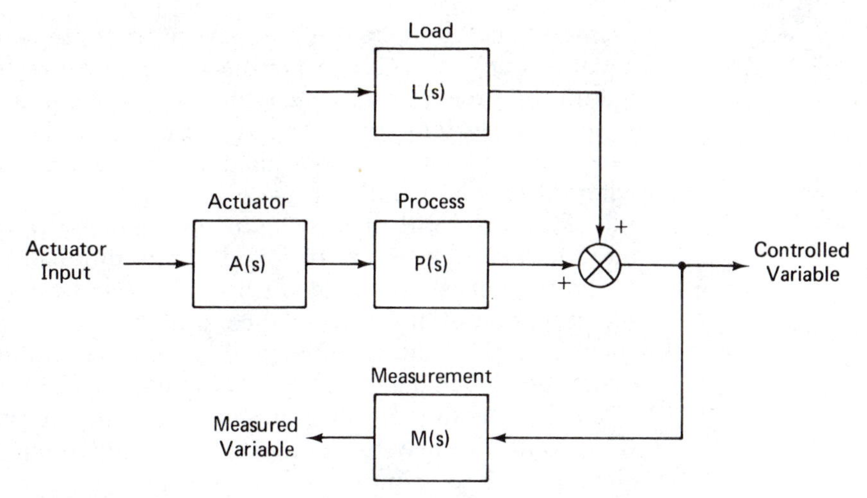

FIGURE 4-11. Relationship between process, load, actuation, and measurement elements in modeling a system.

controller is to maintain the controlled variable. The controller is designed to hold the difference between the measured variable and the set point to as small a value as it can by manipulating the actuation variable.

Figure 4-12 illustrates the simplest form of a single *process control loop* block diagram which also includes straightforward load, measurement, and actuation effects. In real, practical process control systems, there are likely to be many such effects present. Many meas-

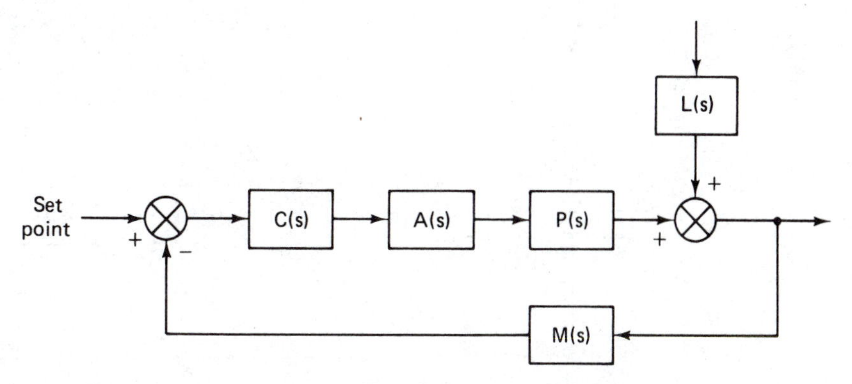

FIGURE 4-12. A single process control loop block diagram with set-point input, measurement feedback, and controller added to the elements of Figure 4-11.

urements may be taken, there may be multiple actuation elements, and the effects of loads may be complex. These effects may be coupled within the process itself in complex ways. When it is possible to do so, control designs are arranged to provide independent control of each process loop, with coordination at a higher, "supervisory" level where it may be handled more easily.

Figure 4-13 illustrates the incorporation of *feedforward control* into the control scheme. In this diagram, the load is also measured. An additional control element uses this measurement to adjust the actuation output of the original controller, in order to compensate for changes in the load. In essence, this added controller anticipates the effects of load changes and "feeds them forward" to the output of the original controller. In some arrangements, the output of the feedforward controller may be applied to the effective output of an actuator element (e.g., by controlling additional valves that add to the output of the valve controlled by the feedback controller). In such an arrangement, the summing point would appear on the right side of the actuation element in the diagram and would have its own actuation transfer function.

The diagrams shown so far are of the form usually employed to illustrate the elements of a process control system. Recall now the way in which process control systems were diagrammed in Chapter 1. The

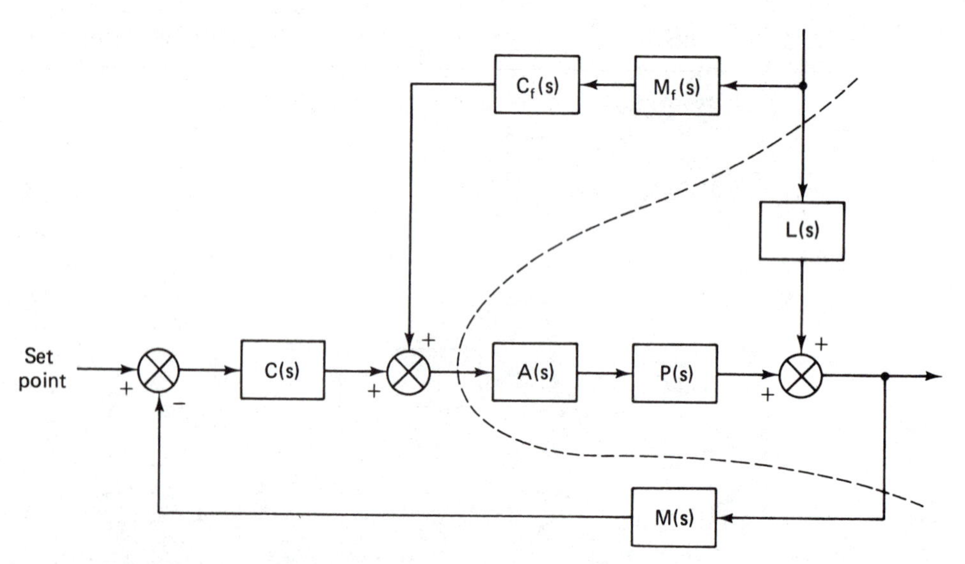

FIGURE 4-13. Process control loop with feedforward elements added. Dashed lines are the boundary between the process and the controller in the Chapter 1 interpretation.

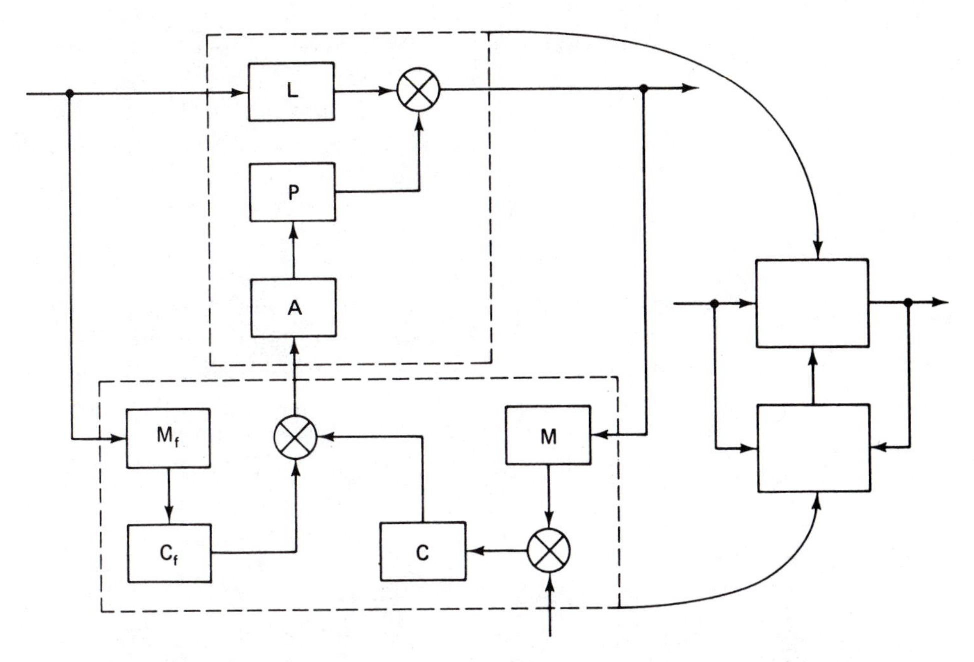

FIGURE 4-14. Figure 4-13 redrawn to show the interpretation of process input and output measurements and instructions to the process.

dashed lines in Fig. 4-13 illustrate the boundary between the process and the controller as interpreted in Chapter 1. The complete block diagram is redrawn in Fig. 4-14 to show this correspondence clearly. The measurement of load by the feedforward controller corresponds to the measurement of the process input by the controller. The measurement of the controlled variable corresponds to the measurement of the process output by the controller. Control is exerted over the behavior of the process through the actuation variable, which corresponds to the controller's "instructions" to the process.

In addition to feedforward modification to the controller structure, a form of control known as *cascade, nested,* or *master-slave* control is sometimes used. This is shown in Fig. 4-15 in two topological arrangements. In a cascade controller, an added controller is used to provide the set point for the original feedback controller. The original controller is called the "slave controller" and the added controller is called the "master controller."

Usually, the master controller measures the output of a second process element (P_M in the diagram) which is affected by the process directed by the slave controller (P_S in the diagram). Its control action is applied by using it to control the set point of the slave controller.

Such an arrangement could be successively extended from left to right in Fig. 4-15(a) in "cascade" fashion (hence the name of this form of control). Figure 4-15(b) is a topological rearrangement of Fig. 4-15(a), which shows that the slave loop can also be thought of as being "nested" within the master control loop.

Modes of Control

We come now to the subject of the internal structure of the controller block in the diagrams—how it is organized inside.

Most continuous process controllers belong to one of two structural categories, of which the first is *on-off control.* Controllers belonging to the second category are based on variations of a single basic control equation and are known under several names, among them *PID control* or *three-mode control.*

On-off Control

On-off controllers are, in fact, nonlinear and not subject to rigorous mathematical analysis using the techniques presented earlier in this chapter. However, in their simplest forms, they are easily understood.

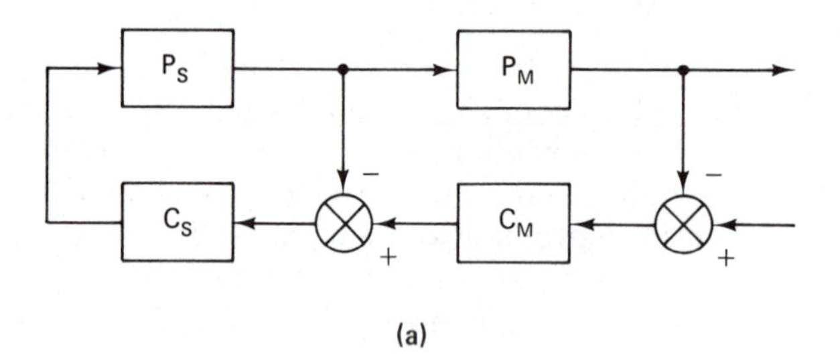

(a)

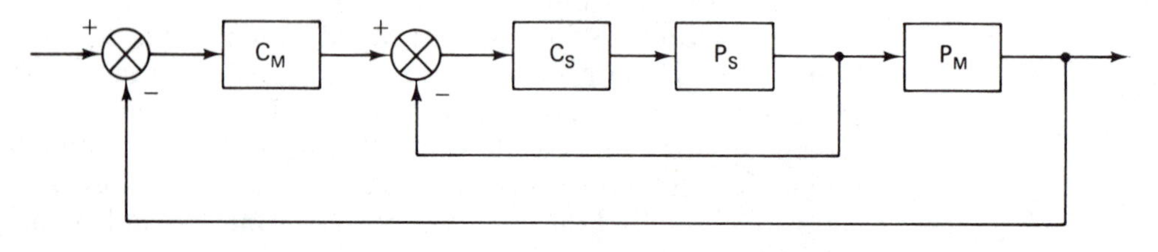

(b)

FIGURE 4-15. Cascade control: (a) drawn to emphasize the cascade arrangement, (b) drawn to emphasize the nesting of the slave loop within the master loop.

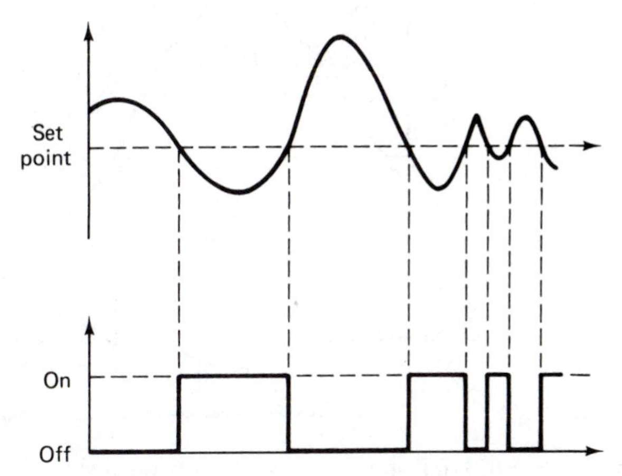

FIGURE 4-16. Measured and actuated variable behavior in a typical basic on-off controller. Note the tendency of the measured variable to oscillate and the frequent controller switching.

A basic on-off controller has two states: "on" and "off." When the measured variable is below the set point, say, the controller turns the actuating element on; when it is above the set point, the controller turns it off. Basic on-off controllers are typified by simple thermostatically controlled furnaces such as those found in home heating systems. Figure 4-16 illustrates the behavior of the measured and actuated variables in a typical basic on-off controller.

Notice that, in general, the controller is unable to maintain the measured variable at the set point. Perfect control can be attained only when load effects exactly balance the effects of the actuated variable. Usually, the measured (and, therefore, the controlled) variable tends to oscillate about the set point and the controller switches on and off frequently.

Frequent on-off control action can be wearing on control mechanisms. Also, when the measured variable is near the set point, the controller may switch several times in rapid succession (due, for example, to noise variations in the measured variable). The imposition of a *differential gap* in the control action can attenuate this effect. The use of a differential gap is illustrated in Fig. 4-17 and acts as follows: When the measured variable is outside and above a gap region centered around the set point, the controller switches off. When the measured variable falls outside and below the gap region, the controller switches on. This behavior is similar to that of a device with hysteresis.

Another on-off controller variation is the *three-position controller,* which acts as shown in Fig. 4-18. Three-position controllers are equipped with an additional state which produces a control action

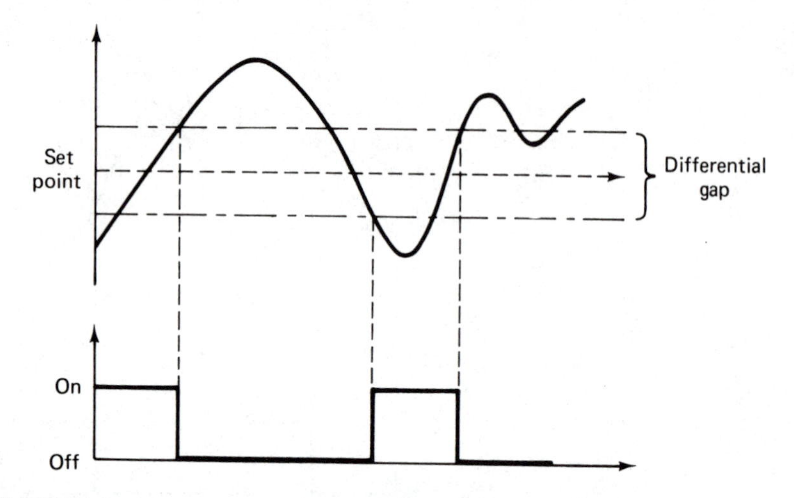

FIGURE 4-17. Use of a differential gap in an on-off controller. The controller switches only at the boundaries of the region defined by the gap.

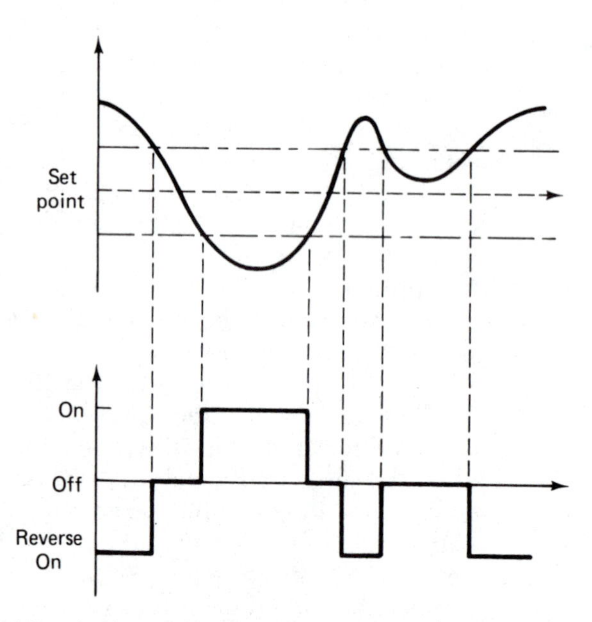

FIGURE 4-18. A three-position controller adds a "reverse" control state to drive the variable into the desired region from above or below.

counter to the on state. As shown in the figure, the original on state is now supplemented with what is often termed a "reverse on" state. When the measured variable is within the gap region, the control is off. When the variable is above the gap region, the control is "reverse on" (e.g., cooling). When the variable is below the gap region, the control is "on" (e.g., heating). Sometimes, the controlled variable is ramped up or down to a limited state (as shown in Fig. 4-19). This type of control is called *floating control* and, although not purely of the "on-off" variety, is frequently considered within this class.

Although seemingly simple, on-off control can become quite complex in more advanced forms. It is sometimes called *bang-bang control* and is equipped with a rich apparatus of mathematical analysis techniques extended to consideration of such items as the time duration of the control action. Examples of this form of advanced control are the steering and attitude control systems for space vehicles, in which the principal control mechanism is the amount of time the thrusters are fired.

PID Control

PID stands for "Proportional-Integral-Derivative." These correspond to the three principal terms in the equation governing the behavior of a PID controller. This form of control is often called *three-mode control.*

To simplify the discussion of PID controllers, we use the revised form of the complete single-loop feedback control system block dia-

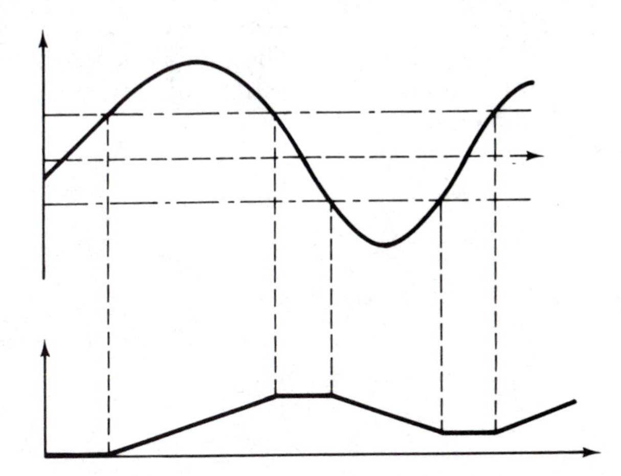

FIGURE 4-19. In floating control, the controlled variable is ramped gradually to drive the measured variable into the desired region. Note that this is not a purely on-off control behavior.

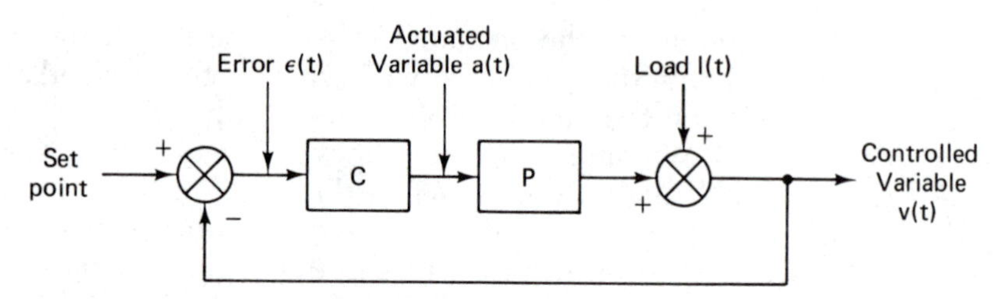

FIGURE 4-20. Simplified single-loop feedback control system block diagram defining symbols for PID control.

gram shown in Fig. 4-20. This figure assumes direct measurement of the controlled variable and direct actuation of the process by the controller; assumptions that are not realistic in practice, but are useful in elementary treatments of the way in which these controllers operate. The figure also shows the symbols used for the variables involved.

The general equation governing open-loop behavior of a PID controller has the following form:

$$a(t) = G_1\epsilon(t) + G_2 \int_0^t \epsilon(t)\, dt + G_3 \frac{d\epsilon(t)}{dt} \qquad (4.17)$$

The first term of this equation produces a control action which is simply proportional ("P") to the difference between the set point and the measured variable, the *error* or *deviation*. The second term produces a control action proportional to the integral ("I") of the error. The third term produces a control action proportional to the derivative ("D") of the error. The block diagram of a complete PID controller, with the transfer functions of each component part, is then drawn as shown in Fig. 4-21.

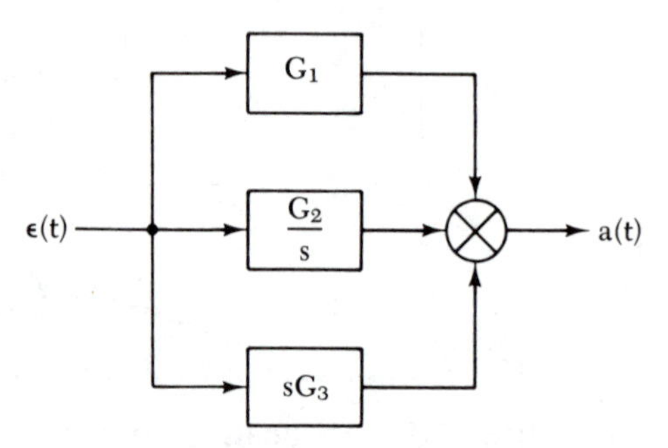

FIGURE 4-21. Block diagram showing the transfer functions and relationships of the three control terms of a PID controller.

Depending on the nature of the process and the load and measurement effects expected, a practical controller may be composed of combinations of these terms. Controllers that include only the first term are called *proportional controllers.* Controllers that include only the second term are called *integral controllers.* Controllers that include only the first two terms are called *proportional-integral controllers* or *PI controllers.* Controllers that include all three terms are called *proportional-integral-derivative (PID) controllers.*

Proportional Control. Proportional controllers are characterized by their *proportional band,* a term expressed in percentage and related to the reciprocal of the constant gain G_1:

$$\text{Proportional band (PB)} = \frac{100}{G_1}\ \% \qquad (4.18)$$

The transfer function between the load and the controlled variable of the system in Fig. 4-20 has the form

$$\frac{\Delta V}{\Delta L} = \frac{P}{1 + PG_1} \qquad (4.19)$$

where P is the transfer function of the controlled process. Since this transfer function is almost everywhere nonzero, proportional control has the effect of failing to converge exactly to the set point except at certain set point values that are related to the process action. The difference between the set point and the value to which the controlled variable does converge is called the *proportional offset.* Proportional control provides fast-acting control but is sensitive to high frequency components in disturbances (changes in load or set point).

Integral and PI Control. The second term in the PID equation can enhance the degree of control by correcting proportional offset effects and attenuating response to high frequency components in disturbances. Because it acts to "reset" the proportional offset, integral control is sometimes called *reset action,* and the term G_2 is termed the *reset rate.*

Reset action alone is not fast-acting, but it does attenuate high frequencies (acting as a low-pass filter). It is often used in conjunction with proportional control to form a PI controller, in an arrangement in which the I term is "tuned" to balance the proportional offset of the P term. PI controllers are, perhaps, the most common form of continuous process control.

Derivative and PID Control. The third term in the PID equation provides control action in response to *changes* in the error. That is, it responds to the time derivative of the error. The purpose of the third term is to obtain a quick response to load or set point changes, in order to shorten the time required to close on the intended value of the controlled variable. Because it is proportional to the rate of change of the error, this part of the control scheme is sometimes called *rate action*. It is obviously even more sensitive to high-frequency components of the error signal (especially noise) than the proportional term. For this reason, rate action is used only in conjunction with proportional or integral control or both (in the full-blown PID controller).

Comparison of Control Action and Tuning. Figure 4-22 illustrates the typical behavior of the actuation (or manipulated) variable for the three most common combinations of these three control modes. Curve *A* is the typical response of a simple proportional controller. Curve *B* is the typical response obtained when integral control is added. Curve *C* is the typical response of a complete PID control. PD control is not shown, because it is not so commonly used as the other combinations. The oscillations in well-tuned PI and PID controllers would not be as great as those shown in the figure. They have been exaggerated for emphasis.

The stability of the complete system (including control, process, load, measurement, and set point change effects) is an important issue in the design and tuning of process controllers. In practical cases, the values of terms which determine controller behavior must be carefully chosen, after substantial measurement and experimentation, to ensure that the complete system will be stable and, at the same time, provide a high degree of control quality. A discussion of the theory of control

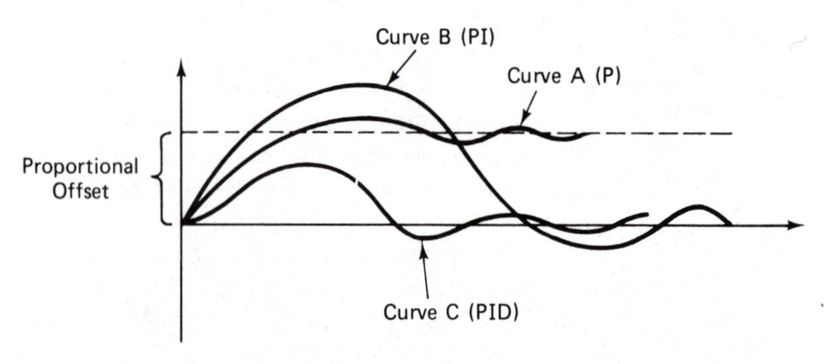

FIGURE 4-22. Actuation variable behavior for a simple proportional controller (P), when integral control is added (PI), and when derivative control is added (PID).

system stability and tuning is beyond our scope here. However, useful discussions of various methods of analytically predicting system stability and control quality will be found in several of the references listed in the bibliography for Part I.

DISCRETE CONTROL

Discrete control is exemplified by the control of processes, machinery, and equipment which are essentially on-off in their behavior. The design and analysis of discrete control systems is based on Boolean logic, and the principal issues are the complexity of the logic and the reliability of the control components used in these systems.

Discrete control systems are fabricated from relays, contactors, and switches of many types (and from their solid-state electronic, pneumatic, and fluidic analogs). These are interconnected with wiring which largely determines their logical behavior. More modern discrete control systems use computer- or electronic logic-based programmable systems called *programmable logic controllers* (PLCs) or *programmable controllers* (PCs), the "logic" having been dropped by some manufacturers in recent years as their products came to be equipped with features which permitted them to perform some elementary arithmetic, data handling, and continuous process control functions.

Discrete controllers are obviously related to on-off controllers, but distinct in that the "process" they typically control is also an "on-off" process. They deal with equipment having "machine states" which are *binary*. That is, they are representable as "on-off," "two-valued," or "discrete." In some extensions of discrete control, machine states which are describable with numbers are included in the theory. Such extensions include the ability to maintain simple counts of machine events and to compare such counts to preset limits. At the furthest extensions, discrete control becomes like numerical control (the subject of the next section).

Discrete controllers respond to changes in certain combinations of discrete machine states, either immediately (i.e., as promptly as they can within the limits of their reaction time) or after the specified delays which are intended parts of the control scheme. Certain types of discrete controllers may act to control sequences of events in the machine. These are sometimes called *sequence controllers.*

Discrete System Components

The fundamental sensing and control component of a discrete system is the *contact*. This may be a part of the controller or the machine it controls. The contact may be a limit switch, pushbutton, toggle switch, or any other electromechanical, electro-optical, pneumatic, or fluidic device having two states: "closed"—meaning that it conducts current (or its analog), or "open"—meaning that it does not.

Contacts may be opened or closed by control action, by mechanical action (as with a limit switch or cam-operated switch), by human action (as with a pushbutton), by the presence of an object (as with an optical or proximity sensor), and so on. A contact may be considered to be "normally open" or "normally closed," where the meaning of "normally" is taken to be with respect to a logically convenient state of the machine or the actuating or sensing element involved.

The fundamental actuating element is the *relay*, generally, and (where it is necessary to switch large voltages or currents) the *contactor*, which is logically equivalent to the relay. Discrete controller logic acts by energizing the coils of its control relays and contactors. Figure 4-23 depicts a typical control relay, which is commonly equipped with two contacts. One contact is open when the coil is not energized and is called the *normally open* or *NO* contact. The other is closed when the coil is not energized and is called the *normally closed* or *NC* contact.

Figure 4-24 depicts the symbols used to represent some of these components in the logic of a discrete controller and the machine that it controls.

Discrete Control Logic

Controller and machine logic may be represented in several ways. Many may be adequately represented for the purposes of logical anal-

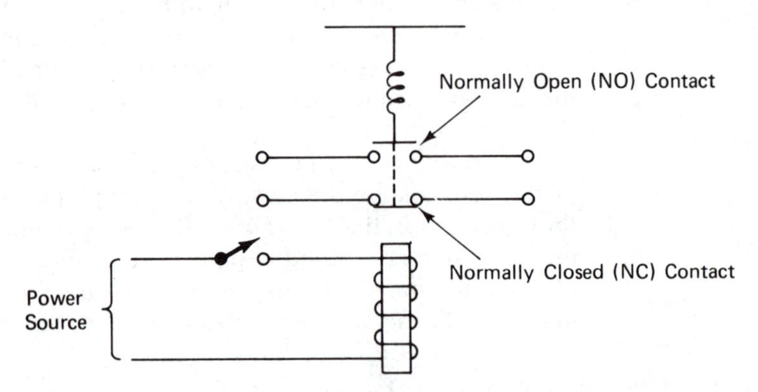

FIGURE 4-23. Organization of a typical control relay equipped with normally open and normally closed contacts.

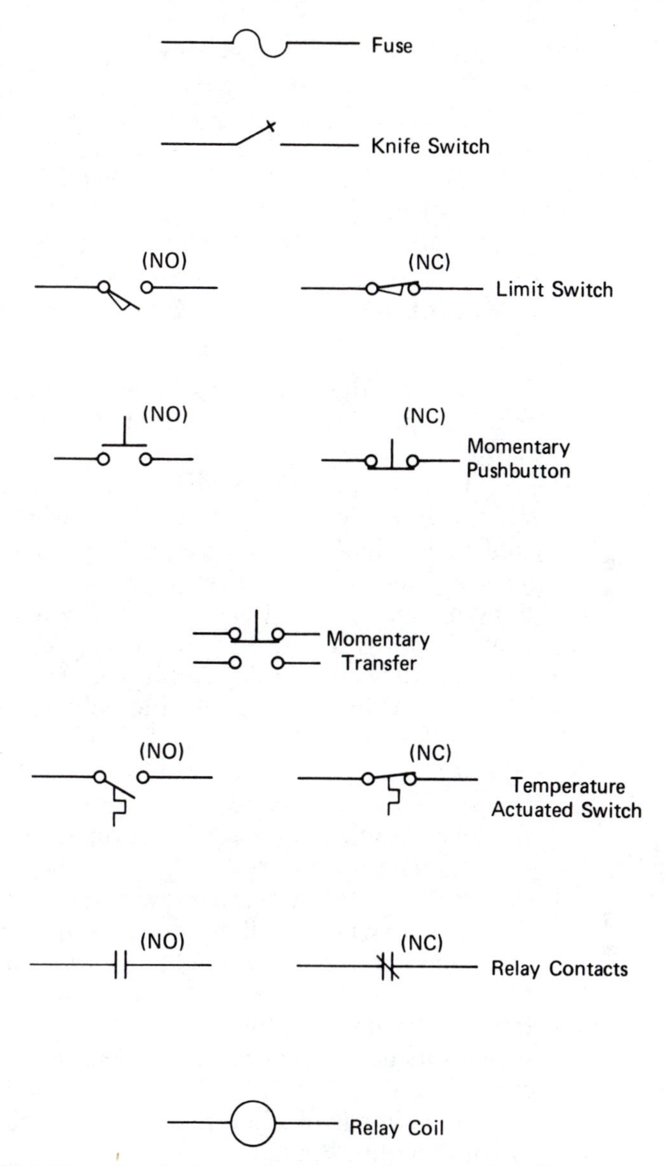

FIGURE 4-24. Symbols used to represent common discrete control and machine logic components.

ysis, if not too complex, by *wiring diagrams* used in their fabrication. These diagrams are sometimes called *elementary diagrams*. Usually, the initial analysis and design of a discrete controller employs either ladder diagrams or Boolean equations (explained below), but we begin by discussing an example using a simple wiring diagram. Figure 4-

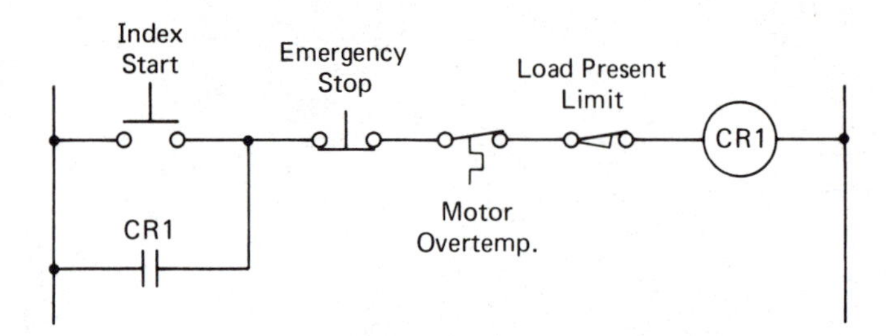

FIGURE 4-25. Wiring diagram for conveyor controller example.

25 is a wiring diagram for a simple control circuit used to index a load from one position on a conveyor belt to the next position. It operates in the following way.

Power is provided to the circuit by applying a voltage across the vertical lines, which represent the power distribution portion of the controller. The conveyor is driven by a motor, which is turned on by a control relay. The coil of this relay is shown as a circle symbol in the diagram, in which the name of the relay (CR1) is entered. If the contacts across the horizontal lines are closed in the proper way, the circuit to the control relay coil is completed and the coil is energized. This relay drives a contactor (not shown in the diagram) which energizes the motor windings and starts the conveyor belt. The relay is equipped with a normally open contact. This is shown as a relay contact symbol, also with the label CR1 to indicate its association with the relay which actuates it. This contact is wired in parallel with the momentary Start pushbutton.

Between the start pushbutton and control relay contact there are three other switches, all of which are normally closed. The first is an emergency stop pushbutton, the second is an overtemperature switch (which opens if the motor temperature becomes too high), and the third is a limit switch (which is opened when the load on the conveyor reaches its next position). In the "normal" condition, there is no load at the receiving end of the conveyor (so that the limit switch is closed), and the motor is off. Let us also assume that the temperature of the motor is within limits.

The operator, wishing to index the next load to the receiving end, depresses the start pushbutton. This completes the circuit to the coils of CR1, which starts the motor and closes control contact CR1. It is no longer necessary for the operator to hold the pushbutton closed, since the CR1 contact now carries current around it. The motor, driving the conveyor belt, moves the next load. When this load arrives at the receiving station, it contacts the limit switch and opens it. This interrupts current to the relay coil, causing the motor to stop and the CR1 contact to open.

So long as the load remains on the belt and in contact with the limit switch, the circuit will not operate again. This prevents reactuation of the circuit if the previous load has not been removed. When the operator at the receiving end of the belt removes the load, the limit switch closes, and the controller is ready to initiate the next index cycle.

If a load falls from the belt or is skewed, or if some other reason arises, the operator may command an emergency stop by depressing the stop pushbutton. In a more sophisticated version of this controller, other features may be added. For example, optical sensors may be used to tell whether the load is skewed. The controller may also be designed to wait until a load is present at the input end of the conveyor belt.

This simple example was easily analyzed through an examination of its wiring diagram. Complex discrete controllers may be difficult to design and analyze when each switch and control element is depicted explicitly with its circuit and electromechanical symbol, as required in a diagram used as an aid to electricians for circuit wiring. For this reason, simplified diagrams (called *ladder diagrams*) are often used in the design of more complex controllers. Ladder diagrams get their name from their similarity to ladders. In fact, this is taken advantage of and extended in that the horizontally arranged portions of the diagram are commonly called "rungs."

Figure 4-26 shows the ladder diagram equivalent of the wiring diagram for the example conveyor controller. Circuit control elements are shown as simple normally open and normally closed contacts and relay coils, without details regarding their specific types, since these do not affect the way they operate logically. Combinations of letters and numbers are used to form convenient mnemonic symbols to identify the components.

Boolean equations are also used to describe the logical behavior of discrete controllers. The Boolean equation for the example con-

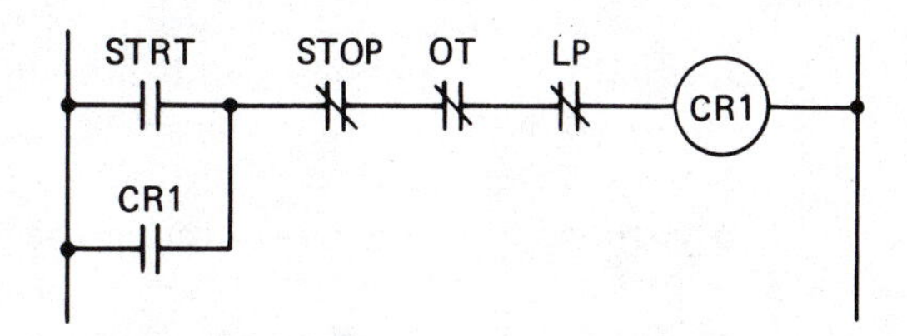

FIGURE 4-26. Ladder diagram for conveyor controller example.

veyor controller is the following:

$$\text{IDX} = (\overline{\text{STOP}}\cdot\overline{\text{OT}}\cdot\overline{\text{LP}})\cdot(\text{STRT} + \text{CR1}) \qquad (4.20)$$

where IDX stands for "index," STRT for "start," STOP for "stop," OT for "overtemperature," and LP for "load present" at the receiving end of the conveyor. Logical AND of two Boolean variables is indicated by the "·" symbol and logical OR by "+". Logical "negation" or "complement" of a variable is indicated by overscoring. If the logical condition of a variable is "true," it is given binary value "1"; if "false," it is given binary value "0".

Thus, when the controller is inactive (not indexing), the equation evaluates as

$$(1\cdot1\cdot1)\cdot(0 + 0) = 0 \text{ or "false" (i.e., off)}$$

When the start pushbutton is depressed, we have

$$(1\cdot1\cdot1)\cdot(1 + 0) = 1 \text{ or "true" (i.e., on)}$$

As the motor starts, CR1 closes, and the pushbutton is released; we have

$$(1\cdot1\cdot1)\cdot(0 + 1) = 1 \text{ continuing "true" (i.e., on)}$$

When the load reaches the receiving end and the limit switch is opened, we have

$$(1\cdot1\cdot0)\cdot(0 + 1) = 0 \text{ or "false" (the motor stops)}$$

Programmable Logic Controllers

Programmable logic controllers (PLCs) are generally designed to accept either Boolean equations or ladder diagrams as programming instructions. In the typical case, a PLC is actually a small computer system which simulates the behavior of a relay-based controller. Some PLCs are programmed by typing Boolean equations on a terminal equipped with a keyboard and cathode ray tube display (described in Chapter 10). Others are programmed by entering ladder diagram symbols on a terminal designed to display them in graphic form. Generally, these are entered in much the same form as depicted in Fig. 4-26.

Because PLCs are made from solid-state electronic components which are not subject to mechanical wear, they have much higher expectations of sustained reliability than discrete control systems made from electromechanical components. PLCs are also smaller and often less costly to use than relay-based discrete control systems. One of the principal reasons for this is that their "wiring" (and therefore the logic by which they operate) is represented in their programs. This makes them easier to install and to change if modifications to the operation of the controlled machine are required.

It seems to be a general fact of life that few controllers are installed and then left completely alone to do their work. The usual rule is that changes will be made from time to time. Thus, flexibility is an important design objective. This objective is more easily met with a programmable device than one which must be rewired in order to alter its logic.

Because most PLCs are based on small computers, which execute their instructions one at a time, the "rungs" of the ladder do not operate in strict simultaneity, as is the situation with a wired, relay-based controller. The computer examines and performs the logical function implied by each rung, one rung at a time. This process can sometimes cause confusion and errors among control system designers and technicians who are more familiar with wired logic.

This process can also lead to difficulties with the speed with which the controller operates. The more rungs there are in the ladder diagram program, the longer it takes the PLC to execute them all and the longer it takes it to get back to any given rung of the ladder. Thus, if the logic represented by that rung is intended to act on an event of short duration, there is a chance that the event will be over by the time the rung is executed once more and that it will be missed.

Programmable logic controllers use optical coupling on their input circuits (to isolate their more delicate electronic components from potentially damaging voltages and currents), and they use power transistors and triacs as output devices. The electrical behavior of these solid-state devices is not identical to that of a control relay contact. For example, control relay contacts literally open up (disconnect) when they are switched off. Solid-state output devices "switch" to a very high (but far from "infinite") impedance when they are switched off. Solid-state circuits are not as rugged as relay circuits and are, therefore, more easily damaged by accidental miswiring and short circuits. As individual components, they are sometimes more expensive than relays.

NUMERICAL CONTROL

Numerical control (abbreviated "NC," but not to be confused with "normally closed") is exemplified by the machining of complex parts.

In most cases, NC systems are used to automate machine tools which turn, grind, mill, cut, and drill parts. These parts may be components of engines and other machines, turbine vanes, aircraft wing components, propeller blades, cams, gears, or virtually any mechanical part imaginable. NC-controlled systems are also used to cut cloth for yacht sails, to weld plates for the hulls of ships, to automatically insert components into printed circuit boards for electronic systems, and to wire these systems.

Numerical control can be thought of as a mixture of continuous and discrete control to which substantial measures of plane, solid, and differential geometry have been added. NC controllers use a series of discrete numerical data points, interspersed with codes which command particular functions, to direct the operation of the machine they control. The numerical data describe the shape of the part to be made. The codes describe particular machine actions (e.g., rates of cutting, tool changes, raising and lowering of the cutting tool).

Starting with the early NC controllers and continuing even today, these codes and numbers were punched into data processing cards or paper or mylar tape. These media were used because they were available as output from the computers used to prepare data for all but the simplest parts. Tape is used often because it is inexpensive, easy to store, and can be corrected or changed, when necessary, by using simple equipment.

Figure 4-27 is a stylistic sketch of the main operating components of a typical NC-controlled machine and NC controller. The tool drive motor, with tool (a milling bit, for example), is attached to a leadscrew assembly. As the leadscrew turns, the bit moves along the Y-axis as shown in the figure. The *workpiece* (the part being milled) is clamped to a table whose position is moved by a second leadscrew, but along the X-axis. By coordination of the number of turns and rates of turning of these leadscrews, the tool may be moved along virtually any path over the workpiece.

The data on the NC *program tape* specify this path. Usually, the path is divided into a series of straight line segments of incremental motion for each axis. If a closely controlled curve is to be traced by the tool, each segment would be quite small. This method is illustrated in Fig. 4-28. The path of the tool must be offset from the desired finished curve on the workpiece in order to account for the diameter of the tool.

The rate at which the tool moves along the workpiece (the *feed rate*) must not exceed the ability of the tool to cut the material. It may also be necessary to withdraw the tool from the surface of the workpiece (to skip over areas which are not to be milled) or to stop and raise the tool head so that the bit can be changed (as when switching

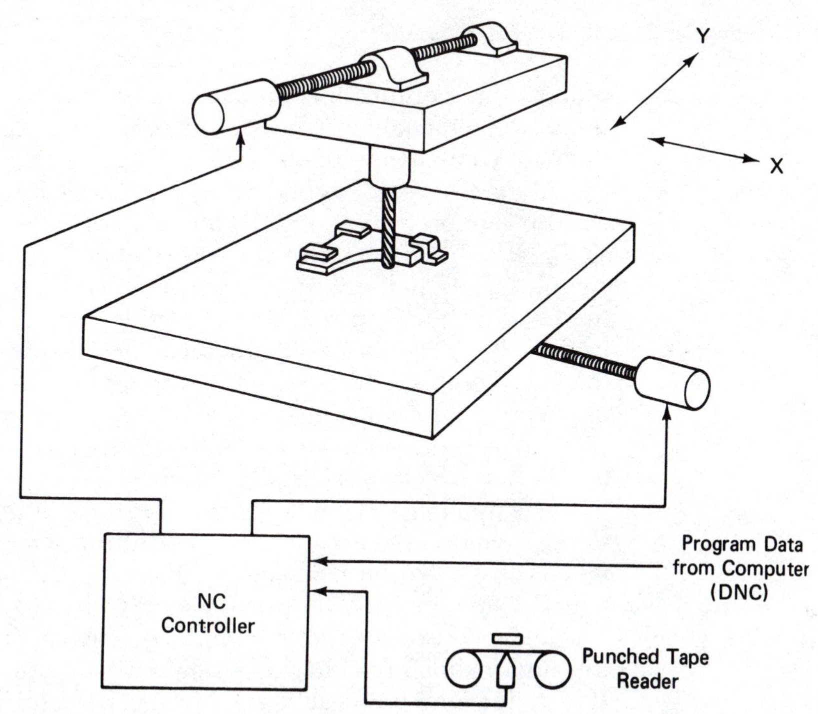

FIGURE 4-27. Main operating components of a typical NC machine and controller.

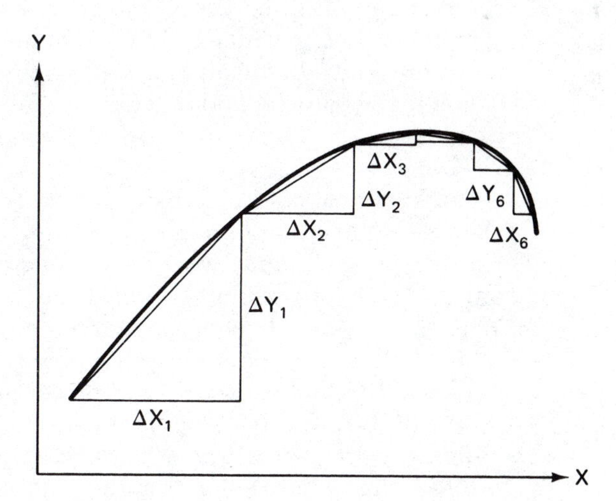

FIGURE 4-28. Division of a curved path into a series of straight line segments.

from a rough milling operation to a smoother one). Functions such as these are commanded by separate codes that are interspersed with the path data at appropriate points along the tape.

Before NC systems became available, the leadscrews that carried out this motion were turned by a human operator using hand cranks, as directed by drawings of the part. It required great skill to make a complex part, and any mistakes usually cost a great deal in wasted time and material. In some cases, finely controlled machining could be done by using cams to control the operation of motors which turned the leadscrews. A "library" of cams could then be used to make complex parts more efficiently, and with fewer errors, but this was still slow and expensive in cases where the required cam was not in the library and had to be specially made.

With numerical control, the NC controller performs these functions. It controls tool motion by controlling servomotors or stepping motors that turn the lead screws. It uses feedback information (obtained from rotary motion transducers attached to the leadscrews) and the pitch of the leadscrew to infer the position of the tool, or its obtains this information from linear position tranducers attached to the tables. It turns the leadscrews at rates which keep the feed rate within proper limits.

The controller gets the information it needs to perform these functions from the NC program tape. In many cases, this tape is read with a photo-optical tape reader. In more modern systems, the NC data may be transmitted directly from a central computer (which is also used to generate and store the data) to the NC controllers over a serial communication link. Arrangements such as these are called *direct numerical control* (or DNC) systems. In still more advanced systems, the controller itself may be computer-based and have local provision for storing the program data (and even for generating it). In this case, the system would be called a *computer numerical control* (or CNC) system.

Numerical control work is generally divided into two broad categories: *point-to-point* and *contouring.*

In point-to-point work, tool operations are usually performed only at separated points. A typical sequence would be as follows: (1) With the tool raised from the workpiece, move to coordinates x_1, y_1. (2) Perform a tool operation at the new coordinates (e.g., drill a hole). (3) Raise the tool from the workpiece. (4) Move to new coordinates x_2, y_2. (5) Lower the tool and repeat from step (2). Examples of point-to-point work are drilling, automatic insertion of components in printed circuit boards, and automatic wiring of electronic systems. "Straight cut" operations are a variation on point-to-point work in which the tool is left in contact with the workpiece as it moves along a straight line.

In contouring work, curves are to be cut in the workpiece. The tool is brought into contact with the workpiece, then moved along the curve prescribed by the NC program. The variety of things that must be considered in contouring work grows geometrically, especially when the manufactured part is essentially three-dimensional.

Usually, computers are used in the preparation of NC programs. Although simple point-to-point programs can be figured out manually, computers are of great help in preparing more complex point-to-point programs, because the computations needed to direct incremental tool motion can be tedious and error-prone. For contouring, computers are a virtual necessity.

Many standard computer programs have been developed to perform the work of translating machined part designs into NC program tape data. Of these, perhaps the most prominent is a software system called *APT* (which stands for Automatically Programmed Tools).

The NC programmer, using APT and working from drawings of the part, prepares instructions which describe the operations to be performed in tractable "English-like" terms. For example, a circle might be defined with a formula-like statement such as

C3 = CIRCLE/5,10,1

representing a circle located at coordinates $x = 5$ in. and $y = 10$ in. with a radius of one inch. In a similar way, points and lines may be defined. The programmer then specifies the path of the tool with respect to these geometric entities with words like GOLEFT (move the tool to the left), GOFWD (move the tool forward), and so on.

The APT software (which is technically a "compiler," a term we discuss in Chapter 13) translates these definitions and instructions into a series of numerical data points (called *CL information*, for "cutter location information") which represent the tool motion. This information is then fed to another program, called a *post-processor*, which adjusts the information to conform to the codes used by the specific type of NC controller to be used. In a DNC system, this information is then transmitted directly to the NC controller via a communication link.

Because the information is often prepared by using computers, and because these computers are also often used in preparing original part designs, the integration of these design and manufacturing activities into a coordinated and comprehensive approach would seem to be generally beneficial. This is, in fact, what has led to *CAD/CAM* (Computer-aided Design/Computer-aided Manufacturing). This term encompasses very nearly the complete scope of using computers in modern automated factories, from design through automated inspection of the final manufactured part.

CHAPTER 5

Actuation

"... suit the action to the word,
the word to the action; ..."

William Shakespeare

The representative signal resulting from a control decision, though it may be certain in its resolve, is usually physically weak. The decision that a valve should be opened, that a motor should be started, that a temperature should be increased, is not enough. The idea must be converted into the reality. The idea is usually represented by a small voltage or current within the logic and circuits of the control and decision-making portions of the system, a signal having little power to cause the desired effects: to produce a change in the reality. Thus, the signal that represents the control decision must be amplified in some way.

Usually the amplified signal must also be converted to a different form. The controller may wish to increase pressure, or temperature, or rate of flow. The signal must, therefore, undergo a transduction which has the required result, preserving the intended "meaning" of the signal.

Thus we have a situation like that depicted in Fig. 5-1, in which the actuator may be thought of as a series amplifier/transducer device. In some cases, it may be useful to have several such internal stages and even to think of the actuator as a device having internal control mechanisms.

Just as inference was important in making measurements, we find an analogous "indirection" in aspects of actuation. For ex-

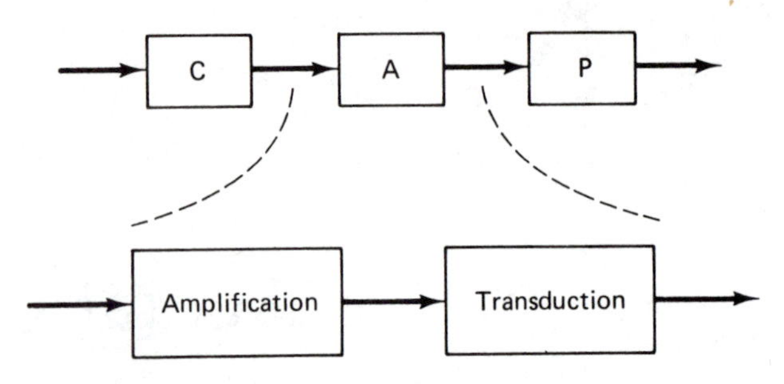

FIGURE 5-1. Actuators may be thought of as combined amplification and transduction devices.

ample, if we wish to raise the temperature of a container, we do not apply heat directly. Instead, we may increase the current flow in a heating element or permit more preheated steam to pass through coils in the walls of the container. If we are controlling a drying operation, we may vary the amount of time that the material is exposed to drying heat by controlling the speed of the motors which convey the material through the drying chambers.

As noted in Chapter 4, actuation mechanisms have transfer functions, which must be considered in the design of the control system. Typical mechanisms are heaters, valves, and motors. These convert input energy into different forms (e.g., current to heat, electrical power to motion). They may, in practical systems, be controlled in many different ways (e.g., electrically, hydraulically, pneumatically). Because electrically controlled mechanisms are most commonly associated with the computer-based systems in which we are most interested, we devote our attention exclusively to these.

The mechanisms are variously called *actuators, final correcting elements,* (or *devices*), *control devices,* and *final controlling elements.* They may be divided generally into two classes according to their action: *on-off* and *continuous.*

ON-OFF ACTUATION

In electrical, electronic, and computer-based control systems, *electromechanical relays* (EMRs) and *solid-state relays* (SSRs) are the principal initial on-off actuation devices. Depending upon the amount of electrical energy to be switched, one or more "stages" of relays may be required. EMRs are typified by the control relays discussed in the previous chapter. SSRs are approximately equivalent devices (similar in logical behavior but not quite identical in electrical characteristics) which are made from semiconductor materials.

Solid-state Relays

The representative signal from the controller is usually a low-level logic signal, with electrical characteristics determined by the type of semiconductor devices used in the logic circuits of the controller. Typical examples of this logic are Transistor-Transistor-Logic (TTL, of which there are several varieties) and Complementary Metal Oxide Semiconductor (CMOS) logic. We shall discuss the electrical characteristics of these semiconductor logic types in more detail in Part II. For now, let it suffice to say that they are characterized by essentially low power handling capabilities. TTL, for example, operates at levels of somewhat less than 5 volts and handles currents that are generally less than 20 milliamperes.

When the initial stages of an actuator are driven with TTL, it is common practice to use circuits in which the output transistor has an uncommitted (or "open") collector, as shown in Fig. 5-2. In this configuration, some TTL devices can handle up to 30 volts. This output transistor acts as a "switch" which is controlled by the logic level. As the logic level changes states, the transistor switches between its "conducting" and "nonconducting" states. When connected as shown in Fig. 5-3, it controls the completion of a circuit which applies direct current power to the load. Because transistors do not switch off completely, a small "leakage" current always flows, even when the logic level control signal is off. This is one way in which these devices differ from electromechanical relays.

If the load is inductive, as would be the case if it were the coil of a control relay, contactor, or small solenoid, protection is required to guard against the effects of the discharge of energy stored in the inductor's fields when the transistor is switched off. This "inductive kick" may exceed the breakdown voltage of the transistor and result in permanent damage. One method used in such cases is to provide a reverse-biased diode in parallel with the relay coil or solenoid, as

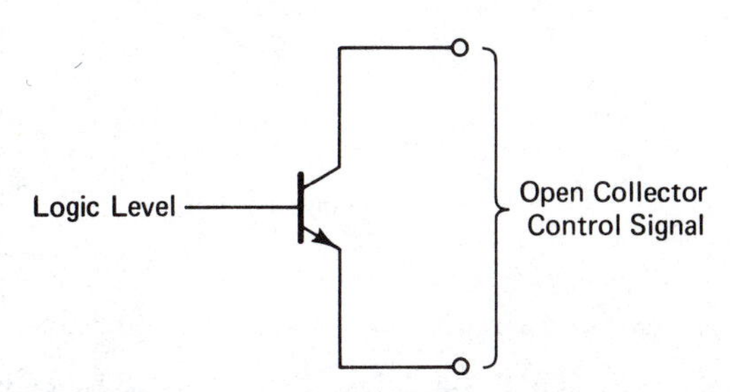

FIGURE 5-2. Open collector output transistor.

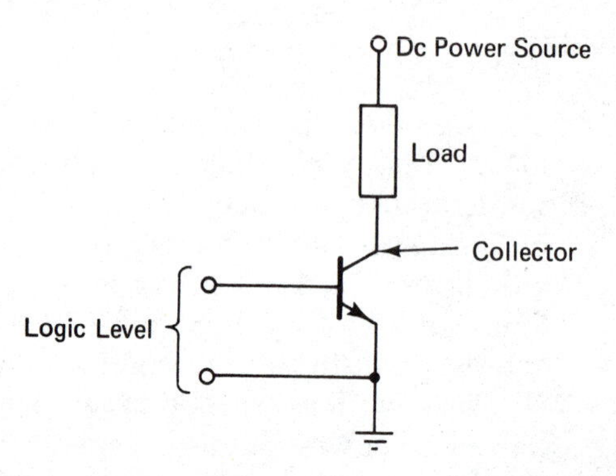

FIGURE 5-3. Output transistor acts as a solid-state switch to control completion of the load circuit.

shown in Fig. 5-4. The diode clamps the voltage at the collector and provides for a more controlled energy discharge.

There are many variations on these circuits which offer tradeoffs between desirable operating characteristics and undesirable side effects, such as improved switching speed versus noise generated by switching transients. Several of the books listed in the bibliography for Part I contain comprehensive treatments of this subject.

At this stage of actuation, the transistor has provided a certain degree of amplification through its ability to switch higher voltages and conduct greater currents than the semiconductor devices used in the control logic (which are usually selected on the basis of different criteria, such as higher switching speeds and lower power consumption).

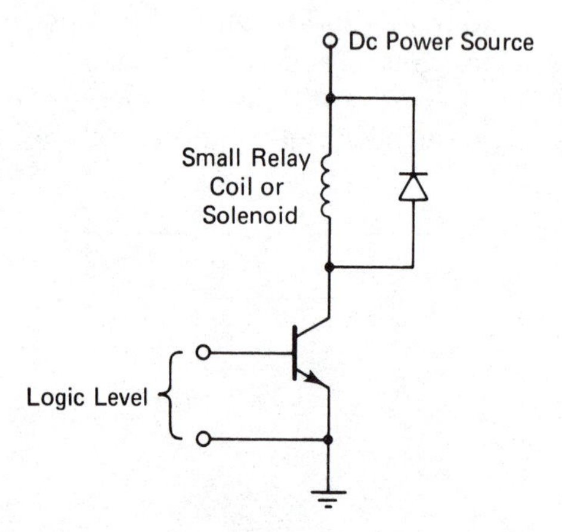

FIGURE 5-4. Use of a reverse-biased diode to clamp and control discharge of energy stored in an inductive load.

The transistor circuits of Figs. 5-2 through 5-4 are simple solid-state "relays." Various classes and combinations of such devices may be used to switch voltages up to a few hundred volts and to handle currents up to a few amperes. Such circuits are commercially available as modules for use with electronic systems and are usually equipped with optically isolated logic level inputs, as illustrated in Fig. 5-5.

The relative cost of SSRs, as opposed to EMRs, depends upon the nature of the control application. Generally, the initial cost of an SSR may be greater, but its long-term cost may be less because of its potentially longer life and greater reliability when it is properly applied.

Electromechanical Relays

When the levels to be handled exceed the capabilities of SSRs, EMRs are often used. Figure 5-6 illustrates the actuation of an EMR used to switch a high power load. Here, a small transistor circuit actuates a relay or contactor which, in turn, acts to switch the heavy load. Each "stage" provides added amplification: the transistor for control of the relay coil, the relay contacts for control of the high power load. Added stages may be used in order ultimately to reach virtually any power

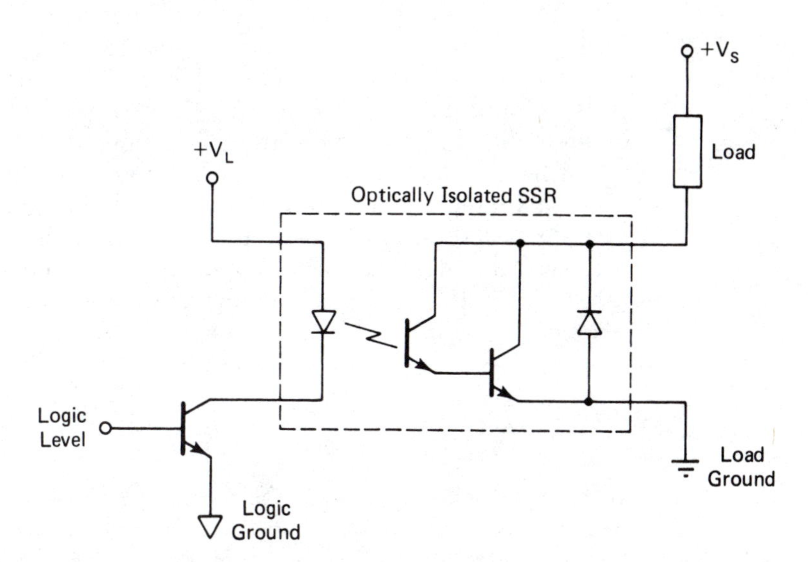

FIGURE 5-5. Organization of a typical optically isolated solid-state relay module. The phototransistor drives the base of the output transistor in order to provide better performance. The output diode protects the device from reversed load polarities.

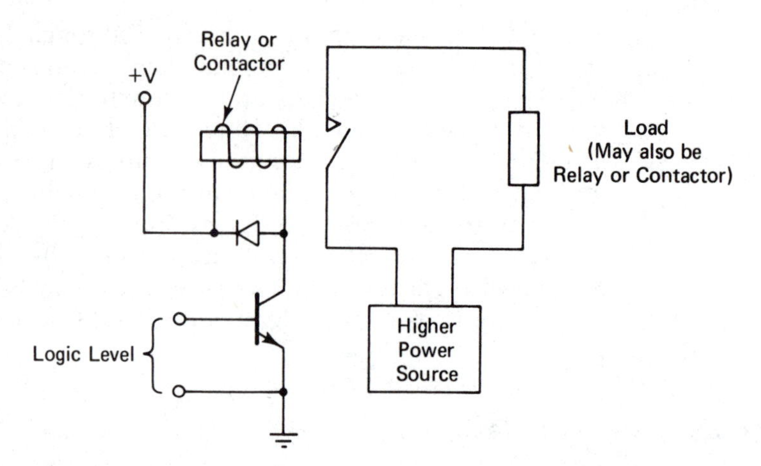

FIGURE 5-6. Use of a small transistor to actuate an electromechanical relay that switches a heavy load circuit.

handling capability (e.g., thousands of amperes for some metal processing applications).

Devices such as these are satisfactory for on-off actuation of process control direct current and voltage loads up to their practical power handling limits. They may be used to increase process temperature through resistive heating elements in cases where simple on-off control provides satisfactory results. They may also be applied to on-off pressure and flow control through the actuation of valves which are opened and closed by the action of a solenoid.

These circuits, and somewhat more sophisticated versions (though similar in principle), are commonly found in discrete control systems based on electronic logic (e.g., in PLCs) as well as electromechanical logic. Examples are conveyor controllers (such as the one discussed in the previous chapter) and automated material handling systems.

CONTINUOUS ACTUATION

When it is necessary to set the value of an actuated variable over a continuous range, on-off final controlling elements will not suffice. If the controller must raise the temperature of a container by a certain percentage, it must be able to increase the average current to its heating elements by an equivalent percentage. If the controller must reduce a pressure by a certain percentage, it must be able to increase the opening of an escape valve by an equivalent percentage If the controller must increase the amount of time that material spends in a drying chamber, it must reduce the speed of the motors that move the material through the chamber.

Operations such as these require continuously variable final controlling elements. The devices used in the initial stages of continously variable final controlling elements are usually *thyristors* and *motors.* These then act through the final actuation stages (actuation transducers) to produce the desired process effects.

Thyristors

Thyristor is a general term for a class of semiconductor power handling devices of which *silicon controlled rectifiers* (*SCR*s) and *triacs* are commonly used members.

The use of thyristors as components of continuously variable final control elements represents an interestingly complementary situation. It is this. The transistor, which is essentially a *continuously* variable semiconductor device, is used as the principal ingredient in solid-state *on-off* actuation devices, as we have just seen. Now thyristors have, as a general class characteristic, what may be described as essentially *on-off* behavior. Yet they are the principal ingredient in solid-state *continuously* variable actuation devices.

Thyristors *fire* (that is, they turn on) at the onset of a particular set of conditions. They then remain on until the onset of a different, but also particular, set of conditions. Generally, their on condition can be thought of as a highly conductive (low resistance) state that exists between a pair of the device's terminals. When off, this state is one of low conductivity (high resistance) between the terminal pair.

SCRs and triacs are, perhaps, the most frequently used thyristors with electronic and computer-based controllers, so we concentrate on them. Other types of thyristors are also used as actuation components and in circuits that fire, or *trigger,* SCRs and triacs. Examples of such devices are *unijunction transistors* (*UJT*s) and *diacs.* We do not cover these in this chapter. Readers interested in these devices will find details in several of the books listed in the bibliography for Part I.

Silicon Controlled Rectifiers. The SCR circuit symbol is shown in Fig. 5-7. It can be thought of as a diode with the addition of a controlling gate. Like a conventional diode, when reverse-biased (cathode voltage more positive than the anode voltage), the SCR does not conduct. Actually, like most semiconductor devices, it is not truly an open circuit, but it appears to be a high-value resistor, so that a small "reverse" current flows, generally small enough to be negligible for practical purposes.

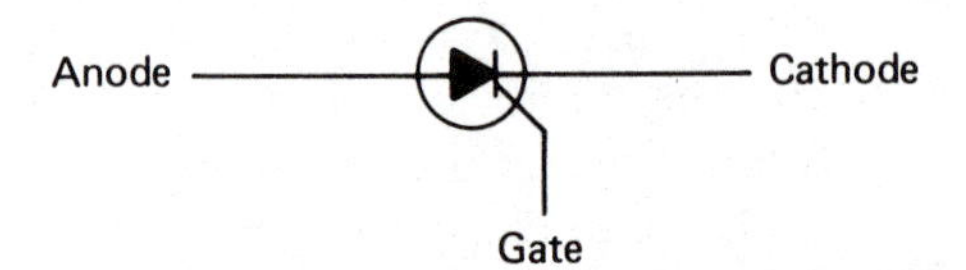

FIGURE 5-7. Silicon controlled rectifier (SCR) circuit symbol.

Until it has been fired, the SCR also does not conduct when forward-biased (anode voltage more positive than the cathode voltage). The SCR is fired by a pulse of relatively small current applied to its gate terminal. This pulse causes the SCR to switch to its conducting state. It then continues to conduct until the voltage across its anode-cathode terminal pair is essentially zero (below the cutoff threshold of the device) or negative (reverse-biased). The SCR then remains nonconductive (even though it may become forward-biased) until it has been fired once more.

The use of an SCR in a simple circuit is shown in Fig. 5-8. Here, current to a resistive load (e.g., a heating element) is being controlled. Power is delivered to the circuit from an alternating current source. By controlling the timing at which the SCR is fired, average alternating currents and voltages may be controlled.

Figure 5-9 shows somewhat idealized voltage and current waveforms across the SCR and the load. The power source voltage sine wave is shown as dashed lines in Fig. 5-9(a). The average current delivered to the load is a function of the *firing angle* or the complementary *conduction angle.* The firing angle is the phase delay between the start of the half wave during which the SCR is forward-biased and the time at which the SCR is fired. The conduction angle is the phase delay between the time at which the SCR is fired and the start of the half wave during which the SCR is reverse-biased. By varying the firing angle, the controller can regulate the average current to the load.

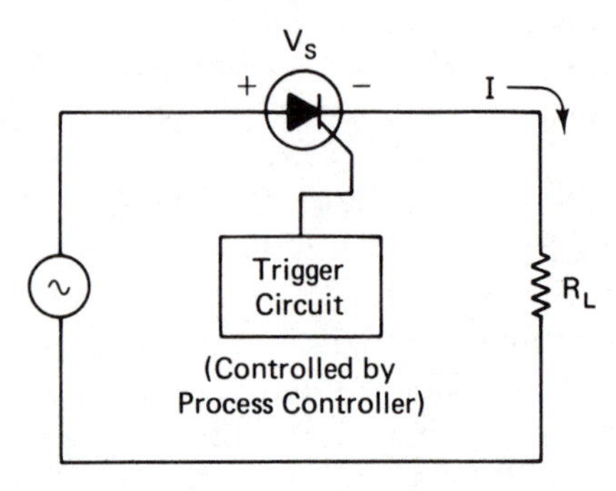

FIGURE 5-8. A simple SCR circuit for the control of a resistive load.

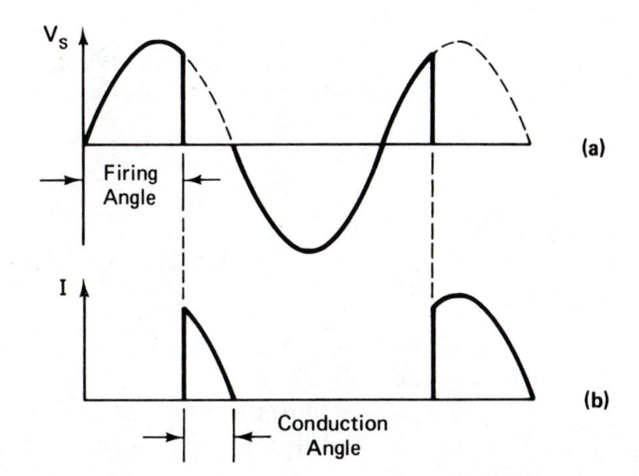

FIGURE 5-9. Idealized SCR voltage and load current waveforms. Dashed lines are the power source voltage. Note that the SCR stops conducting when the voltage reaches zero and must be fired again in the next cycle.

At a firing angle of 0 deg, the circuit of Fig. 5-8 delivers as much current to the load as it is capable of delivering. However, this is only half of what is available from the power source, since only one half wave of the alternating current is used. The circuit in Fig. 5-10(a) is an improvement on this in that it makes the full wave current potentially available. During the positive half wave, SCR1 can deliver cur-

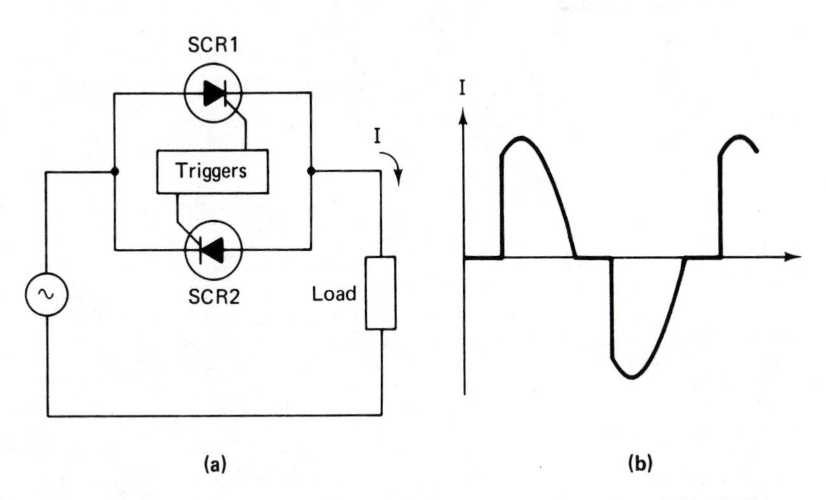

FIGURE 5-10. Use of a pair of SCRs to apply potentially full wave current to the load: (a) circuit, (b) typical load current.

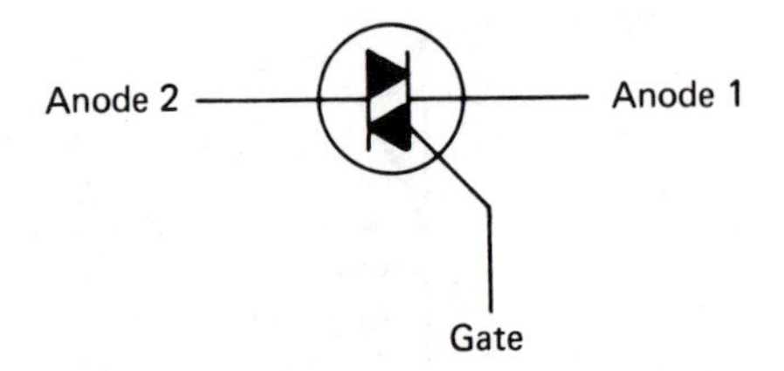

FIGURE 5-11. Triac circuit symbol.

rent. During the negative half wave, SCR2 can deliver current, as shown in Fig. 5-10(b).

When SCRs are used to control direct currents, it is necessary to interrupt current to the SCR (or to shunt current around it) in order to turn it off.

Triacs. The circuit symbol for the triac is shown in Fig. 5-11. This is reminiscent of the arrangement of SCRs in Fig. 5-10. In fact, the behavior of the triac is similar to the behavior of such a pair of back-to-back SCRs, but with a different method of firing.

Like the SCR, the triac does not conduct until it has been fired. When the voltage at terminal A_2 is more positive than the voltage at terminal A_1, the triac may be fired by a pulse of relatively small current flowing into its gate terminal. From this point, it will continue to conduct through the remainder of the half wave. During the succeeding negative half wave, the triac may be fired by a current pulse flowing out of its gate terminal and will then conduct until the onset of the succeeding positive half wave.

The use of a triac in a simple circuit is shown in Fig. 5-12, and the voltage waveforms across the triac and the load in this circuit are illustrated in Fig. 5-13.

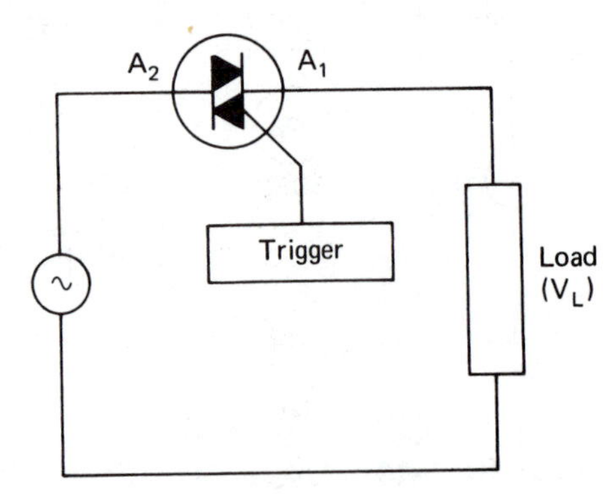

FIGURE 5-12. Simple triac circuit.

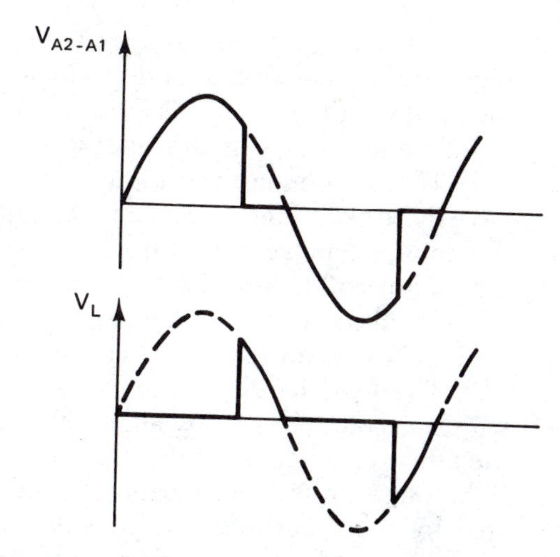

FIGURE 5-13. Idealized triac and load voltage waveforms.

Small triac-based circuits are frequently used with computer-based discrete control systems to switch alternating current signals. They are generally classified as ac solid-state relays in such applications, are optically isolated from their control input pins (which are designed to be electrically compatible with logic-type signals), and are designed to switch at zero-crossings of the alternating current in order to minimize the electrical switching noise generated by their action.

Motors

Motors can be thought of as devices which transduce electrical energy into a change of position (mechanical energy). There are many different types of motors, each suited to particular applications by its unique characteristics. In this section, we briefly discuss three representative types.

Split-phase AC Motors. Figure 5-14(a) is a stylistic sketch of the basic structure of a simple split-phase alternating current motor. Figure 5-14(b) is an electrical schematic for this device. It consists of a *rotor*, which is free to rotate in either direction on an axis perpendicular to the plane of the page. The rotor is surrounded by a *stator*, a stationary magnetic material core on which two windings are arranged. These windings are called *field windings*. The rotor (also called the *armature*) is also provided with a closed arrangement of winding loops.

These motors operate by inducing currents in the armature windings through electromagnetic fields created by current variations in the field windings. For this reason, this type of motor is said to belong to the general class of alternating current *induction* motors.

There are many variations in the designs of such motors. Some have more poles and windings than are shown in the sketch. In some, the armature is fabricated from a core of magnetic material in which aluminum conducting bars are inlaid, with rings at each end of the cylindrical rotor to form a closed conducting path which has the appearance of a squirrel cage. These are called *squirrel cage* motors.

The field windings are excited by separate alternating currents which are out of phase by about 90 deg. This phase difference causes the effective magnetic field surrounding the armature to rotate. As the field rotates, it induces currents in the armature windings. Fields created by these currents act to cause opposing magnetic poles in the armature rotor to try to align themselves with the stator fields. By this action, electromagnetic energy is converted into mechanical energy in the form of armature rotation. Under certain conditions of rotor conductivity, speed, and load, it is possible for the motor to be driven by excitation of only one of the windings. Operation under these conditions is called *single-phasing*. Motors which are capable of single-phasing require less power when operating in this condition.

The direction of armature rotation depends on the lead-lag phase relationship between field winding currents. Its rotational speed de-

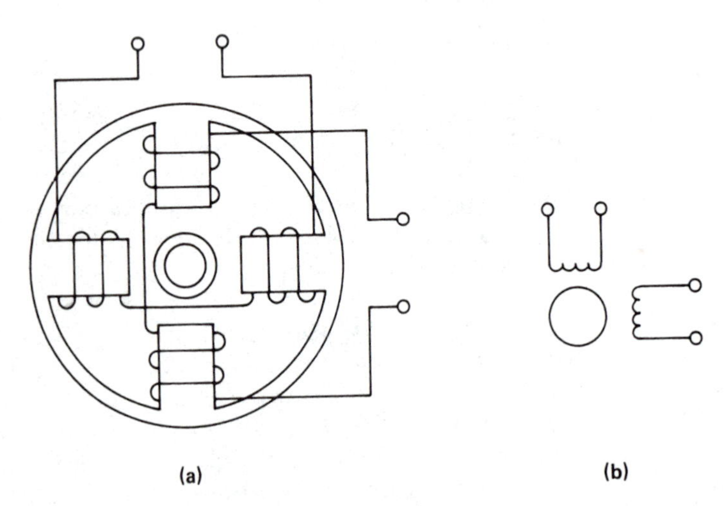

FIGURE 5-14. Simple split-phase alternating current motor: (a) basic structure (note separate stator windings), (b) electrical schematic diagram.

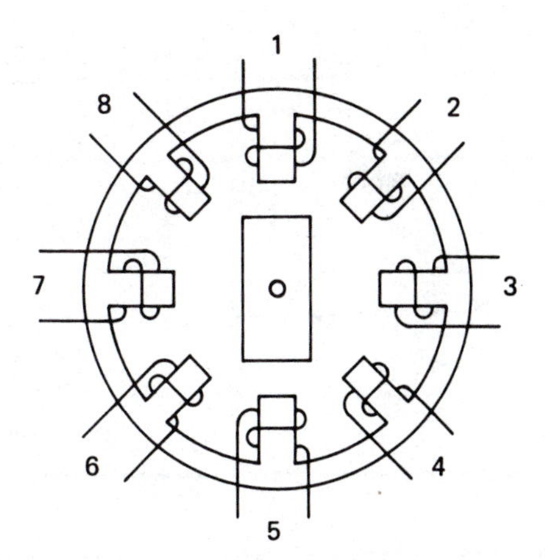

FIGURE 5-15. Basic structure of a simple stepping motor. Note the separate windings for each stator pole.

pends principally on the speed of stator field rotation, which depends, in turn, on the frequency of the excitation currents. The speed also depends on the torque load that the armature must overcome and on the stator field strength, which depends, in turn, on the amplitude of the current delivered to the stator windings. Thus, motor speed can be regulated by controlling the amplitude and frequency of the excitation currents. Further control of the motion of the load can be obtained through mechanical and electromechanical gearing, clutch, and brake arrangements.

Stepping Motors. Figure 5-15 illustrates the basic structure of a simple stepping motor. It differs from the split-phase ac motor in having a separate winding for each pole on the stator core. Its rotor also differs in that a permanent magnet is used.

Simple stepping motors such as this operate by exciting each pole winding individually with a direct current. If the sequence of pole winding excitation is carried out in the proper order, the rotor will seek to align itself with the energized stator pole. This results in an incremental rotation of the rotor. In the motor shown in the figure, since there are eight equally spaced stator poles, each increment of rotation would be one-eighth of a revolution, or 45 deg.

Figure 5-16 shows the waveforms used to control this motor. Say the rotor is positioned initially as shown in the figure and pole 1 is energized. To cause the rotor to take one step in the clockwise direction, pole 2 is energized and pole 1 is de-energized. To take the succeeding clockwise step, pole 3 is energized and pole 2 is de-en-

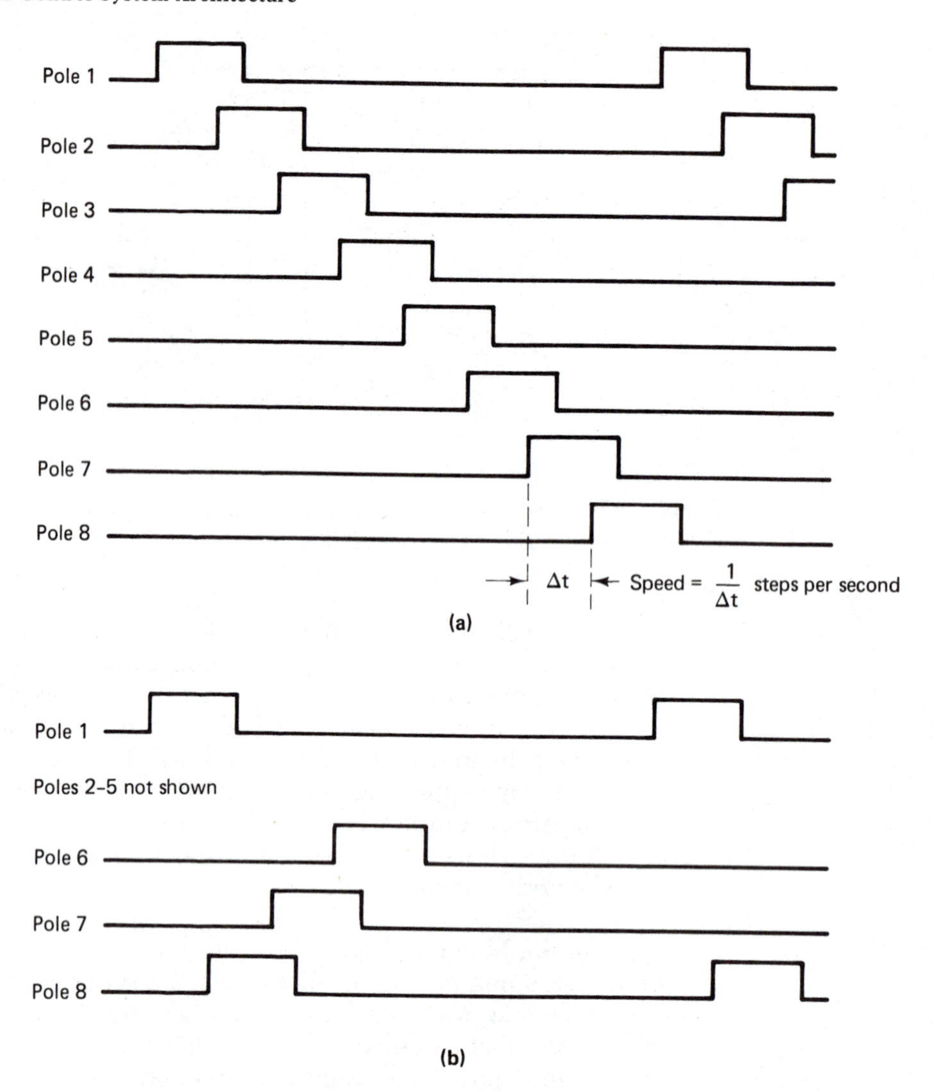

FIGURE 5-16. Stepping motor control waveforms: (a) clockwise rotation, (b) counterclockwise rotation. Note the slight overlap of the excitation of successive poles.

ergized. Continuing this sequence causes the clockwise rotation to continue. The sequence of pole excitation for clockwise motion is shown in Fig. 5-16(a). Counterclockwise rotation is effected by the excitation sequence shown in Fig. 5-16(b).

The holding torque of the motor, its ability to keep its load stationary, depends on the holding strength of the pole/rotor electromagnet arrangement. If the motor is starting its rotation in the clockwise direction, say, against a counterclockwise torque load, it is necessary to maintain some degree of holding torque in order to overcome the load effect. The overlap of the excitation waveforms provides this during the initial portions of each step. Motor speed is controlled by varying the rate of the pole excitation sequence.

Because stepping motor motion is incremental, and therefore digital, in nature, it is easily controlled by digital electronic systems such as computers. Furthermore, when properly applied and not overloaded, the stepping motor can be relied on to make the incremental motions demanded of it. Thus, it is sometimes possible to omit position feedback sensors in positioning systems driven by these motors. This can reduce cost and improve reliability in such systems.

On the negative side, stepping motors are not generally capable of driving very heavy loads, so their application tends to be limited to relatively light-duty work. However, stepping motors have been used to control the pilot valves of hydraulic motors in order to amplify their load-handling capability while retaining the incremental positioning characteristic.

By using variable reluctance rotors combined with permanent magnet elements and additional stator poles, stepping motors have been designed for step sizes down to 1.8 deg, or 200 steps per revolution. Although stepping motor motion is not truly continuous, its resolution (when properly geared) is sufficient to provide the degree of actuated variable resolution required in many applications.

Servomotors. Figure 5-17 shows the basic elements of an alternating current servomotor in simplified form. One winding is arranged in the same way as for a split-phase ac motor. This winding is commonly called the *fixed winding* or the *fixed phase*. The servomotor differs from the split-phase motor, however, in several important respects.

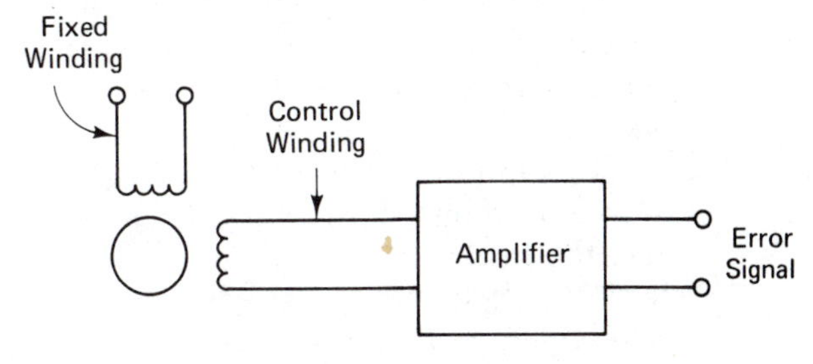

FIGURE 5-17. Simplified alternating current servomotor schematic diagram.

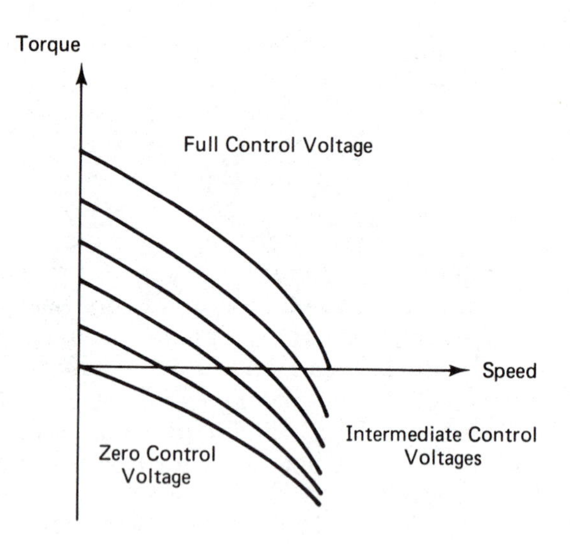

FIGURE 5-18. Characteristic torque-speed curve family for a servomotor.

First, the amplitude of the current applied to the second winding is made a function of an error signal. This winding is called the *control winding* or the *control phase*. The signal on the control winding is an amplified ac signal with its peak value proportional to the error signal magnitude (if the error signal is dc) or peak value (if the error signal is ac). As with the split-phase ac motor, the control winding current must be about 90 deg out of phase with the fixed winding current. The direction in which the motor turns depends on the lead-lag relationship between the two currents.

The net effect is that the motor speed increases as the magnitude of the error signal increases and decreases as the error becomes smaller, the amount of change depending upon the torque load on the motor. Since the motor is intended to stop when the control winding current is zero, it is important to prevent single-phasing behavior. This is accomplished by using higher resistance rotor conductors, the second important difference between split-phase motors and servomotors.

In addition to these characteristics, servomotors are designed to deliver maximum torque at their lowest speeds. This helps to accelerate the load quickly when the error signal is large and the motor is starting; it helps to hold the load in position when it is at and nearing its destination. If load momentum causes the motor to turn more rapidly than directed by the error signal, the motor begins to act like a generator. This effect creates a reverse torque which helps to decelerate the load and minimize the possibility that the destination will be overshot.

Figure 5-18 shows a family of characteristic torque-speed curves for a servomotor as the amplitude of the control winding signal varies and illustrates the effects just described.

TRANSFER FUNCTION/ TRANSDUCTION CONSIDERATIONS

In most cases, the actuated or manipulated variables are not electrical or electronic. Usually, the electrical or electronic controller output signal is intended to change a position, a temperature, or the flow or pressure of a gas or fluid. Therefore, most actuators involve transduction. Both the early (amplification) stages and later (transduction) stages of the actuation elements have transfer functions which must be considered in the design of the system. In this section, we briefly discuss two representative examples.

Consider first the triac-controlled heater shown in Fig. 5-19. In simplest terms, the fluid being heated in the tank has a certain thermal capacity which is a function of its mass and specific heat. The area of the tank and the radiative, conductive, and convective heat transfer coefficients across the boundaries between the tank and its environment govern the amount of heat lost to the environment.

Thus, there is a balance between heat delivered to the tank by the heater (H_d), heat stored in the tank and its fluid (H_s), and heat lost to the environment (H_e).

$$\begin{aligned} H_d &= H_s + H_e \\ &= K_F \frac{d}{dt}(\theta_T - \theta_A) + K_T(\theta_T - \theta_A) \end{aligned} \tag{5.1}$$

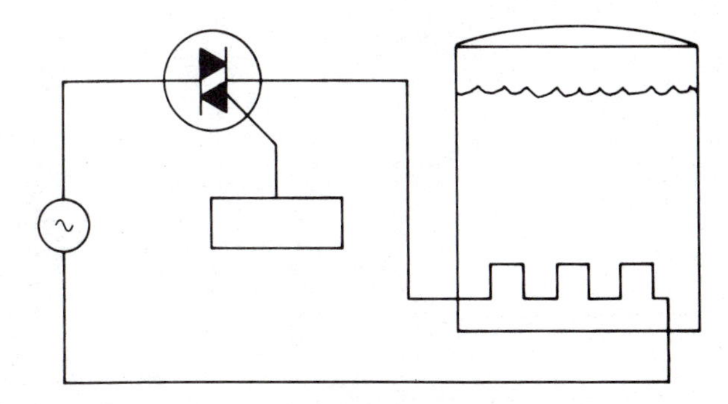

FIGURE 5-19. Triac-controlled tank heater example.

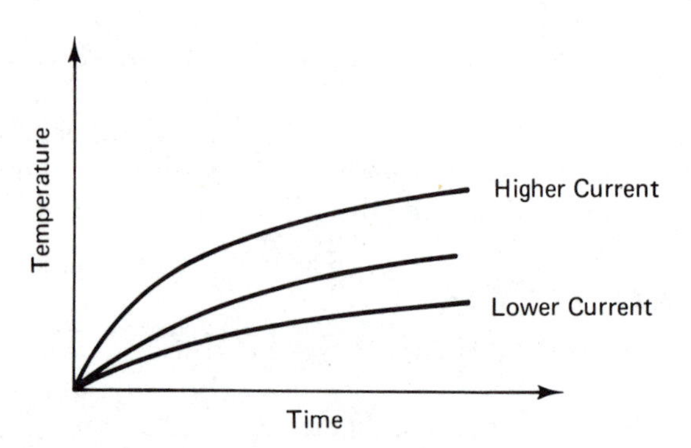

FIGURE 5-20. Idealized behavior of fluid temperature as a function of heater current.

where K_F is the product of the mass and specific heat of the fluid, K_T is proportional to the surface area involved in loss to the environment and combined heat transfer coefficients, θ_T is the tank temperature, and θ_A is the surrounding (ambient) environment temperature. The heat delivered to the tank is proportional to the square of the electrical current delivered to the heating elements by the triac circuit.

Overall, even a simple heater such as this represents a complex transduction which is a function of several variables. Equation (5.1) indicates that the temperature difference between the fluid and the environment has first-order behavior, such as that shown in Fig. 5-20, but the details of this behavior may vary under the influence of other variables. For example, coefficients of heat transfer are affected by temperature. Thus, Eq. (5.1) does not have constant coefficients. The temperature behavior of the fluid is also a function of the amount of fluid in the tank.

This example shows how nonlinearities and variations in the

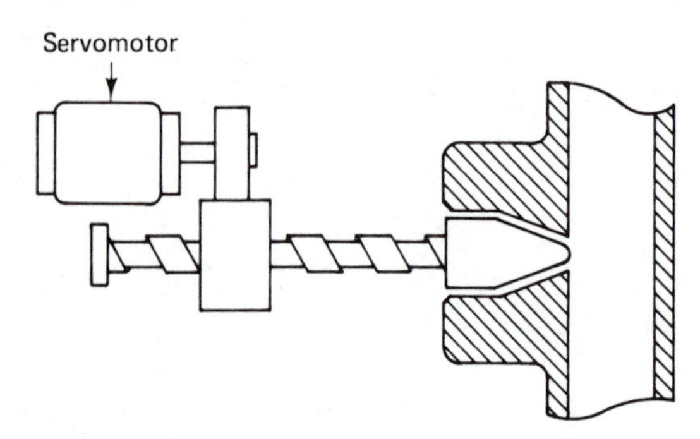

FIGURE 5-21. Servomotor-controlled valve.

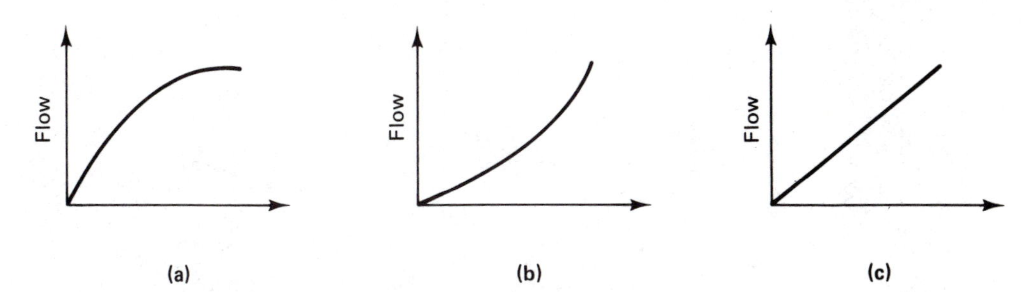

FIGURE 5-22. Flow characteristics: (a) linear valve, (b) nonlinear valve positioning, (c) linear flow obtained from valve designed to compensate for nonlinear flow characteristic.

coefficients of the differential equations involved may complicate the control problem. The next example shows how they can sometimes help.

Figure 5-21 is a stylistic sketch of a valve which is positioned by a small servomotor. Fluid flow through the valve is a function of the difference in pressure on either side of the valve. In the ideal situation, it would be best (from the viewpoint of control simplicity) if the flow were strictly proportional to the valve opening. That is, if the valve were 25 percent open, we would hope for 25 percent of the flow potentially available through the valve; at 80 percent of the valve opening we would hope for 80 percent of the flow, and so on.

Because flow system differential pressure changes as the valve is opened, if the valve is strictly linear, we would actually experience a flow characteristic such as the one shown in Fig. 5-22(a). Here, the slope of the flow versus valve opening curve is reduced at greater opening values.

By designing the valve and its positioning device so that it has *nonlinear* behavior, we may compensate for this effect and produce an overall flow system characteristic which is more nearly linear. This is done by designing the valve and valve positioning device so that, when the pressure drop across the valve is constant, the flow versus opening curve is as shown in Fig. 5-22(b). Here, the slope of the curve is increased at greater valve opening values.

Thus, when exposed to the actual effects of differential pressure drop in the real flow system, the net effect is the flow versus valve opening behavior shown in Fig. 5-22(c); behavior that is more nearly the ideal linear behavior which helps to simplify the control system.

CHAPTER 6

Distributed Control Systems

"Teach us delight in simple things."

Rudyard Kipling

"When in the course of human events, it becomes necessary for one people to dissolve the political bonds which have connected them with another, and to assume among the powers of the earth the separate and equal station to which the laws of nature and of Nature's God entitle them, a decent respect to the opinions of mankind requires that they should declare the causes which impel them to the separation."

Preamble to the Declaration of Independence

Much of what has been said in the preceding chapters had to do with singular elements: the measurement of a single variable (a pressure, a temperature), the control of a single process by the action of the controller through a single actuation mechanism. But we saw also, in Chapter 1, that things are not always so simple (a word used with some reservation, for the control of even a single loop cannot fairly be called "simple"—it has its own kind of sophistication and complexity upon which we have barely touched).

Few systems are composed only of a small number of simply related elements. In many process/manufacturing facilities, there may be many control loops, numbering even in the thousands (in a large refining and chemical manufacturing complex, for example). Although most facilities have loop counts of fewer than a hundred, roughly one-third of the process/manufacturing facilities in the United States have loop counts ranging from one hundred to one thousand per facility. In such facilities, although the aggregate number of loops is not all related to the control of a single process, there are relationships between them which must be considered in managing the facility.

To illustrate this with an example, consider a facility having perhaps over 100 loops in all (Fig. 6-1). This facility might be dedicated to the manufacture of four different chemical products, the process for any one of which might require differing numbers of loops. For reasons of needed process capacity and immunity to equipment breakdown, two more or less identical process systems are set up for each product and operate in parallel.

The manufacture of each product requires raw materials and energy. Some raw materials are used in the manufacture of more than one product, and some are unique to particular products. As the finished product emerges from each process, it must be packaged and

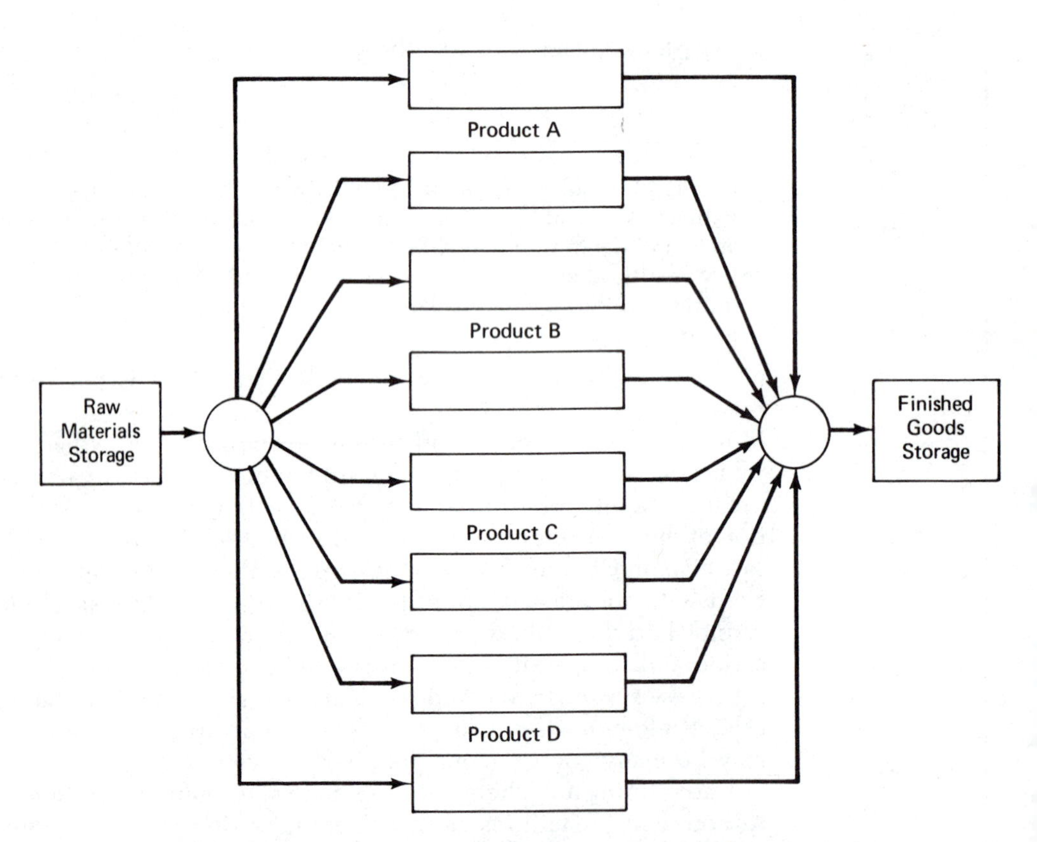

FIGURE 6-1. Example of a facility manufacturing four different products with parallel process systems.

held to await transportation to regional warehouses and distribution centers.

Overall management of a facility such as this is complex. It requires coordination of consumer demand with production schedules and availability of transportation, control of each process operation, coordination of raw material demands by each process operation, inventory management of raw material and finished products, procurement of raw materials, repair and maintenance of equipment, management of the personnel required to run the facility, and research and development activities to improve quality and productivity and to develop new products.

In many cases, the complexity of a modern process/manufacturing facility exceeds this example by orders of magnitude and encompasses many other factors, such as dealing with local, state, and federal government regulations and agencies, dealing with foreign governments for raw materials and export distribution, dealing with sources of raw material supplies to ensure availability and quality and, perhaps, becoming vertically integrated by acting as its own source of raw materials and by becoming a consumer of its own previous products in the manufacture of new products that it may develop.

So we see that assemblies of many things can become quite complex, however simple each may be and no matter how we may delight in their individual simplicity. This chapter is about some ways to deal with this complexity.

COMPLEXITY FACTORS

First, let us consider what factors make for complexity.

Perhaps the most troublesome of these is having many things to control. Closely allied to this are the factors of having limited resources and having to deal with the controlled things quickly. This is analogous to the situation of a control loop having little gain and a narrow bandwidth simultaneously. It cannot respond quickly and has little power to effect a change when it *can* act.

Another factor is reliability. If the cost of failure to control properly is high, then controller reliability becomes more important. If succeeding or parallel stages of the process depend on the success of this controller, and it fails, the whole "structure" may fall like a wall of dominoes.

Fundamental physical limits are still another factor. The speed at which a circuit can operate depends ultimately on the speed at which electrons can travel through it. We know of no "hyper-space warp" which makes it possible to improve this situation.

When efficiency, accuracy, and timeliness are needed, having many things to control can complicate things beyond the limits of tractability. When controlled elements are separated by substantial physical distances, their control becomes more complex. Distance doubly complicates the problem when the controller must have feedback. And "distance" need not be measured only in meters and kilometers. Sometimes it can be the result of other aspects. Extremes of temperature can complicate the control problem, just as physical distance can, because both create conditions of inaccessibility.

Finally, but certainly not least among these factors, are the limits of human cognition. Some things are "simply" too complex for our minds to grasp in their entirety—there are too many aspects, too many interactions happening too quickly, accumulations of more data than we can hold and process in our minds, either as individuals or as teams, working together.

SOURCES OF POSSIBLE SOLUTIONS

Complexity has been an issue for mankind for thousands of years. Well before the rise of modern industry and process/manufacturing systems, we had to deal with factors which make for complex control problems. The earliest examples are found in ancient governments; once human organizations exceeded the scope of the local village or tribe, new ways of ensuring proper operation of the "system" had to be found.

We find the earliest well-studied examples of these solutions in the Mesopotamian civilizations. Representatives and agents of the central government appeared in each village to regulate its activities. With the development of writing, it became possible to communicate regulations widely and uniformly. In a sense, the Code of Hammurabi may be the most prominent example of mankind's early attempts to "program" itself for mutually beneficial concentrated and coordinated action.

The underlying principles of these ancient regulatory systems continue to operate in modern times. Basically, they operate in the following way:

First, a division of labor is made. It may be geographical, as in most governments, where we find federal, state, and municipal governmental organizations. We often find businesses organized along geographical lines, as well. There are other possible divisions. Organizations may be divided according to certain subject areas; for example, commerce, education, defense, research and development, sales, manufacturing, and so on.

A group is then made responsible for governing each division. This group operates according to regulations and policies promulgated by the higher level organization of which it is a part. In turn, it produces more detailed regulations and policies according to which the division for which it is responsible will be operated.

Within these structures, analogs to process/manufacturing control are to be found. Just as organizations, agencies, and departments are responsible for regulating certain areas, so sets of controllers are responsible for controlling certain aspects of the process/manufacturing operation. Just as agencies operate according to policies laid down by higher organization levels, so these controllers operate according to regulations laid down by their designers and policies laid down by managers of that part of the facility. All of these operate through the processing of information, obtained from measurements made on their respective subject areas.

This kind of management has been evolving for thousands of years. With the coming of the Industrial Revolution and the rise of large manufacturing operations, we had to learn about the management of machines also (the principles of which are the measurement and control theories discussed in previous chapters). In more recent times, the digital computer has become a player in this game. In some ways, the evolution of the use of computers in control parallels that of human governmental systems and sheds some metaphorical light on the ways computers are used now in dealing with complexity in the management of large process/manufacturing facilities.

Central Computer Control

Before computers came to be used widely in process/manufacturing system control, control loop status displays were provided in a central control room, along with manual controls for tuning, setpoint changes, and takeover by a human operator in the event of a controller malfunction. The central control room, supervised by humans, was the equivalent (figuratively speaking) of the "brain" that directed the operation of the facility. For particularly complex facilities, many supervisors were required—each responsible for a particular aspect (subject area) of the facility. Higher-level supervisors were responsible for overseeing the work of groups of first-level supervisors. In effect, the control room staff was like a small government, organized into hierarchical levels.

Similar organizational structures continued to be used when the first computers began to be applied to managing the operation of process/manufacturing facilities. In the first supervisory control systems, computers replaced certain lower-level human supervisors (although not always in actuality—it was, and remains, common practice to keep people around to watch the computers, even though this may not be the most effective use of their talents). As larger and more powerful computers became available, they took over more and more of this work. Because the dominant trends in early computer evolution were centered about growth in speed and memory capacity, and because coordination of process information display and control was the principal aim of these computerization activities, it was natural to invest the supervision of as many loops as possible in a single, increasingly larger and more powerful computer.

This tendency was reinforced by economic considerations. The basic cost of early computers was rather large. Because the incremental cost of adding to the memory and input/output of a computer was less than that of adding another computer, it was more economical to attempt to make one computer do as much work as possible. The tendency was toward investing total monitoring and control in a single immense computer. This approach dealt with complexity through a strategy of overwhelming the process with raw computing power, not unlike investing total governmental control in a single, totalitarian monarch or chieftain.

There are several potential problems with such an approach. As Homer has given us insight into the complexity of early Iron Age monarchs, so the experiences of early computer-based control system projects have given us insight into the complexity of large, single-computer systems. They were, in some cases, technical and managerial nightmares; nightmares recorded mainly in myths and legends—official records of such experiences rarely convey their true flavor.

The development of programs for such computer systems required coordinating the activities of many programmers working on different aspects of the design. Operating systems had to be developed to ensure allocation of computer memory and operating time to each part of the system on a basis that was equitable and, at the same time, provided for proper treatment of priorities. The overall measurement and control programs, complex challenges in themselves, had to be developed also. Generally, the project schedule required more or less parallel development of the principal portions of the system. Because of the many interactions between these parts, design and programming errors often had effects which were widespread and difficult to trace.

Another problem with the single-large-computer approach was the potential effect of a malfunction. It was possible for the failure of a single component to bring the computer to a halt or cause improper operation. So it was not prudent, after all, to invest everything in one computer.

Clearly, two would be needed—one more to stand by so it could step in if the first computer showed any signs of trouble. Since a quick changeover would be important, it was necessary that all of the required wiring to the backup computer be in place and that a mechanism be provided for switching everything over to the backup computer. Then there was the issue of maintaining the backup computer in a ready condition. It was possible that simultaneous failures might occur in both computers.

There was also the question of whether the switchover should be done manually or automatically. If manually, a decision by a human operator would be required. If this decision took too long, further problems might arise. Automatic switchover could be done by having the backup computer operate in parallel with the first computer and compare its results to those of the first. If there was any difference, the second computer could command the switchover automatically. But then how would we know *which* computer was right? Well, a *third* computer might watch them both, or a majority vote of two-out-of-three might be used to make the decision. Thus, we end up with an arrangement like the one pictured in Fig. 6-2.

Obviously, it was far from a trivial task to form a computer-based tripartite "government" that was reliable and effective. Nevertheless, such systems were built, and they stand as monuments to marvelous achievements in computer engineering and programming. They were also enormously costly and could be justified only for very complex tasks which could not be accomplished by other means or which provided high returns on the investment in them. Large systems such as these were beyond the economic reach of many operations which might have benefited from computer control.

Small Computer Control

Reductionist methods of solving problems depend on dividing complex problems into more tractable parcels which can be dealt with individually. In government, this means creating subagencies, with their own Assistant Secretaries and Ministers, and assigning to them the more easily discharged responsibilities of governing areas of limited scope.

Large process/manufacturing facilities are usually amenable to several factorings, the most common being first with respect to the products manufactured and then with respect to various stages of manufacture. The problem, until relatively recently, was what to use to

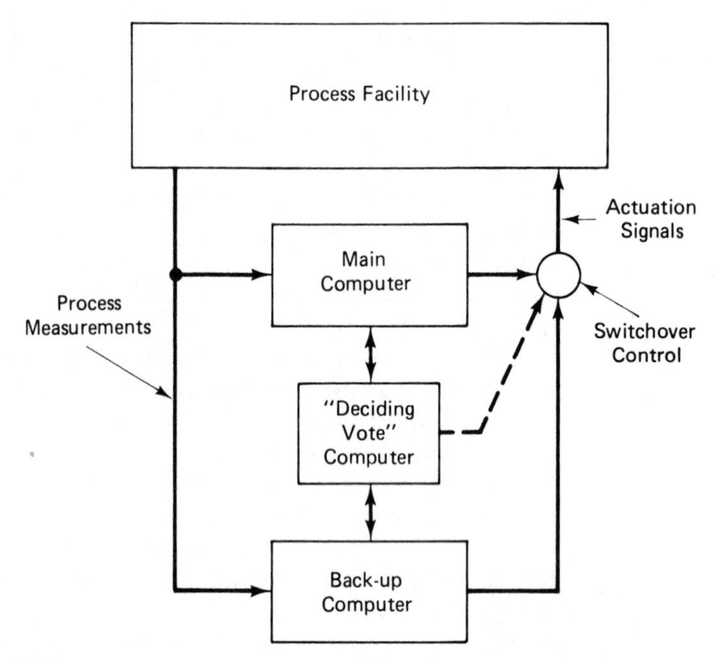

FIGURE 6-2. Central computer control with backup and majority voting for switchover.

monitor and control each factor. To dedicate a single computer to each was rather costly because, at that time, computers were expensive.

The solution to this economic impasse, of course, was to develop a less expensive computer, a task made easier by the assumption that it would not be required to do as much work as its larger predecessor. This led to the development of the minicomputer. Substantially less expensive than large computers, minicomputers could now be considered as candidates for control functions in factorizations of complex facilities. Each could discharge its control responsibilities effectively so long as they remained reasonably limited in scope.

The addition of a few more minicomputers to coordinate the activities of the frontier minicomputers would not add too much to the overall cost of the complete government. Of course, they had to be connected to one another so that they could exchange the information necessary to the coordination activities, but this was also economical, because they could communicate among themselves using only a few pairs of wires by transmitting digital data serially.

Now the *microcomputer* has entered upon the scene. A simplistic view of the microcomputer might see it as an extrapolation of the difference between large computers (which have come to be called "maxicomputers" in modern jargon) and minicomputers: an even less expensive, but also less powerful, computer which simply carried the trendline further. This simple view is crude, but effective. It explains much of the modern interest in microcomputers and many of the reasons why they have become so popular and widespread (even to the extent of making incredibly sophisticated "intelligent" toys common in playrooms around the country).

For now, we leave it at that. There are other reasons for the value of microcomputers, other economic and technical factors behind their utility, which we explore in later chapters. For the purposes of this chapter, let us say that we have come to our present position because we had a complex problem to solve. We considered some of the things which make for complexity and found some tools to use in dealing with this problem (small computers—either mini- or micro-, depending on the magnitude of each factor in the decomposed problem).

We now consider some aspects of their use, bearing in mind that we do not yet fully understand them. We are just interested in "looking over the territory" to get a better feeling for what it might be like.

CONTROL AND DISTRIBUTED TERRITORY

We have decided that our complex problem can be dealt with by dividing it into parts. To the control of each part, a small computer is assigned. Each small computer supervises the operation of a local subject area. It may act as the direct digital control element within one or more control loops in this area, or it may establish setpoints for analog loop controllers in this area. It may measure the process condition in its area and use this information in establishing setpoints and in direct digital control action. It may keep historical records of this information for later study.

Because the operation of each area must be coordinated with that of other areas, a means of exchanging information with the other computers must be established. There may be many such areas. Therefore, we have an arrangement at each area which is as depicted in Fig. 6-3: the small computer, with inputs for monitoring conditions in its area, outputs for controlling those conditions, and one or more communication channels for the purpose of exchanging information with other computers.

This is similar to what we had with the single large computer, but on a much reduced scale. It is, however, different in an important respect: the presence of the communication channels—these are new elements in the architecture of the system.

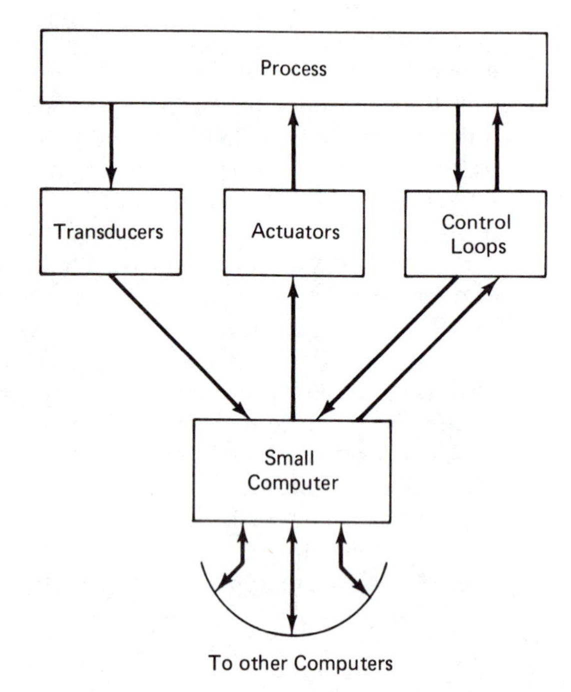

FIGURE 6-3. A small computer responsible for a particular area of the overall facility.

Calling each small computer a *node* in this network of small computers, let us examine how they might be connected.

It is certainly possible that any one node may have reason to communicate with any other node in the network. To retain this generality, there must be some logical path for exchanging messages between these nodes. Suppose it is a direct path with wires strung between each pair of nodes.

If there are only two nodes in the network, there would be only two such paths, as shown in Fig. 6-4(a): one for outgoing messages to the other node and one for incoming messages. With some means of ensuring that both nodes do not "speak" at the same time, perhaps a single bidirectional path may be used (this is possible, but we save it for later treatment). If there are three nodes in the network, there would be six paths, as shown in Fig. 6-4(b). If there are four nodes, there would be twelve paths [Fig. 6-4(c)]; and if five nodes, twenty paths [Fig. 6-4(d)].

The number of paths is clearly growing much more rapidly than the number of nodes. In fact, in a network with n nodes, each node would have $n - 1$ paths entering it and $n - 1$ paths leaving it, $2(n$

− 1) paths at each node. Since a path which leaves one node enters another, it would not do to count each path twice. The actual effect is that each node contributes $n - 1$ paths to the network. Since there are n nodes in all, this makes for $n(n - 1)$ paths in the complete network—growth which is almost quadratic and can quickly lead to an alarming number of paths, as shown by the following table.

Nodes	Unidirectional Paths
2	2
3	6
4	12
5	20
6	30
7	42
8	56
9	72
10	90
15	210
20	380
30	870
40	1560

It is not inconceivable that a large facility might have as many as 40 areas over which we would wish to distribute control in the form of small computers. We can take some comfort from the idea that using bidirectional paths would reduce the number of "only" 780, but we would still be required to equip each node with 39 communication interfaces.

It seems that we have traded one kind of complexity for another. The question is whether we are better off than before and, if so, whether any further improvements are available. The answer to the first part of this question is a qualified "yes." The answer to the second part is a definite "yes."

The reason for the qualification is that the degree of improvement obtained from control distribution depends on the application and the ways in which the original problem can be factored. If the problem is divided too finely, and in such a way as to require communication between every pair of nodes, the logical complexity of the message traffic can be overwhelming (as the table shows). If direct communication is not required between *every* pair of nodes, there are other ways of arranging the network that are less complex (as we see shortly).

Even if communication *is* required between every node, there are interconnection arrangements which at least offer the benefit of

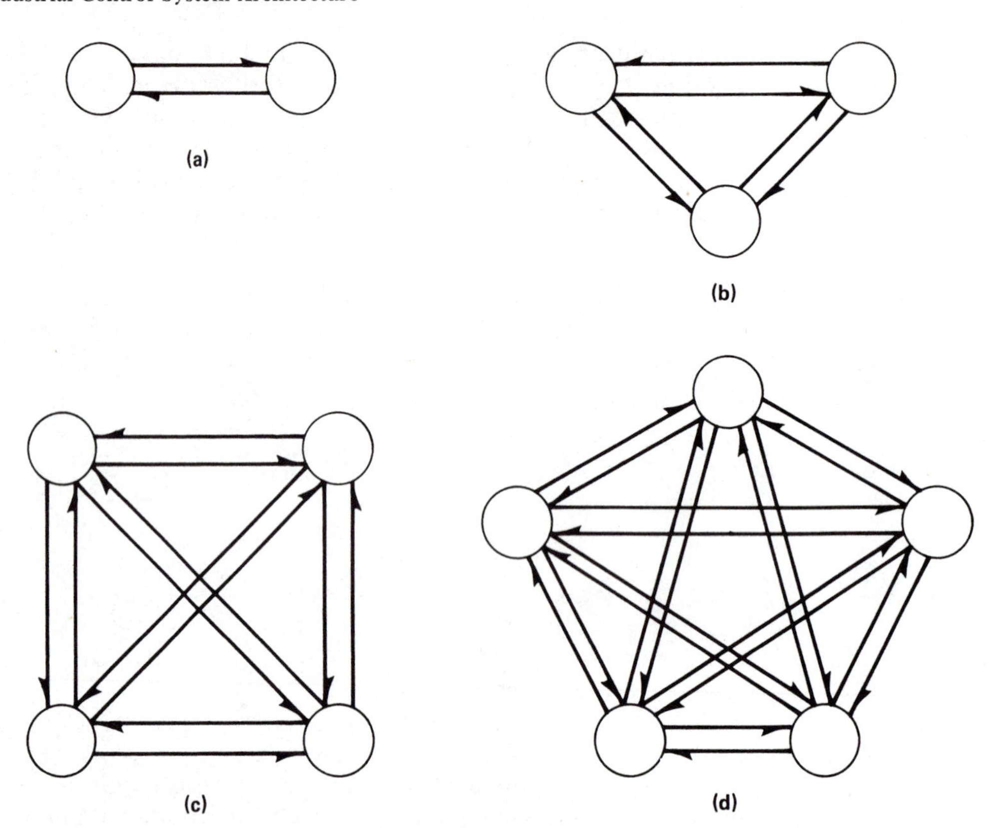

FIGURE 6-4. Growth of the number of communication paths as nodes are added to the network of small computers.

fewer wires. These are called *busses*. With busses, the nodes share a single set of communication lines, as shown in Fig. 6-5. Each node gets to take its turn at using the bus under a set of agreed-upon rules called a *protocol*. We discuss busses in some detail in later chapters. However, busses add their own degree of complexity, because now the protocol must be observed.

We also seem to have found indications of a trend: Each simplification brings with it a new kind of complexity. So long as each new complexity is less troublesome than its predecessor, the strategy is a winning one of increased net simplification. But we may wonder whether there is a stage at which the return on this investment will have diminished to the vanishing point. (This is an unanswered question about modern technology.)

Now let us examine another way of organizing the network under the condition that direct communication is not required between every pair of nodes. This condition holds for many process/manufacturing facilities. It holds for most reasonable factorizations of the example shown in Fig. 6-1. For example, it seems that there would be little need for direct communication between the small computers responsible for manufacturing Product A and those responsible for Product D.

However, there is at least one point at which some coordination between them might be required. Suppose, for example, that both processes use the same raw materials and there is a temporary shortage of one of these materials. If both computers request material from the computer system that manages distribution from the raw materials storage area and there is not enough to go around, which process gets the material?

With direct communication, the Product A and Product D computers may work this out between themselves (this may require some clever programming—the concept of "courtesy" between computers is intriguing). But it is possible that neither computer has enough information to make the decision. It may be that there is, at the moment, little demand for Product A. This would indicate that the Product A computers, in the interests of overall profitability, should defer to the Product D computers.

How are the Product A computers to know this? One way would be to make inquiries of the computers responsible for the finished

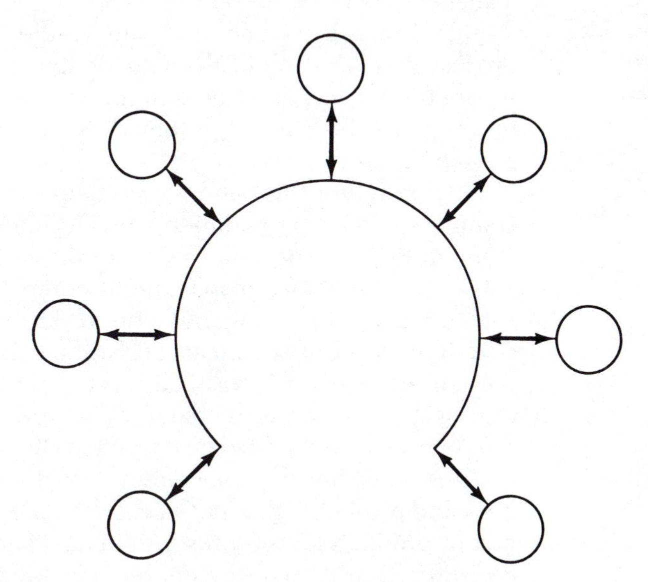

FIGURE 6-5. Use of a bus to reduce the number of interconnecting wires.

goods storage. They might know whether there seemed to be an oversupply of Product A, but they might *not* know about pending orders for the product, some of which might well be urgent.

This is becoming more and more complex and seems not to offer any real simplification. If these were people instead of computers, they might look around and ask who was in charge. People seem to know, more or less instinctively, that someone should be put in charge of coordinating activities in such a situation and would do so. The same principle applies to the computers—put some different computers in charge of supervision and coordination. And just as human managers must have the information necessary to make decisions when given such responsibilities, so must computers. This is the basic idea behind the *hierarchical distributed computer system.*

Figure 6-6 illustrates this concept by showing one possibility for a hierarchical distributed arrangement of our example. Suppose that, in addition to the mutual dependence of Products A and D for raw materials, Products B and C are also mutually dependent. The Product A and D computers are put under the supervision of one additional computer, and the Product B and C computers are put under the supervision of another. Computers responsible for raw materials and finished goods storage are also put under the supervision of an inventory management computer. The added computers are then placed under the direct supervision of still one more computer—the "general manager," as it were.

In Fig. 6-1, we had ten nodes in the system (computers for each process pair for each of the four products, and for raw materials and finished goods). Direct communication among these ten nodes calls for 90 paths. In Fig. 6-4, although we now have 14 computers, there are only 26 links.

Here is how this computer system works: The "general manager" computer, which is assumed to have information about product demand and the amount of raw material on hand (obtained by interrogating the inventory management computer), decides how much of each product is to be manufactured during an ensuing planning period. This kind of work can be handled in a straightforward way by the corporation's business data processing system. It might do this during the night shift, laying out the work to be done the next day by the process/manufacturing computers. By comparing demand for each product, raw materials on hand, production capability of each process area, and profitability of each product, it determines the most favorable mix of products in the present circumstances. These instructions are transmitted to the computers responsible for each area at the beginning of the day.

During the day, these computers keep records of what they did,

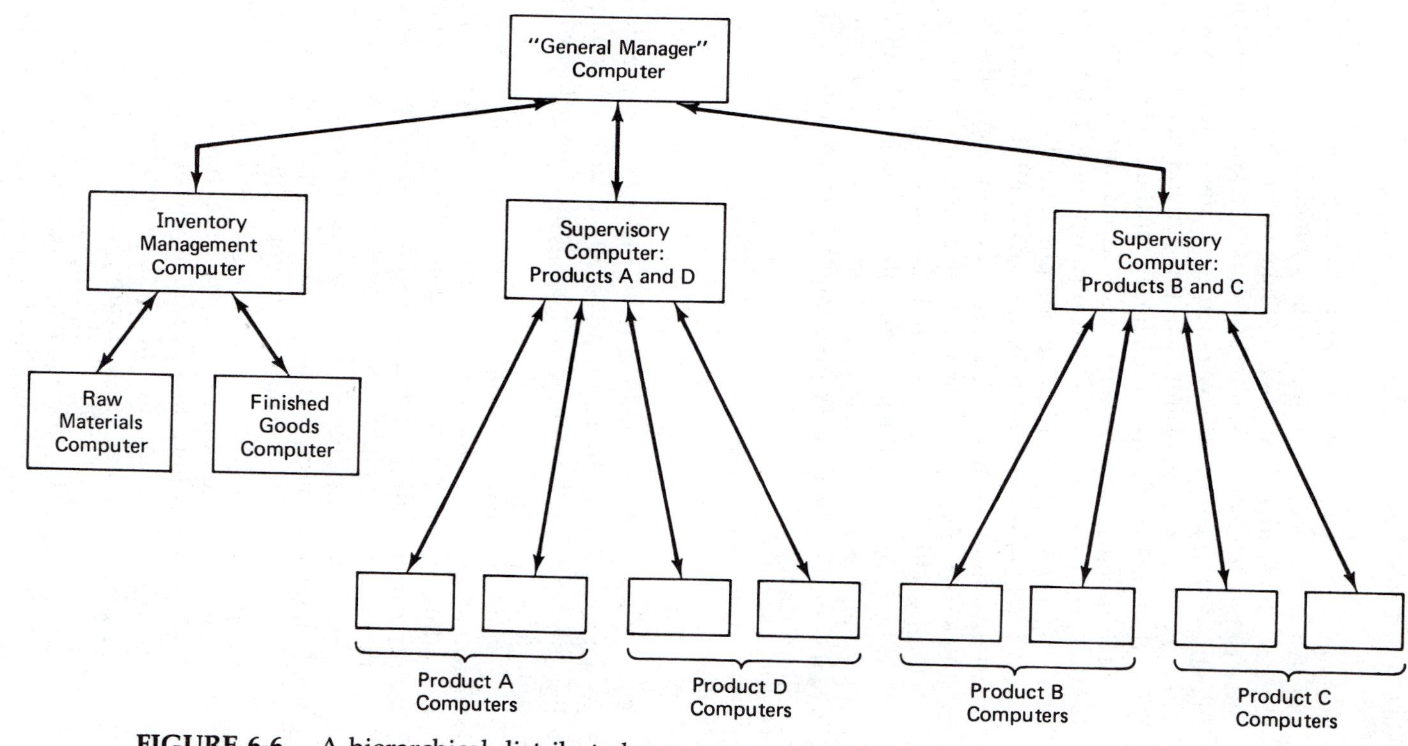

FIGURE 6-6. A hierarchical distributed computer system for the facility shown in Figure 6-1.

how much of each product was manufactured successfully, how much was lost to unacceptable quality, how much was stored away in the finished goods area, how much was taken from the finished goods area and shipped to customers, how much of each kind of raw material was actually consumed, and how much new raw material was received. This information is transmitted to the general manager computer at the end of the day for use in planning for the next day.

Hierarchical arrangements such as this might also be found at levels both above and below those shown in Fig. 6-6. The corporation's business data processing computer might be a member of a network which interconnects computers at all the offices of the corporation around the world. Each set of computers at the lowest level might consist of smaller networks with forms similar to that shown in the figure. The finished goods and raw materials areas might be operated by several coordinated computers—some operating automatic storage and retrieval machines, others operating conveyors and sortation systems. The process areas might have separate computers controlling each loop.

SUMMARY

The "causes" which impelled us to make this "separation" have to do with dealing with complexity—with a search for simplification. Besides reducing the number of communication paths, what have we gained?

First, the overall operation of the facility is more amenable to coordination. The work to be done on one day can be planned the preceding evening. Information about the general operation of the facility can be collected and analyzed in order to work out ways to operate it more efficiently. Situational priorities can be imposed.

Second, the system is less susceptible to malfunction. With a single large computer, one failure could stop the entire facility. With a distributed system, the immediate impact of a malfunction is limited to the area in which it occurs. It begins to be felt in other areas as they come to place demands on that area, but at least there is some delay—an opportunity to effect a repair or change plans to work around the problem. As individual computers become smaller and less expensive, it becomes economically practical to keep complete computers as spare parts. This helps reduce the time needed to make repairs. The division also benefits repair time because it makes detection of the cause of a malfunction easier. If most of the computers used are very similar (the practicality of which becomes greater, the smaller they are), it becomes easier to stock spare parts and to train maintenance people.

The design and development of a system such as this is also simpler. Once coordination rules and protocols for communication are defined, the work of designing and programming computers for each area becomes more independent of other areas and more easily managed. Schedules for development of operating systems, programs, and interface hardware are less tightly coupled. Smaller teams of engineers can accomplish their work with less need to spend a lot of their time in coordination with others.

The system is more flexible and more easily changed. Hardware and programming improvements can be made to one of the computers with less risk that undetected errors will have an adverse effect on other parts of the facility. We also save a substantial amount of wiring. The small computers may be closer to the areas they control, so that the total length of wiring is much less, even with the added communication links.

These simplifications are not without their costs. We have created a mechanism that can generate a great deal of information. We have many computers with which to deal. In hierarchical systems, we must be able to tolerate the delays involved in passing information up to higher levels and in waiting for the instructions to work their way back down to the levels at which action is taken.

Because each simplification can bring its own new complication, it is important not to get lost in detail; both the detail and the overall scene are important. We must be careful to avoid the state of "not seeing the forest for the trees." Sometimes it may be useful to replant the trees, so that the forest becomes an orchard.

PART II

MICROCOMPUTERS

"We are symbols, and inhabit symbols."

Ralph Waldo Emerson

In the introduction to Part I, we claimed that information is the most important driving and motivating force in industry. In this part, we examine the characteristics of one of the two most important processors of information—the computer.

We concentrate almost all of our attention on the microcomputer. Microcomputers now outnumber their larger cousins and ancestors by orders of magnitude. Though physically small, they are still much like larger computers, so much so that, to have a good understanding of how one class works, one will also have a reasonable understanding of how other classes work. The principal differences are matters of scale. This is why we feel comfortable using the prefixes "micro-," "mini-," "midi-," and "maxi-" to distinguish them, though the boundaries between classes may be indistinct.

The other important processor of information is the human being. Humans and computers are complementary in many ways. To understand how computers process information, it helps to have some idea of how humans process information.

Humans are now believed to process information by means of both conscious and unconscious manipulation of symbols in the neurological materials of which their brains are composed.

Symbols are objects with which meanings are associated. Some symbol-meaning associations appear to be formed unconsciously or to have obscure prehistoric origins or, perhaps, genetic origins. Other symbol-meaning associations are contrived and used by common consent. The letters printed on this page and the words they form are symbols having meanings. In reading them, the reader processes these symbols in his or her mind, and these symbols invoke other symbols, which are processed with them. The reader may agree (recognize that the meanings of the symbols conform to "reality"), disagree (recognize that the meanings of the symbols do not conform to "reality"), accept (store the symbols' meanings for later agreement or disagreement), or simply ignore or not understand them.

Each response is a form of information processing for which there are analogs with respect to computers. The logic, physics, and chemistry of the ways in which the human mind processes information are not well understood. Some believe the left and right hemispheres of the brain process information in very different ways, the left hemisphere being a specialist in the manipulation of contrived symbols (with which we commonly associate the concept of "rationality") and the right hemisphere being a specialist in the manipulation of "uncontrived" (but not necessarily "irrational") symbols and in leaps of intuition.

Computers are also processors of contrived symbols. Because computers are capable of processing such symbols at much greater speeds than the human mind, they are valuable tools in the processing of information. This accelerated symbol manipulation can be used to amplify the ability of humans to process information by using it to do some of the processing work.

In science, it can be used to test hypotheses very quickly by programming the computer to try out all (or a sufficient number of) cases that conform to the axioms on which the hypotheses are based. In business, it can be used to carry out the grinding calculations of accounts and costs and payrolls, doing the work of millions of Bob Cratchits every day. In industry, it can be used to calculate the settings needed to control machinery and process equipment more quickly and accurately than could be done by the engineer or technician.

The foundation of this capability is the manipulation of symbols, but the computer is not conscious of their meanings. Recognition of meaning lies in those who program the computers and in those who use the results of the computers' machinations (their output). It is here that the complementarity of humans and computers is most clearly seen.

In the following five chapters, we examine the concrete, physical nature of microcomputers—"hardware"—so that we will understand what they are made of and how they are organized. Then, in Part III, we examine ways of programming them—"software"—that complete the computer as a useful tool for modern control.

CHAPTER 7

General Organization of Computers

"There once was a man who said, 'Damn!
It is borne in upon me I am
An engine that moves
In predestinate grooves,
I'm not even a bus, I'm a tram.'"

Maurice Evan Hare

In this chapter, we examine ways in which computers are organized and operate. The characteristics discussed are applicable to popular modern computers with few exceptions. Some computers are based on somewhat different principles. They are found either in museums or in advanced or specialized applications.

Those found in museums are worth visiting. For the most part, they are examples of early "computing engines" and the forerunners of modern computers. They display their inventors' outstanding ingenuity and insight. They also offer some perspective on how far we have come and why modern computers, numerous though they have become, are extraordinary feats of technological development. Those found in advanced and specialized applications are also of interest, but mostly for the serious student of computing. Also extraordinary feats of technological development, they are important as forerunners of some characteristics that may be found in future computers.

We begin by discussing how symbols may be represented in a computer and treated as data and programs, then go on to discuss ways of organizing the computer so it can carry out its intended purpose.

REPRESENTATION OF SYMBOLS

Many meanings can be represented by contrived written, drawn, and spoken symbols. These graphic and vocal symbols are used in communicating with others and are perceived through the senses of sight and hearing. Contrived symbols may also be mechanical and perceived by touch; an example is Braille writing.

Computers are generally understood by most reasonably well-read people as devices which manipulate numbers. They understand that computers can add, subtract, multiply, divide, and move numbers about. How this is done and how this capability allows computers to translate languages, test hypotheses, print payroll checks, and control machinery is less well understood. Computers do not actually do what most people believe they do.

To communicate with computers, symbols must be chosen for this purpose. Computers are machines ("engines," to use a term popular in some circles) with no history of uncontrived symbols, but they do have *states*—conditions to which useful contrived symbol-meaning associations may be assigned. These associations will be most useful if they have a regular relationship to symbols with which we are familiar and if they can be manipulated and processed by the computer according to procedures which we can dictate.

Symbols are represented in computers by mapping them onto numbers. For example, the following mapping (or encoding) might be used:

A → 65

B → 66

C → 67

D → 68

and so on through the alphabet. Symbols for numerals might be encoded as

0 → 48

1 → 49

2 → 50

and so on. Punctuation marks, lowercase letters, and other useful symbols such as "&," "(," ")," "?," "$," etc., might be associated with numbers left over from the above encoded sequences.

In fact, this loosely described code is the basis for the American

Standard Code for Information Interchange (ASCII), which is a commonly used code for computer systems. This code uses the integers 0 through 127 to represent the upper- and lowercase letters of the English alphabet, numerals, punctuation marks, and other useful symbols, and certain special characters used to delimit groups of symbols and to control the devices which handle them (computers and their input/output terminals).

It is important to remember that this coding is contrived. That the integer "65" stands for the letter "A" in this code is a convention which depends on the context. It may stand for an entirely different symbol in a different context. The integer 65 is itself rather abstract. By agreed practice, the symbols "6" and "5" are written one after the other to stand for (to be a two-character symbol for) this particular "object" which conforms to certain production rules of number theory. Our "name" (our symbol) for this object is "65." There are other ways of naming this number. For example, "LXV" meant the same thing to an ancient Roman.

The operations of ordinary arithmetic are actually manipulations of symbols according to rules that preserve the mapping between the symbols and the numbers they stand for. "2 + 2 = 22" does not preserve this mapping, but it is an often used joke among primary school children. It is remarkable that they recognize the difference, in an intuitive way, and find humor in it.

Because numbers are not concrete objects, the computer must manipulate symbols that stand for these numbers, just as humans must. When the computer manipulates these symbols, primary school jokes are forbidden. When it performs the manipulation "2 + 2," it should come up with the symbol that stands for "4." Thus, a true computer is a machine that manipulates symbols in a way that preserves the mapping between the symbols and the numbers for which they stand.

By means of other mappings, such as the ASCII code, we may store and manipulate other symbols in a computer—symbols which may have other than numeric meanings.

REGISTERS AND BINARY NUMBERS

Registers are used to hold symbols in the computer. Registers are logical collections, or arrangements, of components which have two or more stable states.

In old mechanical calculators, a register was a collection of wheels driven by gears. Each wheel position was associated with a numeral that was printed on the wheel and visible through an opening on the face of the calculator. Operation of the calculator turned the wheels of this "total" register. Each wheel had ten positions, one for each numeral. The arrangement of wheels, side by side in the register, made up the digits of a complete number.

In electronic computers, wheels are not used. Instead, circuits that have two discrete states are used. That is, the circuit is "on" or "off," or current may be flowing or not flowing, or the voltage on a wire may be above or below a certain value. One of these states is associated with the numeral "0" and the other with the numeral "1." Instead of using the familiar decimal number system to represent numbers, the *binary number system* is used.

Binary numbers are used because they have the required properties of preserving the mappings we have been discussing and because circuits that manipulate the symbols used in the binary number system are easy to build. Binary numbers follow very much the same rules and are manipulated arithmetically according to methods which are equivalent to those used in doing arithmetic in the decimal number system.

By convention, when we are doing arithmetic with decimal numbers, the symbol "234" means the number that is obtained by the operations "two times one hundred plus three times ten plus four." That is,

$$234 = 2 \cdot 100 + 3 \cdot 10 + 4 \qquad (7.1)$$

or

$$234 = 2 \cdot 10^2 + 3 \cdot 10^1 + 4 \cdot 10^0 \qquad (7.2)$$

What we are actually doing is representing the number by its relation to a particular polynomial in integral powers of ten, the coefficients of which are chosen from the ten numbers 0 through 9.

The binary number system uses the same approach, using for its coefficients only the two numbers 0 and 1 and representing numbers as polynomials in integral powers of two. Thus,

$$234 = 1 \cdot 2^7 + 1 \cdot 2^6 + 1 \cdot 2^5 + 0 \cdot 2^4 + 1 \cdot 2^3 + 0 \cdot 2^2 + 1 \cdot 2^1 + 0 \cdot 2^0 \qquad (7.3)$$

written as "11101010" as a binary number. An arrangement of eight circuits, each capable of having two states, may be used to store this number. This would be called an 8-*bit* register. *Bit* is short for "binary digit."

This 8-bit register has 256 possible states: the leftmost bit (usually called the most significant bit—just as the leftmost digit is called the most significant digit in a decimal number) may have two states (on or off, 1 or 0, high or low), the next bit to its right may also have two states, and so on to the last bit (the least significant bit). The number of possible combinations is then

$$2 \cdot 2 \cdot 2 \cdot 2 \cdot 2 \cdot 2 \cdot 2 \cdot 2 = 2^8 = 256 \tag{7.4}$$

In general, an n-bit register can have 2^n states. The meaning of any state in which we find this register depends on what the computer programmer intended it to mean. In the case of 234, or its binary equivalent 11101010, this could mean the integer 234 or it might mean the lowercase "j" from the ASCII code, with the most significant bit being used as a *parity* bit, an error-detecting bit which is set or reset so that there will be a particular, in this case odd, number of "1"s in the number. The programmer might have intended, as another possibility, that the 8-bit register be interpreted as containing two 4-bit binary numbers: 1110 and 1010, or

$$\text{"1110"} = 1 \cdot 2^3 + 1 \cdot 2^2 + 1 \cdot 2^1 + 0 \cdot 2^0 = 14 \tag{7.5}$$

$$\text{"1010"} = 1 \cdot 2^3 + 0 \cdot 2^2 + 1 \cdot 2^1 + 0 \cdot 2^0 = 10 \tag{7.6}$$

In fact, without some knowledge of the computer program that placed this number in this register, we cannot tell what its meaning is.

Registers may be realized physically in many ways. Sets of transistor circuits arranged to make up devices (called *flip-flops*) with two stable states are most commonly used to make up registers, but registers can also be made up of circuits that hold information in the form of charges stored in capacitors or in the gates of certain types of field-effect transistors. Registers may even be sets of switches or selectively blown fuses.

The bits of a register may also have logical interconnections between them that permit certain transformations of the data they contain. For example, it is possible in one type of register to copy bits from one flip-flop to its neighbor to the right or left, in effect shifting the number in one direction or another. Some registers are arranged so that they can count, incrementing or decrementing the number they hold. Such *shift registers* and *counters* are essential components of a computer.

Other manipulations can be performed during the transfer of data from one register to another or in combining data held in a pair of registers and writing the result into a third register or back into one of the registers of the pair. The ability of the computer to perform these manipulations efficiently depends on the number and types of registers with which it is equipped, the logic circuits interposed between them, and how they are connected.

BINARY LOGIC AND ARITHMETIC

To see in more detail how numbers are manipulated in a computer, we will examine some simple operations on binary numbers. For convenience, the distinction between numbers and the symbols that stand for them will not be emphasized to the extent that it has been in the first part of this chapter.

First, we consider some simple operations of what is called *combinatorial logic.* These are based on Boolean logic of the same form used in the discrete controllers discussed in Chapter 4.

Four combinatorial logic operations are commonly found among the functional capabilities of a digital computer and in the logical components and circuits of which it is made. These are the *AND, OR, Exclusive-OR,* and *complement* (or *NOT*) operators. The AND, OR, and Exclusive-OR operators operate on a pair of bits to yield a single bit result. These may also be applied in parallel, to the contents of a pair of registers, each corresponding pair of bits being operated on to produce the result. The NOT operator operates on a single bit only, or in parallel on the contents of a register.

The ways in which these work are sometimes illustrated by means of *truth tables,* a term derived from their origins in Boolean Logic and propositional calculus, which are branches of symbolic logic. Truth tables list the results of the operation for each possible combination of input terms. Truth tables for these four fundamental operators are shown in Fig. 7-1.

The AND operator is often called *logical multiplication* and indicated by the dot ("·") symbol used elsewhere for arithmetic multiplication. Similarly, the OR operation is often called *logical summation* or *logical addition* and is indicated by the plus ("+") symbol from arithmetic addition. The Exclusive-OR operation is sometimes indicated by a circled plus ("⊕") and the NOT operation by a minus symbol ("−") or by overscoring. Combinations of terms in such logical expressions are grouped by using parentheses to indicate the order and terms to which the operations are applied. The distinction between the meanings of these operator symbols and their counterparts in arithmetic operations is usually clear from the context in which they are used.

Circuits that perform these operations are easily built from combinations of transistors, but we refrain from discussing how this is

(a) AND Truth Table

A ·	B =	C
0	0	0
0	1	0
1	0	0
1	1	1

(b) OR Truth Table

A +	B =	C
0	0	0
0	1	1
1	0	1
1	1	1

(c) Exclusive-OR Truth Table

A ⊕	B =	C
0	0	0
0	1	1
1	0	1
1	1	0

(d) Complement (NOT) Truth Table

−A =	C
0	1
1	0

FIGURE 7-1. Truth tables for four fundamental combinatorial logic operations.

done. Of more importance is how they work in the combinations used to form the computer, its memory, and its input/output devices.

Figure 7-2 shows the logic symbols used to represent these in schematic diagrams. Note that *NAND* and *NOR* operators have been added to the ones discussed so far. They differ from the AND and OR symbols in that a small circle appears on the output side of the logic symbol. This indicates a complement operation following the operation indicated by the major component of the symbol. NAND is short for "NOT-AND"; NOR is short for "NOT-OR." These are important additions to the set of logical operators which are found often in logic schematics, in many cases more often than the fundamental operations, because their use can result in simpler circuits.

NAND and NOR circuits are related to the other forms by de

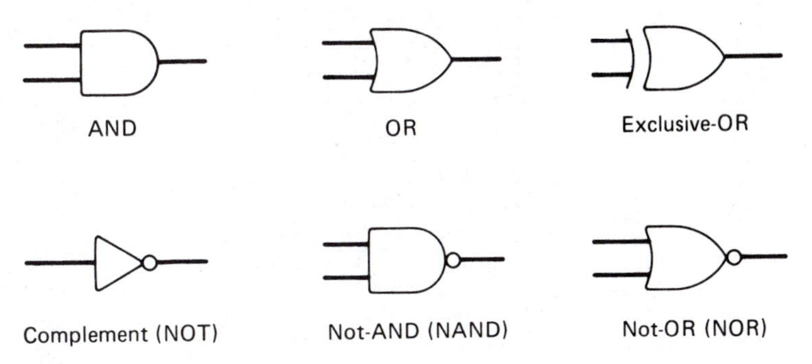

FIGURE 7-2. Schematic diagram symbols for combinatorial logic operators.

Morgan's theorem, which states that

$$-(A \cdot B) = (-A) + (-B) \tag{7.7}$$

and

$$-(A + B) = (-A) \cdot (-B) \tag{7.8}$$

These relations are illustrated further in Fig. 7-3. In this figure, the small circle denoting the complement operator now appears on the inputs to the logic symbols. These are also often used in schematic diagrams.

These operators, in fact, form a more than complete set of logic operators. For example, the AND operation can be performed by a suitable combination of OR and NOT operations. Proof that this is so is left as an exercise for the reader. (Hint: Use de Morgan's theorem.)

The NOT operator as a circuit element is generally called an *inverter*. The other operators are generally called *gates*. For operations on more than two terms, the number of inputs is simply extended. For example, a three-input NAND gate would perform the operation

$$D = -(A \cdot B \cdot C) \tag{7.9}$$

where A, B, and C are input bits and D is the resulting output bit.

Small collections of these elements are commercially available as integrated circuits. This level of integration is called *small scale integration* (*SSI*) as distinguished from *medium* (*MSI*), *large* (*LSI*), and *very large* (*VLSI*) levels of integration in which tens, hundreds, and thousands of such gates are found in single circuit packages, arranged to perform complex functions. At the LSI and VLSI levels we find complete microcomputers in single packages.

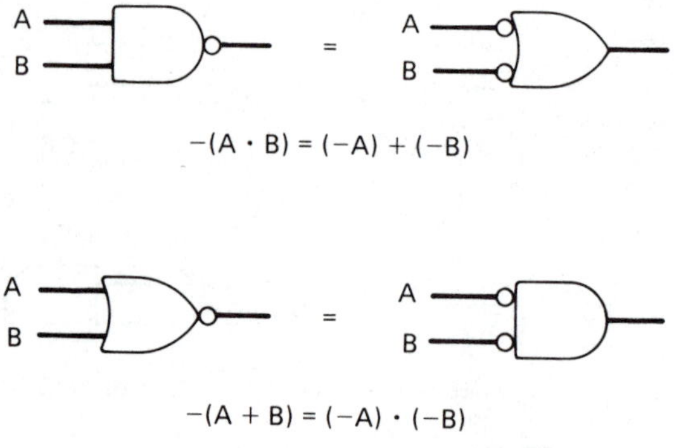

$-(A \cdot B) = (-A) + (-B)$

$-(A + B) = (-A) \cdot (-B)$

FIGURE 7-3. NAND and NOR circuits and de Morgan's theorem.

Gates and inverters can be used to form a circuit which adds two binary numbers. We represent the n-bit numbers A and B and their sum S by concatenated letters, subscripted to indicate the power of two, of which they are coefficients in the binary polynomial representation of the numbers:

$$A = A_{n-1}A_{n-2} \dots A_2A_1A_0 \tag{7.10}$$

$$B = B_{n-1}B_{n-2} \dots B_2B_1B_0 \tag{7.11}$$

$$S = A + B = S_nS_{n-1}S_{n-2} \dots S_2S_1S_0 \tag{7.12}$$

Note that an additional bit may be required to represent the sum S.

Addition of binary numbers follows rules which parallel the familiar addition of decimal numbers. Beginning with the least significant bit of each number (A_0, B_0), the sum of these two binary digits is taken. The least significant bit of this bitwise sum is written down. If their sum is greater than one, a carry is needed. This carry is added to the sum of the next most significant bits, again in bitwise fashion. In this way, we proceed from right to left (least significant to most significant) until each pair of bits has been summed together and with any carries which may have resulted from the summation of their less significant neighbors.

This procedure is illustrated by the following examples of sums of 2-bit binary numbers, with their decimal equivalents shown in parentheses.

A_1A_0	01	(1)	01	(1)	11	(3)	10	(2)
B_1B_0	+10	+(2)	+01	+(1)	+11	+(3)	+10	+(2)
$S_2S_1S_0$	11	(3)	10	(2)	110	(6)	100	(4)

If one were to prepare a truth table for this operation, it could be used to design a logic circuit to carry out this summation by using gates and inverters. Figure 7-4 is a logic schematic for such a 2-bit adder circuit. It is composed of two identical "building block" circuits. Pairs of such building blocks are commercially available, combined in single MSI circuit packages. In the upper section of the circuit, three of the gates are drawn with dashed lines to indicate that they have no function as applied in this example to the least significant bits of the operands. Four such integrated circuit packages may be combined to form a circuit for summing a pair of 8-bit numbers. The input pin labeled C_0 would accept any carries resulting from the summation of lower-order bits, which would come from the output pin corresponding to S_2 in the circuit performing their sum.

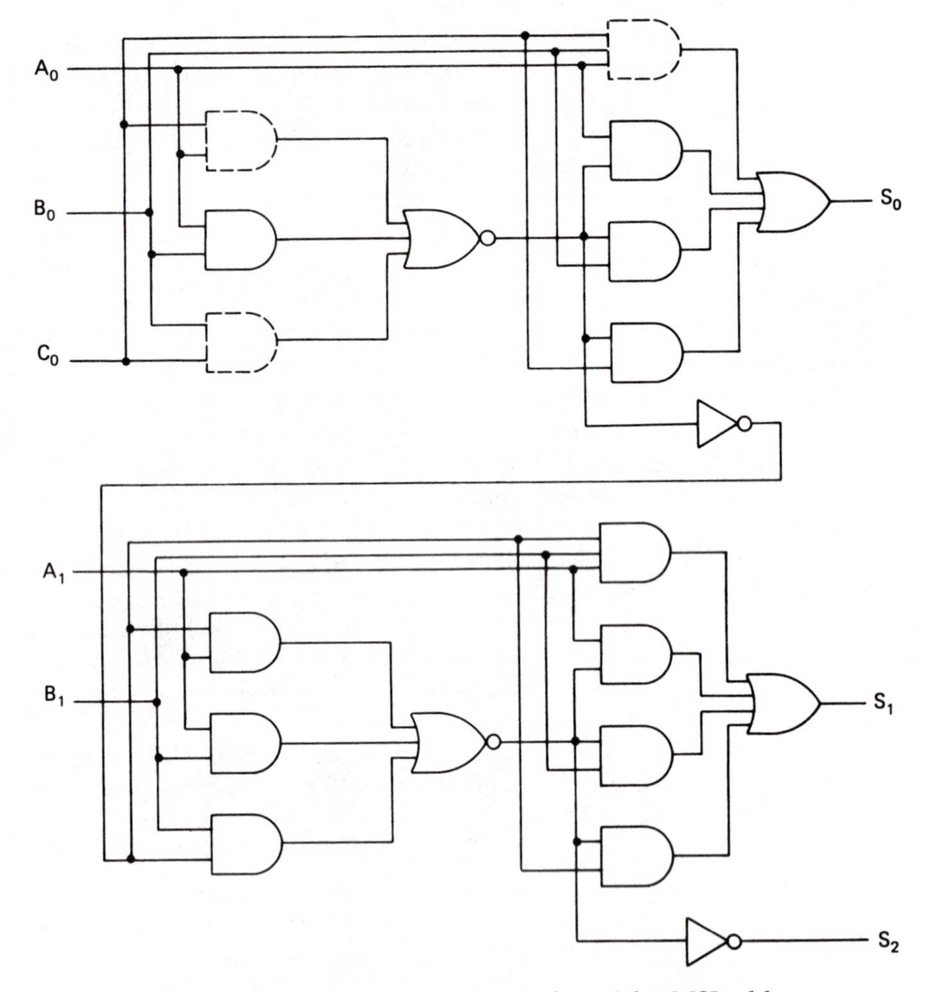

FIGURE 7-4. Logic schematic for a 2-bit MSI adder.

Subtraction, multiplication, and division of binary numbers may also be carried out by circuits that execute the binary analogs of the familiar steps used when these operations are done in decimal arithmetic.

STORED PROGRAMS

We have seen how symbols (to which we have attached meanings in our own minds) can be related to numbers and how symbols for these numbers can be held in computer registers and manipulated by hardware according to rules of Boolean logic and binary arithmetic. We know we can equip the computer with a collection of registers and provide connections and gates between registers so that data can be moved from one register to another, shifted, added, and manipulated logically in many ways. The next step is to consider how we can direct and control the operation of this hardware—how do we "program" it?

This hardware might be equipped with some set of basic capabilities—fundamental operations it can perform on a pair of 8-bit registers, say A and B. Among these might be the following:

- Add B to A, leaving the sum in A.
- Subtract B from A, leaving the difference in A.
- Clear A to "zeros" in each bit.
- Clear B to "zeros" in each bit.
- Increment A (or B) by "one."
- Form A AND B, placing the result in A.
- Form A OR B, placing the result in A.

We may have arranged the logic so that each operation can be invoked by closing a switch or depressing a key, as on a pocket calculator. This would provide us with the capability of directing the operation of the computer. Keys could be used to enter numbers in registers and command the desired sequence of operations. But this could hardly be described as fast, automatic, or even convenient.

We might improve the design by including a mechanism like the one used by Joseph-Marie Jacquard in 1801 to control his loom. He used a card with holes punched in it. Holes punched in particular locations on the card were sensed by mechanical feelers and used to cause particular motions of the loom in order to weave intricate patterns in cloth. This would be somewhat more automatic, but hardly well matched to the electronic speeds of which our collection of registers and logic are capable. It seems cumbersome.

The answer was supplied by John von Neumann in 1947. He proposed that the solution is to represent program steps as encoded numbers and to store the program in the computer itself—put it in registers in the computer, right along with the data. Of course, if the program is lengthy, we will need a large number of registers, but we will concern ourselves with that later.

First, we will see what logical mechanisms we need to implement this *stored program* concept. To keep track of what we are doing, let us call the collection of registers we use for arithmetic and logic the *arithmetic logic unit* and abbreviate it ALU. We will call the collection of registers in which we store the program and the data the *memory*, to distinguish them from the ALU and any other registers we may find necessary.

The program can be expected, for the most part, to be a linear sequence of steps: "add register B to register A," then "divide by 2," then "move the result to register C," and so on. Under some conditions, however, we might want to carry out a different set of steps. For example, if an intermediate result is less than a particular number or if a sum is so large that it cannot be held in the accumulating register (which we can call the *accumulator*)—if it "overflows"—we might want the program to carry out a different sequence of steps.

This will be facilitated if we have some way of specifying particular registers in the memory. We can do this by numbering these registers (we can also call them *locations*) sequentially. In effect, these numbers are *addresses* of locations or *words* in the memory.

The address of the current program step can be kept in another register devoted to this purpose and used to control the circuits that select the next program step (the next *instruction*). By counting up sequentially in this register, we can generate the addresses of program steps in order, one after another. *Program counter* seems to be a reasonable name for this register.

If we also provide for setting the contents of this register to a particular value (which might be selected according to the results of a computation), we can switch to a different sequence of instructions—in effect, we can cause the program to "branch," or "jump," or "go to" a different memory location, at which the beginning of the different sequence of instructions has been stored.

As the program operates, we will want to test certain aspects of the results of instructions. Such things as whether the contents of the accumulator are zero, positive, or negative, whether it has even or odd parity, and whether there has been an overflow can be tested by providing flip-flops which record these conditions following each arithmetic or Boolean logic operation on the accumulator. *Conditional branch* operations may be enabled by the states of these flip-flops. These flip-flops can be considered to be bits in a register which we may call the *program status word*, since it contains certain information about the status of computations. Individual bits in this word can be called *flags* (the "zero flag," the "overflow flag," etc.).

While we are adding new registers, we may include still one more and use it to hold other memory addresses. By loading this *memory address register* with other addresses (which might be results of computations and might even be addresses of other program instructions), we can read from or store into other parts of the memory, independently of the address in the program counter. This allows us to keep data in the same memory with the program and, if the memory is large enough, manipulate even more symbols than we could otherwise handle.

We can represent the program in memory by assigning code numbers to each operation of which the logic is capable. For example, "1" might mean "Add B to A and place the sum back in A," "2" might mean "Subtract B from A and place the difference back in A," and so on. We will need some more logic to "recognize" (to "decode") these numbers and provide the signals to the gates and registers in the ALU that cause the specified operations to take place. It would seem fair to call this logic the *instruction decoder*. Its design will determine the number and types of instructions in the computer's *instruction*

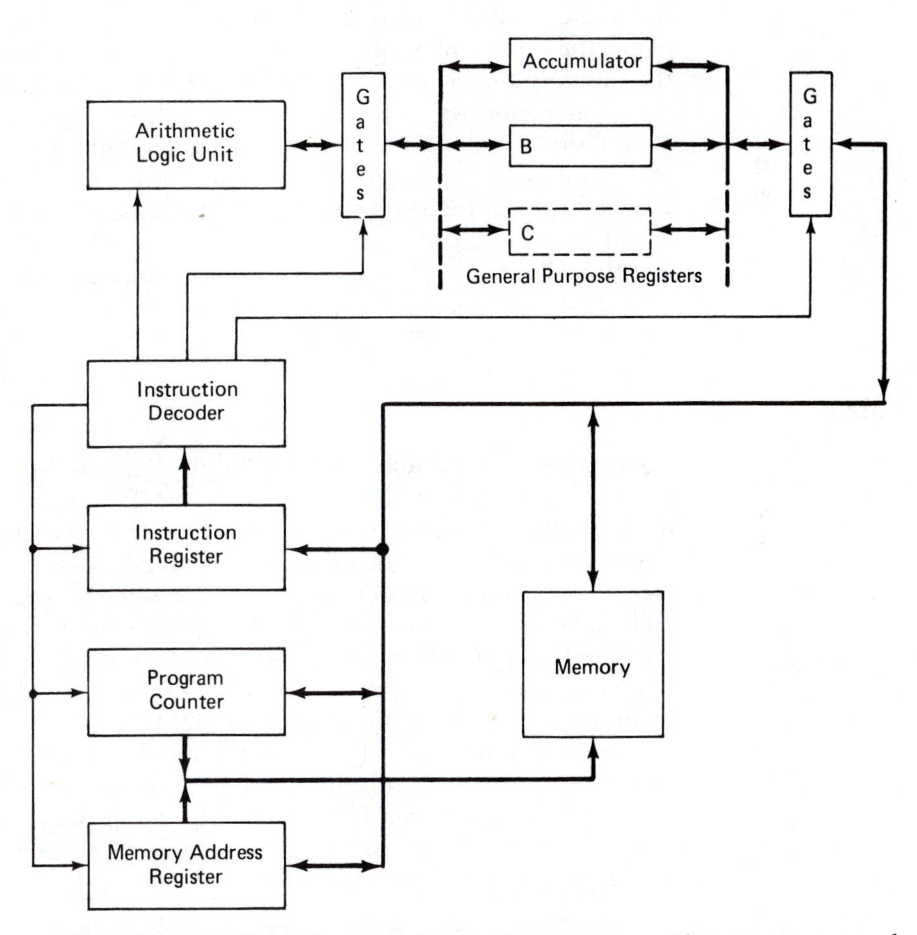

FIGURE 7-5. Organization of a simple computer with memory, general purpose registers, ALU, instruction holding and decoding, program counter, and memory address register.

set. This will also need an *instruction register* in which the instruction fetched from memory can be held while it is being decoded.

Figure 7-5 is a picture of the things we have developed so far and how they might be connected to one another.

Steered by the instruction decoder, numbers may be fetched from the memory and placed in the program counter (to cause the program to branch), in the memory address register (to be used as the addresses of any location in the memory), or in the accumulator or any of the "general-purpose" registers (B, C, etc.) we might include in our complement of hardware. Numbers may also be transferred from registers to the ALU, for arithmetic and logical operations, and back again.

Because it is devoted to processing numbers, logical terms, and instructions (and to distinguish it from the memory and other items of equipment we may add), we will call everything in Fig. 7-5 except the memory the *central processing unit* and abbreviate it "CPU."

The names we chose for the parts of this computer are the names commonly used for them by engineers and programmers. It is evident that many variations and refinements of this arrangement are possible—additional registers might be useful, more powerful instructions might be added, different ways of interfacing between the memory and the CPU might yield a higher performance design, and so on. We see some examples of these in the next chapter, where we examine the designs of several actual microcomputers.

SIZE

Computers are often referred to as "8-bit," "16-bit," and "32-bit" computers. There are also "4-bit," "12-bit," and "24-bit" computers. When used in this way, these terms refer to the number of bits in the principal general-purpose registers and accumulator of the CPU. The most commonly used computers have register sizes that are multiples of eight bits. These are not necessarily representative of the majority of the world population of computers, because most of the microcomputers used in the ubiquitous pocket calculators are of the 4-bit type. Some computers have mixed register sizes.

The optimum size of a register depends on how it is to be used and, most particularly, on the number of *states* needed—that is, the number of numbers that it must be able to represent. A register of n bits has 2^n possible states. The table below lists several common sizes and their number of possible states:

n	*States*	n	*States*
4	16	16	65536
8	256	24	16777216
12	4096	32	4294967296

Clearly, the more bits in a register, the more resolution and dynamic range it has. A 32-bit computer can perform computations to more than nine significant digits. It can count up to the estimated world population as of 1978.

In talking about these numbers, the symbol "K" is used to stand for $2^{10} = 1024$ (very close to the number 1000, for which "k" is sometimes used). Thus, a 16-bit register has

$$2^{16} = 2^{6} \cdot 2^{10} = 64 \cdot 1024 = 64\text{K} \qquad (7.13)$$

possible states. This is a popular size for program counters and memory address registers in many computers, allowing them to specify 65536 distinct memory locations.

The optimum size of a register is a compromise between hardware cost and the degree to which high resolution and wide dynamic range are actually needed. If the computer is to be applied to complex computational problems in which high resolution and wide dynamic range are important (e.g., weather prediction, astrophysics, quantum physics), it will be most convenient for its programmers if it is equipped with "wide" registers. If the computer is to be applied to a simple control problem in which resolution to only about 0.5 percent is needed, then 8-bit registers will generally suffice and will cost less.

If, from time to time, a program must handle numbers that will not "fit" within one of its registers, the programmer can use *multiple-precision* techniques: treating a pair or triplet of registers as if they were concatenated, with an implied factor of the appropriate power of 2 applied to the most significant of the pair. Although there may be added costs for these approaches, they are generally less than the cost of making *every* register in the computer wide enough to handle every number it might encounter.

Most numbers handled by most programs are less than 256. This range allows us to represent the letters of the English alphabet (and also many non-English alphabets), numerals, punctuation marks, and so on, using the ASCII code or similar codes. It allows us to count or total many commonly used quantities (e.g., the number of steps in one revolution of a stepping motor, the degree to which a valve is open to within about 0.5 percent). Thus, 8-bit registers are a reasonable compromise found in many computers.

These 8-bit numbers are called *bytes*. (In an even lighter vein, 4-bit numbers are then called "*nibbles*" by some, although this term has not attained the same official stature in the jargon as has the term "byte.")

Because 8-bit numbers are so common, it has become widespread practice to organize memories around this size. The use of "byte-sized" memory locations has enjoyed the support of the semiconductor manufacturers who supply the integrated circuit packages used in the fabrication of computer memories. Various building block packages are found in *octal* (from the Latin for "eight") forms, meaning that they include sets of eight similar circuits, sometimes arranged with gating and control inputs which act on all of the circuits simultaneously, so that they operate in parallel. An *octal latch*, for example, is a set of eight flip-flops into which an 8-bit number could be set and held ("latched"). For handling pairs and sets of four bits in a similar manner, "dual" and "quad" devices are offered.

The use of a byte-oriented memory does not restrict us from using wider general-purpose registers in the computer. The interface between the CPU and the memory can be arranged to fetch two bytes whenever a 16-bit register is to be loaded. If the instruction set provides for it, instructions may even specify the number of bytes to be moved to the registers (or even from one set of memory locations to another directly in what is, to the program, a single operation).

Thus, it is common to find computers with byte-oriented memories and 16-bit or larger general-purpose registers and accumulators. This is convenient for programmers, because it gives them a "byte" more room in which to maneuver in the registers without wasting lots of bits in memory words which, because of the high relative population of 8-bit quantities, may go unused in a 16-bit word memory organization. It also gives them the ability to address commonly used quantities directly, without having to go to the extra trouble of separating two bytes out of a 16-bit word in which they might have been packed in order to conserve memory locations.

INPUT/OUTPUT

So far, our computer is capable only of introspection. We have made no provision for entering symbols or examining the results of its ruminations, nor even for entering a program. For us to communicate with it, and for it to control machinery and process equipment, we must provide for input and output.

The beginnings of an input/output device interface would be some more registers that could be easily connected to the CPU. Program instructions can place output bytes in such registers, and output devices can then remove the information when ready to accept it. A simple octal latch could hold these data. A device with an input byte for the computer could transfer it to the CPU in the same manner, but in the opposite direction. Since the number of input/output devices would vary from one application to another, it would be most eco-

the details of how output instructions are implemented in the transmitting device. If they are such that the presentation of data can occur in synchronism with the DATA PRESENT signal, the location of (and, indeed, whether there is a need for) the latch depends on the design of the receiving device.

Part (b) of the figure illustrates the sequence of events. While Device B is unable to accept new data, indicated by the BUSY signal being at a "1" ("high" or "logically true") level, which by mutual agreement means that the device is busy—much like the busy tone on a telephone—Device A refrains from placing new data in the latch. Instead, it monitors the BUSY signal until it sees it return to the "0" ("low" or "logically false") level. From this, it concludes that the next data transfer may be carried out. Device A then places the new data in the latch and activates its DATA PRESENT output signal.

In some systems of handshaking logic, the DATA PRESENT signal causes BUSY to go high, and BUSY then feeds back to turn DATA PRESENT off in readiness for the next transfer, as indicated in the figure. In others, DATA PRESENT may simply be a pulse.

Interrupts

This method of transferring data requires a certain amount of "concentration" by Device A (which, let us assume, is the computer) if it is to move along as fast as possible. Device B, if it is a relatively ordinary output device, has little else to do but dispose of the data given to it in the manner prescribed by its purpose (e.g., write it onto magnetic tape, convert it to an analog value, convert it to serial form and transmit it). But the computer, perhaps having other devices to handle and computations to do, might be used more efficiently if its program were not required to monitor the BUSY signal continually.

The program could be designed so that it went about other business and came back periodically to check the status of its input/output devices. At convenient points in the program, it might *poll* each device to see if there were any new input data for it and to see if output devices were ready to accept any data the computer has for them. Polling might be satisfactory in a system that is rather simple, but there are several situations in which polling would be unsatisfactory.

One situation is when input data must be taken at intervals that are precisely separated in time. This would be the case when one is sampling an input signal for digital signal processing. Another is when input data arrive without warning and in a "burst." If the computer does not break away from its other work and accept these data quickly, some may be lost. In order to handle such situations, it is clear that we must be able to *interrupt* the program, so that it can take care of these input/output requirements in a timely fashion. Once they have been taken care of, the program can resume whatever it was doing when it was interrupted.

One way to do this would be to include logic that monitors the BUSY lines on output devices and the DATA PRESENT lines on input devices while the computer is carrying out other operations. If this logic observes any of these lines in an active state, it could interrupt the program by loading a particular address into the program counter. This would have the effect of causing the program to continue execution from the particular address automatically. If the programmer had placed the input/output handling program at that address, it would then be executed, and input/output operations would be taken care of promptly.

By making this interrupt circuitry only a bit more sophisticated, it could provide a unique address for *each* line. Thus, the programmer could provide separate routines for each device, each routine starting at a unique location to which an interrupt would *trap* the program in the event of an active signal from the corresponding device. This *multilevel interrupt* circuit would be more efficient than a circuit which trapped the program to only a single memory location, where the program would then have to poll each device in order to determine which one caused the interrupt.

If more than one device might become active at the same time, we might even arrange for the interrupt logic to accept them according to a defined *priority*, so that the most important devices could be assured of receiving the most prompt attention. For example, an interrupt that indicates that it is time to take a periodic analog input signal reading could be given priority over one that indicates that a communication interface is ready to accept another byte for serial transmission. Most modern computers are equipped with such a *multilevel priority interrupt* structure.

To ensure that a high-priority interrupt handling routine does not become the victim of a lower-priority interrupt while it is running, it will be necessary to provide circuits which can be used to *disable* other interrupts. Interrupts could be disabled automatically when one is accepted by the CPU, then enabled later, using an instruction provided for this purpose, when the routine completes its work. Because it may be inconvenient to accept any interrupt at all during certain parts of the program, it would also be useful to provide an instruction which allows the programmer to disable interrupts. This instruction might be further refined by designing it so that the programmer could specify a priority level below which interrupts would be disabled.

Such an arrangement certainly provides for prompt response,

but, if we are going to permit somewhat random program interruption, we must also provide a way for it to get back "on track" after it has taken care of the requirements of the device that interrupted it. After all, we expect our computing "engine" to "move in predestinate grooves." There must be some means by which the program can resume from the place at which it was interrupted. This leads us to add some more logic to the computer. As it turns out, this logic can be used for other purposes as well.

SUBROUTINES

Subroutines started as a programmer's way of saving work. Programs often contain instruction sequences that are very similar to other sequences. For example, two otherwise identical sequences might differ only in that they work with different sets of data. Others might perform slightly different operations depending on some data characteristic (e.g., if the data are negative, take the absolute value before continuing to process it).

Rather than write different programs to handle each of these similar cases, a clever programmer might write just one, making it a bit more versatile and able to be used whenever such similar processing was to be done. Typically, there are many such opportunities to save programming work, and they also offer the benefit of smaller programs.

An example is converting ASCII-encoded numbers into binary numbers and back again. Such operations are called *BCD* conversions. BCD stands for *binary-coded-decimal*. We noted earlier that ASCII was the most popular form of such encoding. Almost all of the widely used methods share the characteristic of a simple relationship between the four least significant bits of the numeral codes and the binary numbers for the ten decimal digits 0–9.

Rather than write out the instructions for this operation each time it has to be performed, could we write *one* such *subroutine* and somehow "call" it into operation whenever it is required? This subroutine would need to know the location of the string of BCD numerals that it was to convert and where it should leave the resulting binary number.

In general, it would be necessary to provide a means of communicating this information to any subroutine we might create, a *calling sequence* for the subroutine. The specific information to be communicated depends on what the subroutine is to do. There are several ways in which this information could be "passed" to the subroutine from the main program. For example, it might be left in prearranged memory locations or held in CPU registers when the program enters the subroutine. A means by which the subroutine can return to the main program will also be required. Because it can be called from many places in the main program (or even from other subroutines), it will be necessary to provide the subroutine with the memory location (the *return address*) at which the calling program is to resume after the subroutine has done its work.

Very different methods of doing this are found in different computers, especially in some older designs. However, a pleasing regularity of methods has emerged in modern designs. In some cases, the method was to use CPU registers to pass parameters (or parameter addresses) and return addresses to subroutines. In others, these were stored in line with calling program instructions, immediately following the *call instruction*, which was provided in order to branch to the subroutine. Both methods are still used by some programmers because they are easy to understand and satisfactory in simple situations. More modern methods use both registers and "stacks" (which we discuss in the next section) to pass this information.

The call instruction differs from an ordinary branch instruction in that it is designed to leave behind some record of its location in memory. In older designs, it saved the contents of the program counter in a special register set aside for this purpose, then branched to the subroutine. The subroutine could locate its list of parameters by addressing memory beginning at the location stored in this register and also could use this address to determine the location to which it was to return. In some old designs, the return address was stored in memory at the start of the subroutine and the first subroutine instruction to be executed was fetched from the following location.

The means by which subroutines find their way back to the main program provides a mechanism for resuming interrupted programs. To take advantage of this mechanism, when interrupted, we save the previous contents of the program counter in the same way it would have been saved by a call instruction—in effect, the interrupt logic simulates a call to a subroutine which starts at the interrupt address. The interrupt handling routine may then return to the main program at the point of interruption by using the same technique an ordinary subroutine would have used.

But there is still another issue to consider. To do its work, the interrupt handling routine will undoubtedly need to use some CPU registers. If it does not leave them in the same condition in which it found them, the interrupted program is not going to operate properly; it might even seem to be yielding random results. This means the interrupt routine must "save" any *machine conditions* that might be altered by its operation and then "restore" them before it returns to the interrupted program. It can do this by writing the contents of any registers it uses into memory locations it has set aside for this purpose. Just before returning, the registers can be reloaded from these locations.

Several modern computer hardware and software developments acted to bring about some basic changes in the ways interrupts and subroutines are handled in modern computer designs.

A dedicated microcomputer in a control application must be able to start operating virtually the moment power is applied to it. To require manual intervention for loading programs is cumbersome and requires peripheral devices which add to cost and decrease reliability. Programs can be left in an unpowered computer in magnetic core memory, but these memories are expensive, difficult to miniaturize, and, although reasonably reliable, do sometimes lose information during the removal and reapplication of power.

Semiconductor and fused-link *read-only memories* (*ROMs*) offer a superior approach. For now, it suffices to say that the most important characteristic of a ROM is that the computer cannot write into it—it can only read from it. (We discuss the technology of these memories in Chapter 9, including how programs and data can be written into them if they are "read-only.")

The use of ROMs created a situation in which the computer memory had to be divided into two sections: one to contain only programs and constant data, and one to contain "volatile" data. This meant that programs could not modify themselves. This practice of changing instructions in a program while it was running, according to results of previous operations, was frowned on (correctly) as error-prone and dangerous. With ROM-based programs, it became physically impossible. It meant that data were now removed from the program to a different section of the memory. Computer designs in which programs and data could still be intermingled came to be called "von Neumann architectures," to distinguish them from their new cousins of more structured design. This change also meant that return addresses could not be stored in the vicinity of subroutines. If calling sequence parameters had to be changed, they could not be placed in line with program instructions.

Another development which led to basic changes was the concept of *recursion* in programming. Recursion is based on repeated applications of the same procedure until some terminating condition comes to pass. Programmers who are interested in taking full advantage of recursive techniques need computers in which a subroutine can call *itself* and in which two routines (called "coroutines") can call one another. This was very difficult to do with computers of older design.

The idea of subroutines calling other subroutines (even themselves), being nested within one another as it were, and the possibility that the same subroutine might appear at several "levels" of this nesting, meant that a moderately large number of return addresses and sets of machine conditions would have to be stored away and saved in what might be described as a carefully controlled and delicate order. A subroutine or interrupt handler could not simply reserve a private set of memory locations for this purpose. Not only would too many be required for a moderately large program, but a subroutine would have to keep a separate set for every level to which it might be nested and somehow keep track of the level at which it was being used. Passing parameters also became complicated. How could a subroutine change a calling sequence used to call itself without becoming irretrievably confused as to what it was doing and what level it was operating on?

Computer people have a remark which they use for sardonic comic relief when wrestling with a seemingly intractable problem: "Don't worry about it—it's only ones and zeroes."

STACKS AND OTHER ADDRESSING

All these issues have to do with "structure"—ways in which programs and data are organized and ways in which the computers that run the programs are put together. It is indeed "ones and zeroes," but it is how they are put together that makes it all work—how they are structured—how they are organized. For a collection of programs and subroutines to have tractable "structure" and for data to be handled in a useful way, there must be orderly relationships between them. The more well-organized these are, the more tractable and useful they will be.

An integral part of the concept of structure is the concept of "place"—"location"—or, to use a "postal" notion, "address." Where is this item of data "located"—what is its "address?" If one part of a program wishes to use another part of a program (call it as a subroutine, for example), how does it "get to it?"

These questions are parts of daily discussions among programmers. They even go so far as to talk (usually unconsciously) as though *they* were the programs. "Where do you leave the data pointer for me?" "After I call subroutine A and branch to you at B, what do you do?" This use of anthropomorphism and identification with the programs they write is not unusual in the least and is indicative of the intensity and involvement with which many programmers pursue their work. It also illustrates the importance of "place"—in a way, the programmer has stepped inside, and become a part of, the structures that are created by the act of programming.

Up to this point, we have been talking about addresses as locations in the computer's memory—this was as sophisticated as we needed to be until we bumped into recursion, ROMs, multilevel priority interrupts, and subroutines. We will have to deal with these again later on, and we will find a few more strange computer science animals lurking in ambush for us as well—things programmers call "processes" (not the same kind we find in manufacturing chemicals) and "semaphores" (not quite the same things Boy Scouts use to send messages).

Addresses are among the basic ingredients in systems of programs and data structures. The full scope of the science of such structures is containable only in a lengthy series of graduate level courses which specialize in this particular branch of computer science. We limit our scope to a few primary concepts which will help us to understand the general organization of computers and appreciate some of the features of the microprocessors, memories, interfaces, and bus structures described in the chapters immediately following.

Recognizing Addresses

At the most primitive level, an *address* is a number that corresponds to a unique location among the registers of the computer. Here we include in the term "registers," all of the memory, CPU, ALU, and input/output device interface points—anything that can receive, hold, or be a source of data, even down to the level of a single bit. Each thing has, at the very least, a "conceptual" address; we can write down strings of numerals which are in a one-to-one correspondence with these physical entities in the hardware.

When our programs refer to an item of data or another program, they use its address. The logic uses this address to select a particular register and to transfer data to or from this register or to perform some operation on it. It makes this selection by *decoding* the address. A *decoder* is simply a collection of gates which has, as its input, an n-bit number which represents the address and has, as its outputs, a set of signal wires—one for each possible state of the n-bit input address. Decoders are common integrated circuits and are usually equipped with one or more additional control input lines which are used to enable them. Figure 7-7 shows the logic diagram and truth table for a simple 2-bit decoder. Pairs of such circuits are commercially available as integrated circuits. An example is the 74LS139. Similar devices which decode 3-bit and 4-bit numbers are also available.

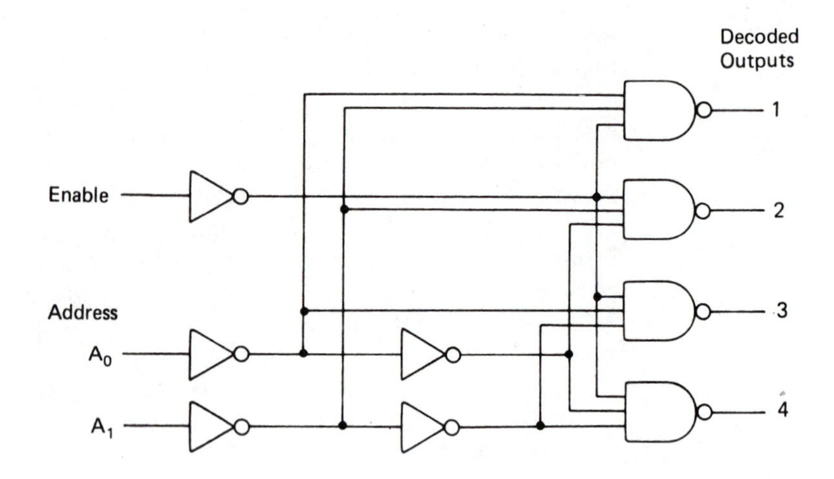

Enable	Address		Decoded Outputs			
	A_1	A_0	1	2	3	4
H	–	–	H	H	H	H
L	L	L	L	H	H	H
L	L	H	H	L	H	H
L	H	L	H	H	L	H
L	H	H	H	H	H	L

FIGURE 7-7. Logic diagram and truth table for a 2-bit decoder.

The output and enable signals are of the *active-low* type—they are considered to be logically "true" when at the lower of the two voltages at which this particular logic family (TTL) operates. "Low" and "high" are symbolized by "L" and "H" in the truth table. When the decoder is not enabled, the address inputs do not matter—all of the outputs remain high, logically false.

The output signals are used to enable other decoders and ultimately a set of gates which transfer the contents of the register selected by this decoding process onto the wires which carry the data to another place. In a similar way, input gates and latch controls on registers that are to receive the data are enabled by their addresses. So we see that, in general, the complete transfer requires two addresses: the address of the *source* of the data and the address of the *destination* of the data.

Using decoders exclusively can be cumbersome, especially if addresses are wider than 4 bits (16-bit addresses are common in modern microcomputers). It can require large numbers of decoders, and output signals may have to run over considerable distances. Mixing decoders with other approaches usually yields a better design.

Digital comparators are designed to recognize particular addresses. They are usually used to recognize "subaddresses" in a field of bits (usually the most significant bits) within a larger address and then to produce enabling signals to other comparators or decoders which are responsible for register selection based on numbers in the remaining address fields.

Figure 7-8 illustrates a form of digital comparator based on Exclusive-OR gates. This circuit tests whether the 4-bit address $A_3A_2A_1A_0$ is equal to 1001. If any bit in the address fails to match, the output of the corresponding gate will be high and cause a high output from the 4-input OR gate. A high output indicates that the addresses are not equal. If they are equal, the output will be low (an active-low signal that can be used to enable decoders equipped with active-low control inputs).

A third method is based on figuratively arranging the registers in array form. A pair of decoders are then used to generate "row" and "column" selection signals. Gates are placed at each row-column intersection. The gate at the intersection produces an enabling signal only if its row *and* column have been selected by the decoders. Figure 7-9 illustrates this scheme for decoding a 4-bit address. This scheme is particularly useful for selecting registers which contain a single bit—a situation often found in semiconductor random access memories (RAMs) which are commonly organized in 1K, 2K, 4K, 16K, or 64K by 1-bit arrangements. A memory that is one byte wide can then be formed with eight such circuits arranged in parallel fashion.

Types of Addressing

If the instruction set of the computer gives the programmer the ability to manipulate addresses as if they were data—to compute their values

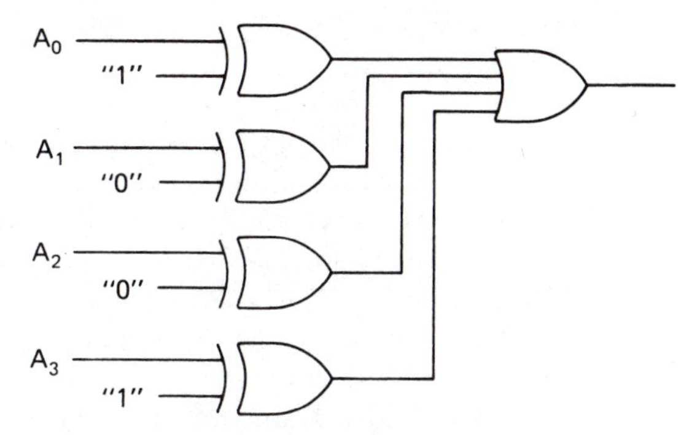

FIGURE 7-8. Recognition of the address "1001" using a digital comparator.

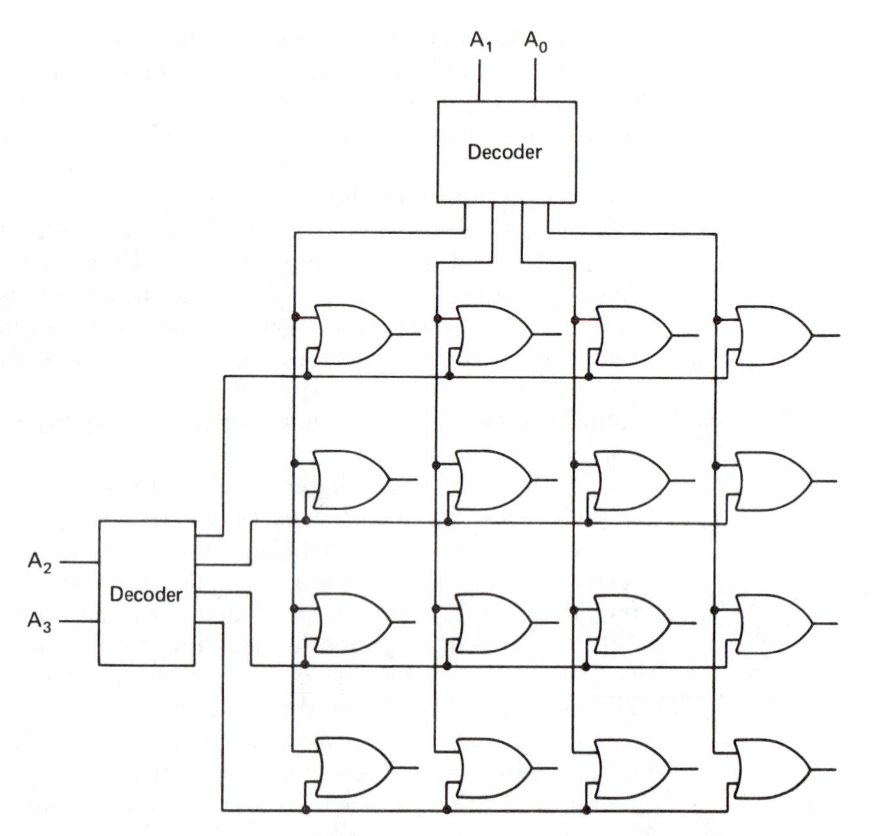

FIGURE 7-9. Decoding of a 4-bit address using 2-bit decoders and an array of gates.

and use them to refer to memory or to registers—then the computer is capable of *indirect addressing*. Virtually every modern computer provides such capability in one form or another and, in many cases, in several forms.

Indirect addressing works in the following way: First, provision is made in the instruction set for recognizing that the instruction intends to use indirect addresses by giving the instruction a unique code number or by providing a bit in the instruction which, if set, means that the address in the instruction is to be treated as an "indirect" address rather than as a "direct" address.

If a *direct address* is specified, the data at that address are used as the operand of the instruction (e.g., moved, incremented, added to the Accumulator). If an *indirect address* is specified, the data at that address are used as the *address* of the operand and *not* as the operand itself. In some computers, the most significant bit of these data was used to indicate that it *also* should be treated as an indirect address. In this way, a "chain" of indirect addresses could be used to reach the location containing the ultimate operand. The address in the indirectly addressed instruction "pointed to" the address of the address of the address . . . of the address of the operand.

Such chained or multilevel indirect addressing is rare in modern designs, having been replaced by more versatile and reliable (less error-prone) methods. One such variation on indirect addressing, sometimes called *indirect register addressing,* provides one or more registers which the programmer can load and manipulate and which, for certain instructions, are interpreted by the hardware as containing addresses of operands.

Another variation is *index register addressing.* Here, one or more registers, called *index registers,* are added to the complement of CPU hardware and used as offsets to the address specified in the instruction or in an indirect register. Thus, the operand address becomes the sum of the contents of either the address field of the instruction or the indirect address register and the contents of the index register. By incrementing the index register, the program can work its way through a list of operands which begins at the address in the instruction or at the address in the indirect address register. Because the address in the index register is "relative" to another address, this method is sometimes called *relative addressing.*

This method can be applied both to instructions that refer to data and to instructions (such as branches and subroutine calls) that refer to other instructions in the program. This makes it possible to set up a list of addresses as branch destinations which can be selected based on a computed result in an indirect address register or index register. Such destination addresses are often called *vectors* because they "point" to particular places in the memory. Usually the term "vector" is used when they are program addresses and the term "pointer" is used when they are data addresses, but this distinction is not formal nor is it observed universally.

One more variation is often called *immediate addressing.* This simply means the contents of the address field of the instruction is taken to be the operand itself. This is often used to include simple constants within the instruction and as the original source of addresses to be loaded into indirect address registers and index registers.

We ended the previous section of this chapter in the middle of a wrestling match with subroutines and nesting. Now that we know a bit more about addressing, we are ready to return to this issue and see how we might resolve it.

First, let us consider how a collection of subroutines might be "nested," one within the other like Russian dolls, and still have a means for the subroutine at each level to return to the proper place in the subroutine at the next level above it.

So long as the same subroutine does not appear more than once in this nesting, the problem is easily handled. A calls B, which saves the return address to A. B calls C, which saves the return address to B. C calls D, which saves the return address to C, and so on, until we come to subroutine X, let us say, which saves the return address to W, which called X.

When X completes its work, it must retrieve the return address to W and branch to that point. W, when it completes its work, must retrieve the return address to V and branch to that point. This sequence continues back up to each succeedingly higher level until B completes its work, retrieves the return address to A and branches to that point, returning to the highest level—the doll has been taken apart and reassembled.

The mechanisms we discussed in the previous section would suffice for this—each routine has only one address to save and could reserve a spot for it in its own private part of memory. But they fail miserably if one of the subroutines appears more than once in the nesting. If, for example, the sequence were A B C D C on the way "down," the second call to C would overwrite the return address to B. As the last call to C finished and started back up, the return sequence would be C D C D C D . . . forever. The program would go into an endless loop, never getting back to B and then to A. Not quite the kind of predestinate groove we had in mind for our computing engine.

The problem arises from overwriting return addresses. Simply providing more places to put them will not work. Computer designers and programmers cannot always predict how many times a subroutine will be called nor how many there will be; nor can the subroutine tell what level it is at (and, therefore, in what unique place to save each address). Even if it had room for all of its possible return addresses, how would it keep track of them?

Perhaps, if we had some means of keeping track of return addresses, we could break the problem into two tractable parts. There may be a clue in the order in which return addresses are saved and retrieved. Retracing the path down into the nesting of unique subroutines and back up again, we see that the order of saving addresses is

A B C D ... V W

The order of retrieving ought to be

W V ... D C B A

The last address saved (that for W) is the first one retrieved on the way back up; the first address saved (that for A) is the last used. This is like soldiers marching single file into a blind alley, facing about, and marching out again. It is like stacking coins into those spring-loaded automobile dashboard coin holders, so that they will be handy for tolls. The last one pushed in is the first one popped out.

If we were to "stack" return addresses like coins, would it work for the A B C D C sequence of nesting? We need four locations for the stack. As we push addresses onto it, it look like this:

A calls B	*B calls C*	*C calls D*	*D calls C*
A	B	C	D
	A	B	C
		A	B
			A

As we pop addresses from the top on the way back up, it looks like this:

C returns to D	*D returns to C*	*C returns to B*	*B returns to A*
C	B	A	(empty stack)
B	A		
A			

It works. It also helps the problem with regard to the amount of memory needed. All we need to do is reserve as many locations for the stack as there are levels of deepest nesting. This can be done easily by the programmer, who should have a rough but reasonable idea of how much depth is needed for the program—it does not have to be precise, just sufficient. Virtually any practical nesting depth can be accommodated within reasonable limits.

Recalling that interrupt handlers are "called" by the same mechanisms used for subroutines and return to the interrupted program in the same fashion, it is clear that this stacking mechanism would also serve that function and would allow one interrupt handler to interrupt another or even itself. Interrupt handlers could simply save machine conditions on the stack with the addresses so that they would also be retrieved and restored in the proper order.

We can provide for manipulating such "stacks" by equipping the CPU with a new special register for this purpose, providing some additional instructions, and slightly changing what the logic does when a subroutine is called or an interrupt occurs.

First, we need "push" and "pop" instructions. "Push" works as follows: (1) The item of data is stored at the address contained in the new register (which we call the *stack pointer*). (2) The stack pointer is then decremented by one, leaving it ready to push the next item of data onto the top of the stack. "Pop" works in opposite fashion: (1) The stack pointer is incremented, to point to the last item placed on the stack. (2) The item at this address is fetched. Decrementing the stack pointer for a "push" makes it easier for programmers to assign the stack to an area of memory with higher-numbered addresses. Programs and data can then be placed in lower-numbered addresses. This permits a certain dynamism in the use of memory. Starting at opposite ends of the memory, as it were, programs and data and the stack can grow toward each other with a postponed likelihood of overlapping.

Instructions for initializing the stack pointer will be needed, as will instructions for fetching the contents of the stack pointer itself so that the program can check whether the stack is about to "overflow" the memory area allocated to it. With these in hand, the programmer could also arrange to have several stacks in the program. It turns out that stacks have other applications as well—for example, they are useful for passing parameters between subroutines and in translating high-level programming languages to the binary codes ("machine language") for the instructions executed by the hardware.

Finally, we redesign our logic for handling subroutine calls and interrupts so that they "push" the contents of the program counter onto the stack automatically, and we provide a "return" instruction which "pops" the stack and loads the address so fetched into the program counter.

The types of addressing of which a computer is capable are usually discussed in its instruction manual under the heading "Addressing Modes." The terms used by manufacturers may vary depending on how they perceive their particular features to be unique or better than those of other computers. Most modern designs provide the types we have just described but with minor variations from case to case.

INSTRUCTION SETS AND FORMATS

As we have gone along, we have equipped the computer with certain instructions that seemed useful or necessary. The time has come to see just how we might do this and whether we might have planned on more instructions than are practical or even achievable.

Instructions are stored in memory, fetched by the CPU logic from locations addressed by the program counter, and decoded in order to tell what operation the instruction is to perform. We have seen that, in general, instructions have "operands"—the data items they use or modify when they operate. In general, operands are identified by their addresses. Operands may be CPU registers, registers associated with intput/output interfaces, or locations in memory. Conceptually at least, an operand might be a word of some number of bits that depends on the "size" of the CPU or its memory, a byte, a bit, perhaps even a "nibble."

We must consider how we arrange the form of the instruction, the parts which make up the instruction, how its operands are specified, and how it is to be decoded.

The usual form of an instruction is one in which it is divided into "fields," each of which specifies a different part of the information required. One field contains the code which specifies the operation (e.g., "add," "subtract," "shift left," "branch," "branch if the accumulator is zero"). This is called the *op-code field.* Fields are also set aside to specify addresses of operands.

Generally, the op-code field must have enough bits to identify each type of operation. A field of four bits, for example, could specify sixteen (2^4) different operation codes, but this seems to be a rather small number for a computer of reasonable capability. For most convenient access to the memory, it would be best to have an operand field (or *address field*) with enough bits to form the direct address of every location. In a computer with a 64K memory, this means the field must be sixteen bits wide. If a second operand from the memory is required, and we want to address it directly, another sixteen bits are needed.

So far, we have imagined a requirement for more than 36 bits, a rather wide instruction, and wasteful of memory, since many instructions would not need that many bits (e.g., a simple branch instruction requires only one operand—its destination address). Just as we saw earlier—that numbers which are wider than eight bits may be divided into bytes and stored in memories which are addressed to the byte level—there is no reason that instructions might not be similarly divided and stored. In fact, this is common practice in modern designs and provides not only economies of memory utilization but also opportunities for designing instruction sets with a greater variety of operations than was possible with older designs in which memory, data, registers, and instructions were all of the same width.

Because most modern microprocessors have addressing schemes which specify a particular byte, we will assume such a "byte-oriented" addressing scheme in the remainder of this discussion. Another common characteristic of such microprocessors is their ability to address 64K bytes directly. This we shall also assume. It means that a 16-bit address will be required.

Next, we consider that there will be, within our instruction set, several possible "classes" of instructions. Membership in one of these classes depends on the type of information needed by the instruction. For example, one class might need only a single memory address. Branch and subroutine calls would be members of this class. Another class might simply form combinations of data in CPU registers or move data from one register to another. These need only the addresses of the registers involved. If we have a limited number of CPU registers, say eight, then three bits are enough to identify the registers. For other classes, the "destination" operand might be implied by the op-code. This would be the case for all arithmetic operations that left their results in the accumulator.

We then design our instruction formats to use only as many bytes as required by the particular class of instruction. If one is required, only one is used. If three bytes are required, then three are used. This permits us to encode many more instructions without wasting memory.

This scheme requires some additional logic in the mechanisms used to fetch instructions from memory. They must recognize the class to which the instruction belongs so that they can fetch the number of bytes required to make up the complete instruction. This can be done by agreeing that the first byte of the instruction will be encoded in such a way as to contain this information. If more bytes are required, the logic can proceed to fetch them before initiating execution of the instruction.

In addition to specifying the number of bytes in the instruction, a portion of the first byte can specify a "context" in which the logic is to decode the remainder of the op-code. For example, if the two most significant bits are "10," it might identify the instruction as belonging to the class of arithmetic operations which involve the accumulator and one general-purpose register. In this context, the next three bits can be interpreted as encoding the specific arithmetic operation, and the least significant three bits can be interpreted as specifying the general-purpose register to be used as the other operand.

Similarly, if the two most significant bits are "11," it might identify the instruction as belonging to the class of "program transfer" instructions which includes branches, subroutine calls, and returns from subroutines. Within this content, other bits and fields of bits can be interpreted to indicate whether the transfer is to be made unconditionally or only if certain conditions are true (e.g., if the contents of the accumulator are zero).

Once the first byte has been fetched, its context determined, and the meanings of the remaining fields decoded, the instruction can be executed immediately, if no more bytes are needed, or the remaining bytes can be fetched and the instruction then executed. Remaining bytes would be fetched to form transfer addresses, to be handled as constants ("immediate" data), or even to be more "detailed" addresses. They might be added to the contents of the program counter or index registers, to form relative addresses, or they might be used to select individual bits within a register or a location specified by the principal address field of the instruction.

Instruction encoding methods such as these are found in the Intel 8085 and the Zilog Z80 microprocessors, among others. These are, perhaps, the two most widely used 8-bit microprocessors in industrial control applications. We examine some of their characteristics in the next chapter.

There are several ways to implement logic that decodes instructions and controls the routing of data and the operations performed on the data. In a modern CPU equipped with a moderately complex instruction set, it is difficult to imagine how its logic can be made to fit within a reasonable volume. Typical modern designs have many more instructions than their predecessors, yet are much smaller physically. Not all of this reduction in size is due to miniaturization realized by integrated semiconductor technology. A great deal of credit must go to the computer "architects" who developed new ways to organize computer structures and more efficient techniques for decoding and combinatorial logic that blended well with the new semiconductor technology.

Among these techniques is the concept of a *logic array*. Logic arrays are highly regular and uniform arrangements of gates. Because of their uniformity, they are more easily integrated into circuits which are quite small physically. The functions performed by these arrays are determined by the ways in which they are connected internally. In general appearance, their logic diagrams are similar to the array shown in Fig. 7-9, which was used as a memory address decoder. They differ in their arrangement of input circuits, which are typically rows and columns of 2-input AND or OR gates, and in their output circuits, which are typically multiple input gates that accept combinations of row and column signals from the central portion of the array.

By selecting combinations of row and column signals, logic arrays can be made to perform a wide variety of decoding and other combinatorial logic operations. They can be made in various forms, with different arrangements of input and output gating for different classes of logical functions. In a fixed design, such as a general-purpose microprocessor, the way in which they are interconnected is selected by the designer and incorporated into the masks used in the manufacture of the circuits. Because they are so generally useful, however, several semiconductor manufacturers have chosen to offer logic arrays in which the interconnections can be selected (programmed) by the user after the circuit has been assembled. These *field-programmable logic arrays* (FPLAs) are programmed by specially designed systems which selectively blow small fuses within the central array in order to remove unwanted connections, leaving only those specified by the user in data entered into the system by means of switches, or a keyboard or punched tape reader.

COORDINATION AND TIMING

Fetching, decoding, and executing instructions, fetching operands and input data, and writing results back to memory and output devices requires a certain amount of coordination. An instruction must not be decoded until it has been fetched. A certain number of combinatorial logic steps must be taken to perform an addition operation. Logic devices require a certain amount of time to switch states. A memory device can be read from and written into only within certain limits of speed. Signals take time to travel from their origin to their destination. It is, therefore, necessary to provide mechanisms for "pacing" the operation of the system. This is the purpose of the system's *clock* circuits.

As a rule, clock circuits are based on a crystal-controlled oscillator which provides a regular timing signal used to step logical operations and data transfers. The overall timing of the computer's operations is based on a number of periods of this clock. To see how this works in a typical design, we retrace the steps involved in fetching and executing an instruction.

At the conclusion of an instruction execution, the program counter contains the address of the next instruction, having been incremented or loaded by a branch, subroutine call, or interrupt. It is at this point that we begin.

Because most operations involve fetching instructions and data and writing results back to the memory, it is useful to imagine the timing sequence as being made up of a series of *memory cycles*, each composed of a number of more detailed steps. *Input/output cycles* would be very similar, if not identical. Fetching a three-byte instruction, for example, might proceed as follows.

First, the contents of the program counter are gated onto the address inputs to the memory. This is followed shortly by activation of a "memory read" signal. This enables memory subsystem decoders, which select the memory elements at the location specified by the address and gate them onto memory output lines.

After a delay to allow time for the memory subsystem to perform this work and for memory data lines to stabilize, the CPU gates these data lines into the instruction register. Until this register has been loaded, decoding cannot begin.

As decoding begins, the CPU decides whether more bytes are needed to form the complete instruction or whether execution can begin. In the case of our example, two more bytes must be fetched. This means that two more memory cycles similar to this one must be executed. For each one, the program counter is incremented so that the byte at the next sequential address is fetched. These bytes must be gated into different registers to be held until the instruction is complete and its execution can be carried out.

The actual execution of the instruction requires further steps, the number of which depends on its specific function. A shift instruction would require steps in which the contents of the specified register were sequentially moved one bit to the right or left until they had been moved the specified number of positions. A multiply instruction would require a number of shifting and adding operations. Branch instructions would require that the program counter be loaded with the fetched address and, if it is a subroutine call, the prior contents of the program counter must be pushed onto the stack. This will require steps in which the stack pointer is decremented and also two more memory cycles for writing the old address bytes onto the stack.

In a typical design, the system clock frequency is selected so that one clock period provides enough time for the longest essentially indivisible step. Each clock signal transition then steps the logic through the required sequence of operations at a rate commensurate with the times they require.

In some more sophisticated designs, the logic may be arranged to perform certain operations in parallel or to "anticipate" subsequent steps. A CPU might, for example, continue immediately into a subsequent instruction fetch cycle while decoding of the previous instruction was still in process. This would make the next instruction available sooner. If it is not required, perhaps by virtue of the previous instruction's having been a branch, nothing would be lost. If it *is* the next instruction to be executed, it would be ready and waiting. In effect, the logic fills up a "pipeline" with instructions waiting to be executed. Such *pipeline architectures* are often found among large computers and now appear in the designs of some of the more sophisticated microcomputer systems.

In some systems, the rate at which data can be transferred to and from the memory is limited by characteristics of the memory—it may not be able to handle data transfers as fast as the CPU. It is also sometimes necessary to have more than one type of memory in the system, and each type may have different speed limitations. This would be true for certain combinations of ROM and RAM, for example. The designer cannot always anticipate the speeds of memories which are slower than the CPU, nor can he dictate that they have certain minimum speeds. To do so might unnecessarily restrict applications of the computer.

To accommodate the possibility that the CPU might have to wait for slow memories, most CPUs are also equipped to accept *wait* input signals. During transfers between the CPU and memory (and, in some designs, input/output transfers also), if the wait signal is logically true, the CPU will suspend operation until the signal returns to the false state. This causes the CPU to wait for the memory or input/output device.

In some designs, the CPU delays for a short while after activating its memory control lines, generally one clock period. If the wait signal is not active by that time, the CPU proceeds, assuming that the data have been placed on its input lines. If the wait signal is necessary, the memory subsystem logic is responsible for recognizing that it is being addressed and activating the wait signal before this time. In other designs, a positive acknowledgment of readiness for the data transfer is required of the subsystem.

Subsystem coordination is one among several issues in the general subject of *bus structures*, a subject which deals with the relationships among processors (of which there may be many in advanced designs), memories, and input/output devices. We cover this subject in Chapter 11.

DESIGN CHOICES

We have been able to touch only briefly on the fundamentals of the general organization of computers. There are many ways to design effective computers, many choices which, taken together, produce designs that may have greatly differing characteristics and "flavor." To a large degree, these characteristics can be analyzed to select the design most suitable for a particular application. Test programs, called *benchmarks*, can be written and tried on different computers to decide which is fastest, which requires the least memory, which is easiest to program, and so on.

Because computer manufacturers intend them to be generally useful in a broad range of applications, some compromises may have been made which favor generality over features which might make them clearly superior for a more specific application. This can tend to make designs somewhat similar and make analytical choices difficult. It has been suggested, only half jokingly, that it might be practical to make a choice by throwing darts at a board on which the names of the computers were written than by going to great lengths to compare them by means of benchmarks and quantitative analyses of their capabilities.

It is in such cases that the flavor of the design becomes important. Although not amenable to quantitative analysis, it can determine, to a significant degree, the extent to which the computer will be an outstanding success. The subjects on which we have concentrated in this chapter are the principal contributors to a computer's flavor. Clarity of design—the degree to which the design is easy to understand, to use, to "keep in one's head"—is sometimes more important than such features as the number of instructions or the number of addressing modes. To equip a computer with several hundred unique and complex instructions is of little value if its programmers have difficulty remembering them and must repeatedly look them up in the reference manual.

As a user of microcomputers, it is unlikely that you will find it necessary to design logic for instruction decoding, memory addressing, stacking, and so on. Most of what you will need in the way of hardware will be found to be commercially available as printed wiring boards, modules, or complete systems. However, you must select the microcomputer systems you plan to use from among the wide variety which is available, and there will be times when you must specify or design interfaces between the computer and the equipment it is to monitor and control. There will also be situations when it is useful to have a general understanding of how they work on a more fundamental level, because this knowledge can help in the diagnosis of system malfunctions and in tracing subtle design errors.

You will be helped in each of these cases if you choose a design with which you feel comfortable—a design with an agreeable "flavor."

CHAPTER **8**

Microprocessors

> "Non aliter, si parva licet componere magnis,
> Cecropias innatus apes amor urget habendi
> Munere quamque suo."
>
> (Just so, if one may compare small things with great, an innate love of getting drives these Attic bees each with his own function.)
>
> Virgil 70-19 B.C.

In this chapter, we discuss microprocessors in particular. We begin with some background information about how they are made, and then discuss general characteristics of different classes of microprocessors.

It should be noted that there is a definite distinction between the terms *microprocessor* and *microcomputer*. Microprocessors are the CPUs of microcomputers. A microcomputer is a computer in which the central processing unit is made by using one or more microprocessor components. A collection of electronic logic does not become a computer until it has been equipped with memory and input/output capability. A microprocessor does not become a microcomputer until it has been surrounded with memory and input/output circuits.

There are many types and models of microprocessors, so many that it would be safe to say that no individual is thoroughly familiar with them all. As of the fall of 1981, there were nearly 200 different models commercially available and buzzing about like so many bees, getting information and performing their functions of processing it in millions of systems ranging from calculators to "supercomputers." Some are doomed to eventual commercial extinction in spite of the technical excellence of their designs. Others are destined for reasonable commercial success in spite of technical defects in their designs. Fewer still will enjoy widespread use in many areas because of their technical excellence and the extraordinary effectiveness with which they are marketed.

Most engineers engaged in the design and use of industrial systems will work with what are called "board-level" products or with "system-level" products. "Board-level" products are completely assembled printed wiring boards on which integrated circuits have been mounted to form CPUs, memory subsystems, and input/output subsystems. They have been designed, assembled, and tested as individual subsystems. "System-level" products are completely assembled and tested systems.

In designing a control or measurement system using board-level products, the user provides (or purchases) card frames, which hold and interconnect the boards, power supplies, and enclosures to contain the complete assembly. The user then writes programs for this microcomputer, interfaces it to the equipment to be monitored and controlled, and tests the operation of the complete system.

In working with system-level products, the user purchases completely assembled and tested systems and is required only to supply programming for them, a task which may be straightforward or complex, depending on the nature of the application and the degree to which the system can be programmed in languages which are easy to use.

Few such engineers will actually be required to design a microcomputer. Nevertheless, it is important to have a good understanding of how they are made and how they work. This is useful in evaluating their capabilities, so that the best of them may be selected. It is useful in interfacing them to equipment, for it is in the area of interfacing that the engineer may be required to execute some original design work. It is useful in designing and writing programs that take best advantage of their capabilities. And it is useful in trying to figure out what is wrong when the system does not work as intended (few do when they are first turned on).

EARLY HISTORY OF MICROPROCESSORS

In the fall of 1971, Intel Corporation introduced the 4004. This was the first commercially available microprocessor, the outgrowth of a custom development project undertaken by Intel for Busicom, a Japanese manufacturer of calculators. In its design, it differed significantly from other efforts at large-scale integration in that it was programmable in essentially the same way that larger computers were programmable.

At the time, modestly integrated semiconductor memories were coming into use with minicomputers and large computers. The manufacture of such memory devices was among the original purposes for which Intel was formed. Recognizing that microprocessors could be the "seed crystals" around which memory-intensive systems might grow, Intel made an arrangement with Busicom by which Intel was permitted to market these microprocessors. Original buyers of these devices were required to agree that they would not use them to make calculators.

The introduction of the 4004 was followed the next year by the 8008. While the 4004 was a 4-bit microprocessor, the 8008 was the first 8-bit microprocessor and was better suited to industrial control and measurement and communications applications. It was the outgrowth of an Intel development project for the Datapoint Corporation and was intended to be the central element in an intelligent terminal, but was not actually used for that purpose because it was eventually considered to be too slow.

Because of rapid advances in LSI semiconductor technology, the years immediately following were characterized by the introductions of many new microprocessors. Some were the products of electronics corporations that had been established for many years. Some were the products of corporations newly formed for the purpose of manufacturing microprocessors and memories.

INTEGRATED SEMICONDUCTOR TECHNOLOGY

Microprocessor performance advances are driven by two major forces. One is the field of computer architecture, which is somewhat abstract in that it deals with nonphysical structures for the processing of information. The other force is the field of solid-state physics, which is manifestly physical in that it deals with the basic physical elements that compose microcomputer systems. Advances in semiconductor technology arising from solid-state physics result in higher and higher levels of performance and integration—the incorporation of greater numbers of functions within faster, smaller, more reliable, and less costly solid-state devices which can then be applied to a broader range of applications safely and economically.

In this section, we examine several aspects of semiconductor technology to see how these devices work and how they are formed into the components of which LSI circuits are made.

Basic Principles

The fundamental issue is the degree to which solid materials conduct electrical current and ways in which this property can be controlled in order to perform logical functions. Just as a control relay contact is used to interrupt and continue current flow in a copper wire, we wish to find a physical mechanism that can be used to control currents and that will switch at greater speeds, consume less power, and be much smaller physically.

Current is the net result of positive and negative charges moving in opposite directions in a material. *Charge* is a bipolar quantum property of atomic particles, the charge on an electron being considered in classical physics to be the smallest amount of charge that exists. An electron carries a negative charge. Atoms in which the number of electrons and protons (which are positively charged) are not equal carry a positive or negative net charge and are called *ions*.

Conduction in solids is the process whereby loosely bound electrons are freed through attraction by the nuclei of neighboring atoms in the crystal lattice structure of the material. Under the influence of electric fields, current resulting from the movement of such electrons is increased to a degree that is proportional to the applied field strength and the number of such free electrons per unit volume, a property of the material. This is the basis of Ohm's law. Insulators have few free electrons and conduct poorly. Good conductors have a high density of such free charge carriers. Between these two extremes are the materials which are *semiconductors*.

Semiconductors

Germanium and silicon are the most used semiconducting materials. Of these, silicon is the basic material of microprocessors. The structure of a silicon crystal is based on covalent bonds in which neighboring silicon atoms share electrons. When provided with sufficient energy, an electron may break this bond and become a free charge carrier, leaving behind a positively charged region in the lattice (called a *hole*) and creating a negatively charged region wherever it is located. As this process continues, free electrons may fill holes left behind by other electrons, and leave holes in their own wakes. The net flow of this process, electrons in one "direction" and holes in the other, forms the basis for conduction in these materials.

Just as the electron can be modeled as a particle having certain charge, mass, and "mobility" in terms of classical physics, holes are modeled in solid-state physics as "particles" having related characteristics of positive charge, mobility, and even mass.

Compared to usual conducting metals, pure semiconductors are not outstanding conductors. Silicon has a resistivity (the reciprocal of conductivity) eleven orders of magnitude greater than that of copper. Nevertheless, it enjoys a valuable property which copper does not. Its conductivity can be controlled and increased by combining it with small amounts of certain impurities.

Silicon and germanium are found in the fourth group of the periodic table of elements. Next to oxygen, silicon is the most abundant element on this planet. When combined with small amounts of certain elements from the adjacent third (trivalent) and fifth (pentavalent) groups of the periodic table, silicon and germanium acquire useful distributions of charge carriers. This process of adding small quantities of trivalent or pentavalent elements to semiconductors is called *doping* and yields a material with charge carrier properties that depend on the type of doping element used.

A trivalent atom (such as boron) has only three valence electrons available with which to form bonds with adjacent silicon atoms in the crystal lattice. This leaves a hole which may be filled by an electron from a neighboring silicon atom, which in turn leaves a mobile hole in its wake. In such a crystal structure, conduction by the motion of positively charged holes is possible. Such materials are called *p-type semiconductors.*

Similarly, a pentavalent atom (such as phosphorus) has five valence electrons—one more than needed to form covalent bonds with neighboring silicon atoms. This remaining electron can become free for conduction by negatively charged carrier motion. Such materials are called *n-type semiconductors.*

pn Junctions

If a silicon crystal is doped nonuniformly, charge carrier density will also be nonuniform. Even in the absence of electric fields, there will be a thermally induced statistical diffusion of charge carriers, away from regions in which they are concentrated. The combined effects of this diffusion and local fields caused by it at junctions between *p*-type and *n*-type semiconductor materials form the basis of microprocessor operation at the physical level.

Put very simply, it works in the following way.

Figure 8-1 depicts the junction between a *p*-type and an *n*-type material. Circles depict bound ions in the lattice with their net charges shown by plus and minus signs inside the circles. A pentavalent atom, in the *n*-type region, is called a *donor* because it can donate an electron to the charge carrier distribution and is marked with a plus sign because, once it has made its donation, it becomes a bound ion with a

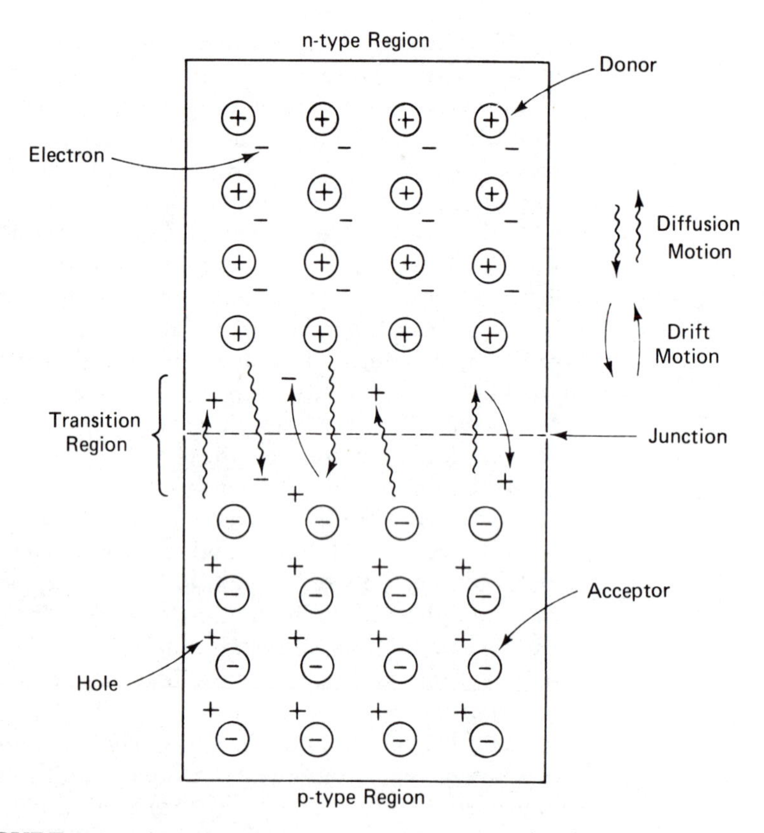

FIGURE 8-1. A *pn* junction. A positive external voltage causes current flow to increase exponentially.

net positive charge. A trivalent atom, in the *p*-type region, is called an *acceptor* and marked with a minus sign for complementary reasons. Plus and minus signs outside the circles depict mobile holes and electrons, respectively.

Charge carrier motion due to thermally induced diffusion is illustrated by wavy lines crossing the junction between the regions. As a carrier crosses this boundary, leaving its native region, it is called a *minority carrier*. In its native region, it is called a *majority carrier*. The region in the vicinity of the junction is called the *transition region* or *depletion region*.

As majority carriers diffuse, they leave behind "uncovered" charges bound in the lattice. These charges cause electric fields in their vicinity that induce minority carriers to return to their native regions. This field-induced carrier motion is called *drift* and is depicted by curved lines associated with holes and electrons in the figure. The net field of these bound charges creates a voltage potential that opposes majority carrier diffusion and enhances minority carrier drift across the junction. With no externally applied potential across the device (from top to bottom in the figure), drift and diffusion effects balance, and there is no net current flow.

For a majority carrier to cross the barrier presented by this potential, it must have sufficient kinetic energy. The probability that a carrier has such energy is inversely exponentially related to the magnitude of the potential. As a result, diffusion current is very sensitive to the magnitude of the barrier. If a positive external voltage is applied from the *p*-region to the *n*-region, it reduces the magnitude of the barrier and a positive current flow results which increases exponentially. Applied with the opposite polarity, an external voltage increases the magnitude of the barrier, and no current flows until the field magnitude becomes great enough to break electrons directly from their covalent bonds, at which point there is a sudden increase in current in the reverse direction. This is known as the *Zener effect.* At still higher voltages, the carriers themselves possess enough kinetic energy to break electrons from their bonds and produce a further increase in current known as the *avalanche effect.*

Figure 8-2 is a graph of the behavior of current with externally applied voltage for this device, which is called a *junction diode* or a *pn diode.* When voltage is applied with polarity which reduces the potential barrier, the diode is said to be *forward-biased.* Applied with the opposite polarity, the diode is said to be *reverse-biased.*

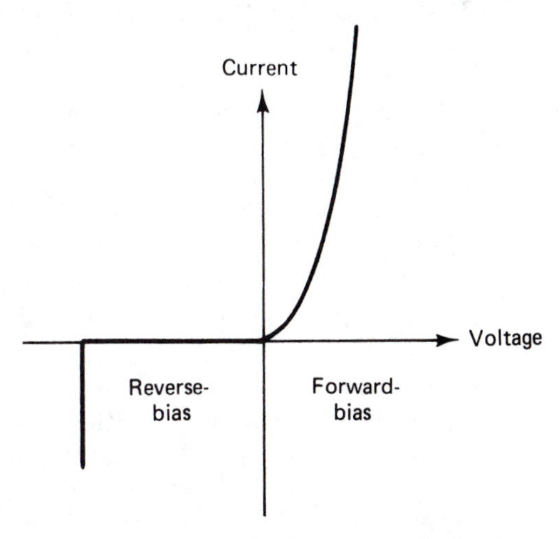

FIGURE 8-2. Junction (or *pn*) diode current behavior as a function of externally applied voltage. Note the rapid current rise when forward-biased and the sharp negative current when sufficiently reverse-biased.

More complex combinations of dopant concentrations and junctions can be used to form transistors of several types, as well as semiconductor equivalents of resistors and capacitors.

Semiconductor Device Fabrication

Several processes are used to manufacture discrete semiconductor components and integrated circuits. The relative merits of each have to do with the degree of process control they offer the manufacturer, the proportion of acceptable devices produced ("yield"), and their ability to form the complex structures required in highly integrated devices.

Semiconductor crystals can be grown from molten silicon by touching its surface with a seed crystal. As the seed crystal is slowly drawn away from the molten surface, atoms from the molten material form extensions of the lattice to yield a larger crystal. Crystal rods so formed are then sliced into *wafers*, which are used as the *substrate* material of the device. By doping the molten material with trivalent or pentavalent elements, the resulting substrates will be either *p*-type or *n*-type materials.

Figure 8-3 illustrates succeeding steps for one type of process used to form *pn* junctions in a substrate region. This is called a "planar" process. A wafer of *n*-type silicon [Fig. 8-3(a)] is heated in the presence of oxygen to produce a thin layer of silicon dioxide on its surface [Fig. 8-3(b)]. (Silicon dioxide is an insulator.) The wafer is then coated with a photosensitive compound. Areas of the compound that have been exposed to light through an optical mask will dissolve in an etching bath. Unexposed areas will not, remaining to protect the silicon dioxide from the etchant. By this photolithographic technique, certain areas of oxide are etched away to expose the underlying substrate [Fig. 8-3(c)].

Next, the wafer is exposed to a dopant gas, in this example a trivalent gas. Atoms of the gas diffuse into the exposed substrate to convert it into a *p*-type region [Fig. 8-3(d)]. Subsequent masking, etching, and diffusion steps are then used to create additional regions, which may also be doped to various concentrations and types. Finally, metal interconnections are made between devices formed by this process by aluminum deposition, the patterns of which are controlled by similar masking and etching operations.

Another manufacturing method creates regions through a deposition process in which added crystal structures of elements from doped vapors form on areas exposed by etching the oxide coatings. This is called an *epitaxial* process.

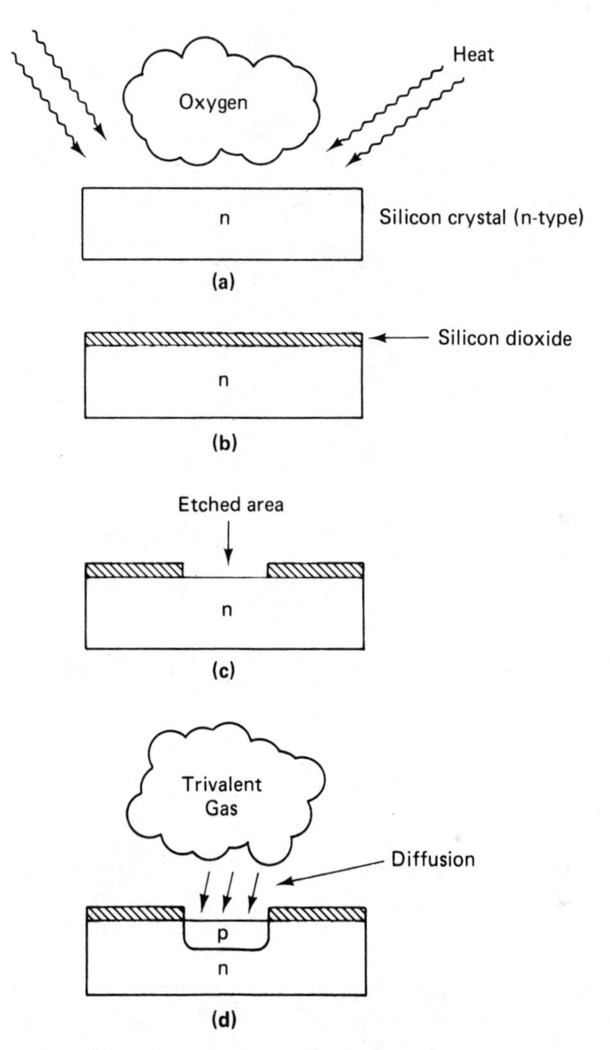

FIGURE 8-3. Semiconductor device fabrication with a planar process: (a) heating in oxygen to produce (b) a silicon dioxide layer, (c) etching to expose substrate, (d) doping to produce a *pn* junction region.

Figure 8-4 illustrates cross-sectional forms of common integrated circuit components. Heavily doped regions are shown with plus signs on their type letters. Inductors cannot be formed in integrated circuits, but resistors, capacitors, and active elements can be formed. In terms of required area, capacitors take up the most room and are followed by resistors. Bipolar transistors and diodes are the next smallest. The

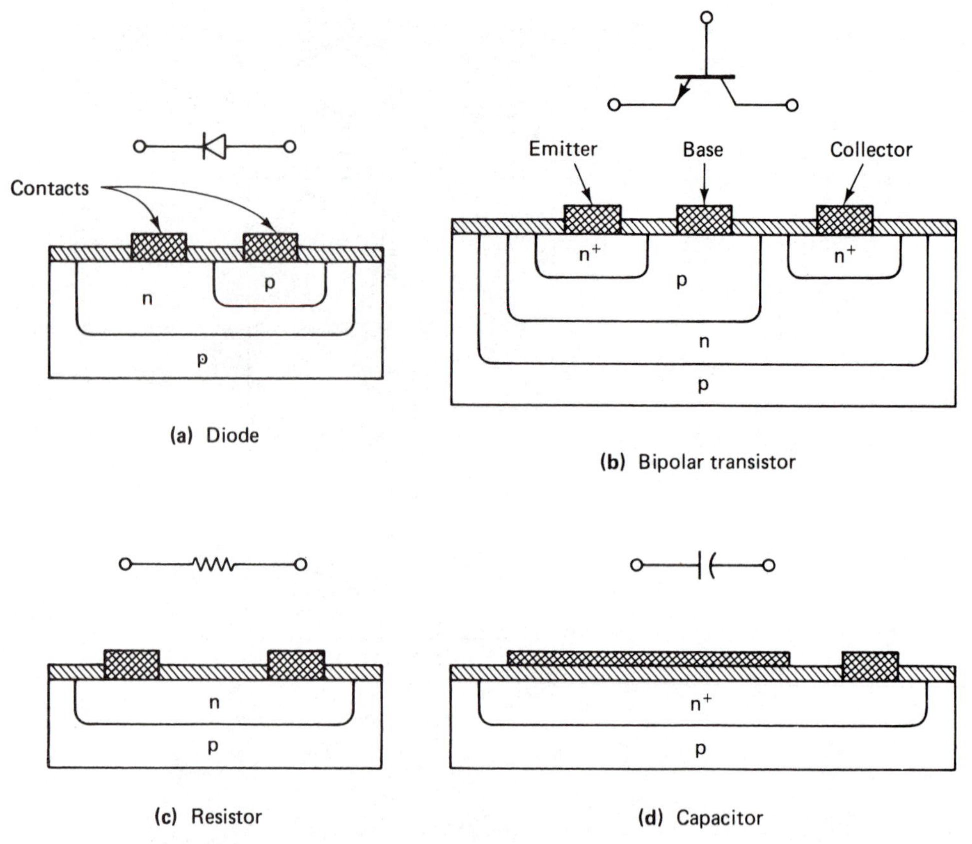

FIGURE 8-4. Cross-sections of common integrated circuit components: (a) diode, (b) bipolar transistor, (c) resistor, (d) capacitor.

metal-oxide-semiconductor field effect transistor (MOSFET) requires even less space. For this reason, it has the greatest utility in highly integrated devices such as microprocessors and memories.

Many copies of these integrated circuits are made on each wafer, laid out in arraylike fashion. The lines separating the circuits are then scribed, and individual circuits are broken from the wafer (Fig. 8-5). These individual blocks of silicon are called "*chips*" informally and *dice* formally. Extremely fine metal wires are then bonded to pads around the periphery of the die. The die is fastened in place in the cavity of an integrated circuit package, the fine wires are bonded to stronger leads (*pins*), and the package is capped to form the complete IC package. The most common form of package is called a *DIP* (which stands for "dual-in-line package"). This is illustrated in Fig. 8-6.

In combining and interconnecting ICs to form CPUs, memories,

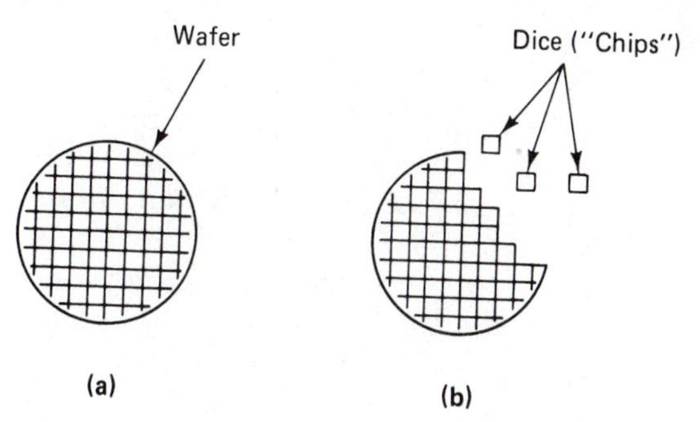

FIGURE 8-5. Mass production of integrated circuits: (a) array of circuits etched on a wafer, (b) dice scribed and broken from the wafer.

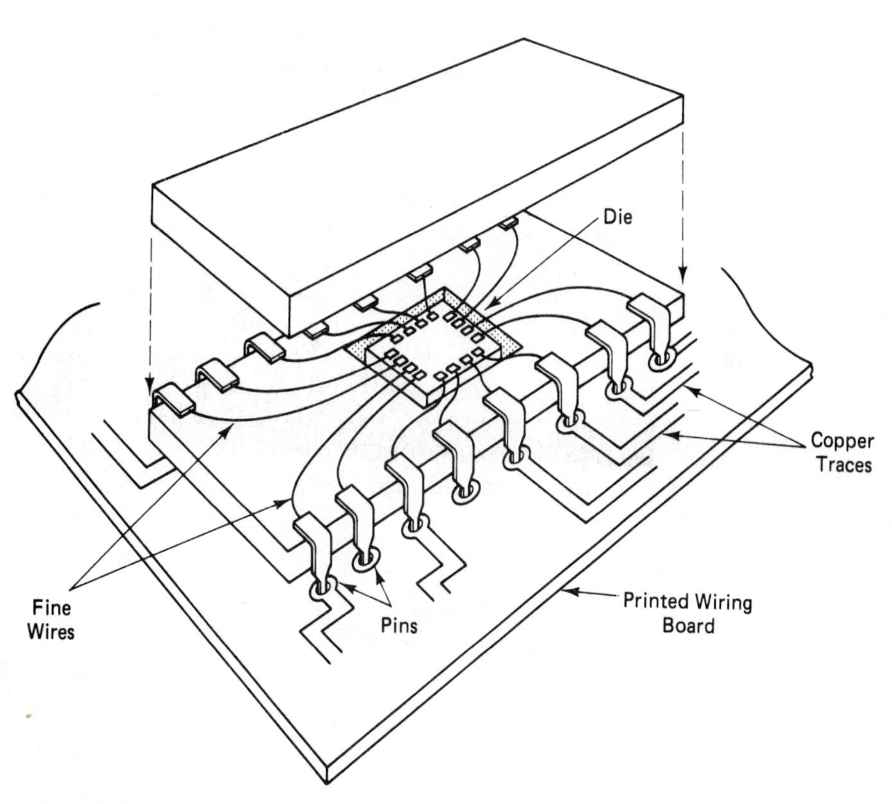

FIGURE 8-6. Assembly of an integrated circuit and its mounting on a printed wiring board.

and I/O subsystems, ICs are mounted (by their pins) in holes drilled in printed wiring boards. *Printed wiring boards* are usually sheets of glass-epoxy laminate on which interconnections between ICs and discrete components are made by copper traces, which are formed into the desired patterns by etching away unwanted areas of copper (in much the same way unwanted areas of silicon dioxide are removed from the surface of the silicon).

MOS Transistors

Early microprocessors and semiconductor memories used *p-channel MOS* transistors. Figure 8-7 is a simplified cross section of this device with its circuit symbol. LSI circuits based on this semiconductor technology are commonly called *PMOS* devices or *p-channel* devices. The substrate is lightly doped *n*-type material. Two heavily doped regions (*p*+), called the *source* and the *drain*, are deposited with aluminum for connection to other circuit elements. A lightly doped *p*-region is placed between the source and the drain, with an insulating layer of silicon dioxide covering it. An aluminum contact for the *gate* is placed here. Bias voltage on the gate affects the conduction properties of the lightly doped *p*-region beneath it, in effect "pinching off" the ability of this "channel" to conduct current between the source and the drain.

Later microprocessors and memories, typical of many types used today, use *n-channel MOS* transistors. Figure 8-8 is a simplified cross section of this *NMOS* device with its circuit symbol. In its simplest form, it is like a PMOS device with the region types reversed.

Because NMOS device performance is theoretically superior to that of a PMOS device, NMOS microprocessors generally exhibit faster instruction speeds and often use less power. They can operate at lower voltage thresholds that are compatible with bipolar TTL devices, which can serve as interface circuits. NMOS manufacturing difficulties forced the use of PMOS technology for early devices. The

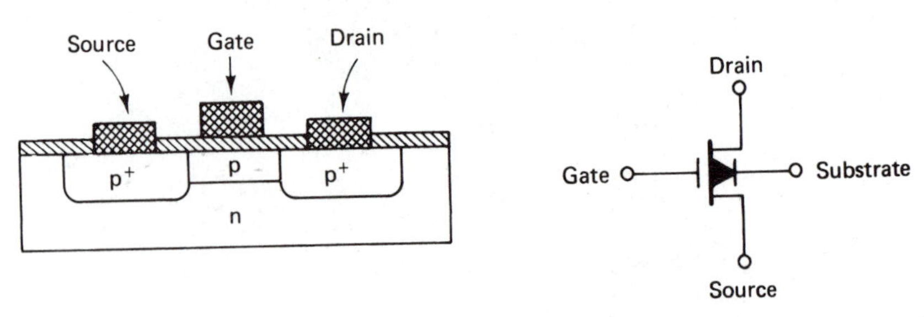

FIGURE 8-7. Cross-section of a *p*-channel MOS (PMOS) transistor and its circuit symbol.

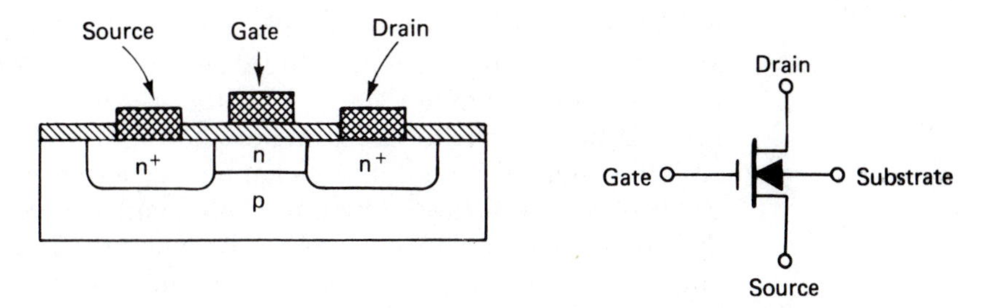

FIGURE 8-8. Cross-section of an *n*-channel MOS (NMOS) transistor and its circuit symbol.

solution of these problems brought NMOS to its present position as the principal microprocessor and memory technology, and continuing improvements strengthen its position and lead some to feel that it may someday rival bipolar technology in performance.

The third principal form of semiconductor technology is *complementary MOS* (*CMOS*), shown in Fig. 8-9 along with its circuit symbol. This device, a CMOS inverter, uses two transistors: a *p*-chan-

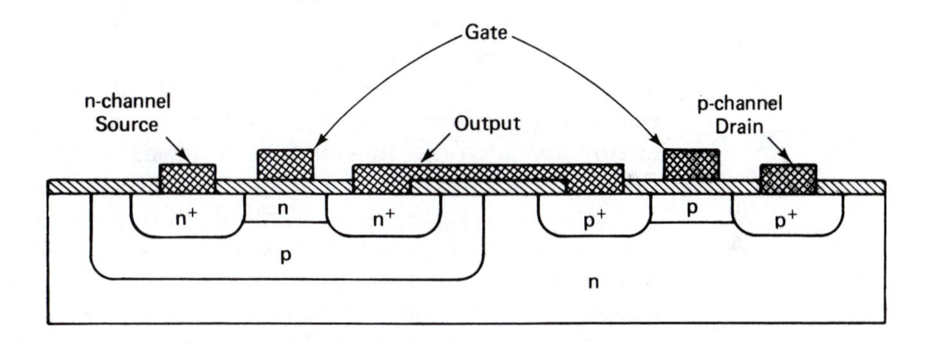

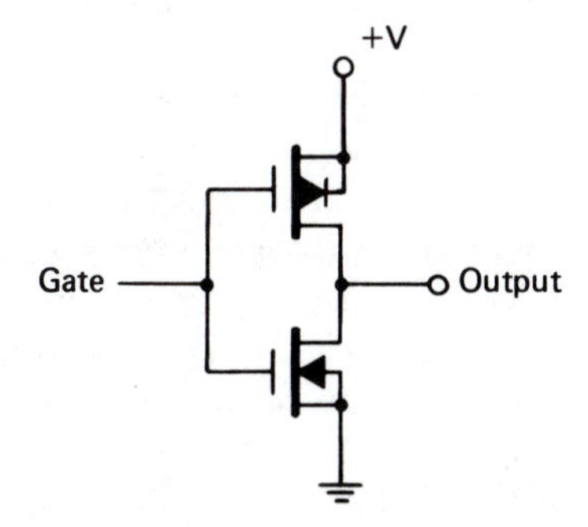

FIGURE 8-9. Cross-section of a complementary MOS (CMOS) inverter and its circuit symbol.

nel transistor, arranged to act as a load, and an *n*-channel transistor, arranged to act as a driver, with its output taken from the connection between the *p*-channel source and the *n*-channel drain. Input is applied to both transistor gates. When the input level is near the power supply voltage ("high"), the *p*-channel device is turned off and the *n*-channel device forms a low-impedance path between the output and the power supply return. Conversely, when the input is "low," the *n*-channel device is "off" and the *p*-channel device forms a low-impedance path between the power supply level and the output.

CMOS devices have several performance advantages, the principal one being that they consume very little power when not actually in the process of switching. When they switch, they do so very quickly, aided by the "active pull-up" effect of the *p*-channel device. They are also very tolerant of the actual power supply voltage, permitting operation with unregulated power supplies, which are smaller and less costly, and with small batteries. A small battery can provide enough power to retain data in a 4K-byte CMOS memory for weeks.

In spite of these advantages, CMOS is used in only two currently popular microprocessors, one newly introduced model, and a few CMOS versions of microprocessors that were introduced originally as NMOS devices. Among the problems with CMOS devices is the need to protect their inputs from static discharge (they must be handled very carefully during manufacturing and care must be taken during their power-on sequences) and the tendency for some types of circuits to "latch-up" in a mode of behavior similar to that of an SCR. Reliable solutions to input protection and latch-up problems have been developed recently and will result in its increased use.

Since most microprocessors are based on MOS technologies, we will not devote any more attention to bipolar technology than to note that it is used principally in SSI and MSI devices, in "bit-slice" central processor elements (which are explained in the next section of this chapter), and in extremely fast (but expensive) memory devices. These memories and central processor elements are used principally in very powerful special-purpose computers and general-purpose large computers and minicomputers.

There are many reasons for the degree to which one semiconductor technology enjoys favor over another. Semiconductor manufacturers try to deliver devices that combine high performance and reliability with reasonable cost. High performance is related to features that can be integrated into the chip and the speed at which the devices switch. The smaller the area required for an element, the more of them that can be integrated into the chip. The desire for speed comes from the need for devices to do more work in less time, but this is countered by the power consumed and the number of integrated devices. The higher the device power and the greater the number of devices, the more heat generated. Heat contributes substantially to reduced reliability. Fundamental physics limits the amount of heat that can be dissipated by a small package. Thus, one principal thrust of semiconductor research is the search for smaller, faster devices that require less power.

In terms of combined speed and power (called the *power-delay product*), CMOS is the best of widely used present technologies, followed by NMOS and the Low-Power Schottky Transistor-Transistor Logic (LSTTL) form of bipolar on a roughly equal footing, then by PMOS. Although LSTTL is faster than NMOS, it consumes more power. In terms of the area required for a device, NMOS is best, followed by PMOS, then by CMOS and LSTTL. Cost is a function of the number of masking and diffusion steps required in manufacturing the device; the fewer there are, the less cost in manufacturing. Here PMOS is least costly, followed by NMOS and CMOS, then LSTTL.

GENERAL CHARACTERISTICS OF MICROPROCESSORS

There are several characteristics by which microprocessors may be categorized and judged fit for particular purposes. The principal ones are semiconductor technology, size, "completeness," ways in which it is interfaced to memory and input/output subsystems, internal organization, addressing modes, instruction set, and the degree to which it is supported by other devices ("support chips") which work with it as interface, logic, and processing function elements. All of these operate in concert to determine the degree to which a selected microprocessor will be an effective tool for an industrial measurement and control application.

Microprocessor performance (speed and processing capability) is not the only issue. It must also be a good economic choice. Of all economic factors, the price paid for the microprocessor itself is often least important. You probably carry enough small bills in your wallet to pay for several of the most common types. Vastly more important are the costs of memory and input/output hardware needed for the complete microcomputer system of which the microprocessor is only a small, albeit critical, part.

Depending on the number of systems over which programming efforts will be amortized, you must also place substantial weight on the ease with which it can be programmed. Programming costs can easily become the principal factor in the total cost of a microcomputer-based industrial measurement and control development project. This is one reason why the availability of high-level languages for microprocessors has become such an important consideration.

If only a modest number of systems are needed, the availability of board-level and system-level products becomes important. A great deal of engineering effort and cost can be saved by making systems from board-level products. Even more can be saved by using completely assembled system-level products if only a small number of systems is required.

In this section, general microprocessor characteristics are reviewed and advantages of different approaches are weighed. Unless you become involved in detailed microcomputer design, you will not be substantially affected by the finer points of semiconductor technology, "completeness" of the microprocessor package itself, and details of interfacing the microprocessor to the circuits immediately surrounding it. These will be treated only briefly. The merits of various semiconductor technologies were reviewed in the previous section, so we begin with the next item, which is the one by which most microprocessors are classified—their "size."

Size

Microprocessors are commercially available in the following "sizes": 1, 2, 4, 8, 12, 16, and 32 bits. The term *size* refers to the number of bits in its principal internal registers generally and, more specifically, to the width of its ALU—the number of bits in each ALU operand handled by a single instruction, e.g., an addition operation. Thus, a microprocessor that adds a pair of 8-bit numbers in one operation is classified as an 8-bit microprocessor; one that performs addition four bits at a time is classified as a 4-bit microprocessor.

Microprocessor size usually characterizes the size of data items transferred between it and its memory and I/O devices, but not always. In some larger types (16- and 32-bits), data and instruction exchanges use bytes to make up larger "words." Being of a specific size does not mean it cannot do arithmetic on larger items. Many 8-bit processors, for example, can perform increment and decrement operations on 16-bit data. Virtually every 8-bit processor provides instructions that make double-precision (16-bit) addition and subtraction possible with only a few program steps.

Microprocessors of the 1-, 2-, and 4-bit sizes encompass three particular application areas: very small logic controllers, bit-slice processors, and calculators. Very small logic controllers are used instead of SSI and MSI logic for relatively simple control problems that use Boolean operations and no computations. An example would be a simple conveyor system sequence controller like the one described in Chapter 4.

Bit-slice processors are used in microprogrammable special- and general-purpose computers. These components can be imagined as cross sections ("slices") from which processors of greater word size can be constructed. The most popular types are 4 bits wide and are members of a collection of devices that can be used to make rather fast and versatile CPUs. A 16-bit ALU can be formed from four of the "central processing element" members of the collection—essentially fast bipolar ICs capable of executing a small number of basic operations and including elements of several registers. Other members of this matched collection serve as instruction fetchers, registers, and decoders. These are assembled to form CPU logic circuits which handle relatively wide instruction words (20 or more bits per instruction is not unusual).

At this stage of assembly we have a fast, but relatively simple, ALU and fast wide instruction handling logic: a simple, but fast, computer. At this point a substantial departure is made from conventional design. Instead of requiring the usual programmer of this computer to use the simple (but wide) instructions, the designer programs it, using these instructions, to *emulate* the instructions of a computer of a different type—instructions better suited to programming for its special purpose (e.g., high-speed digital signal processing) or as a much more versatile general-purpose computer.

The steps required to carry out more powerful instructions (e.g., multiplication, division, index register-based access to memory) are coded in the wide instruction words as a series of more fundamental operations and stored in a very fast bipolar memory of the same width. This program is called a *microprogram*; the memory in which it is kept is called the *microprogram memory* or *control store*. Instructions in the user's program (*macroprogram*) are fetched from main memory under microprogram control and their instruction fields used by the microprogram to select individual microprogram routines, which carry out the operations implied by the *macroinstruction*.

Many minicomputers and large computers are microprogrammed designs and have several useful features. For example, they can be microprogrammed to emulate the instruction sets of older computers. This permits software developed originally for the older computers to be run on the newer, faster computer. If the control store can be changed dynamically or programmed by the user, it becomes possible to change its "personality," depending on the program it is running at the moment, or to add special instructions that make it more effective for particular applications.

Four-bit microprocessors are commonly used for calculators and electronic cash registers. Four-bit registers are used to hold each decimal digit during the computation. By carrying out arithmetic operations in this way, certain problems with representing decimal fractions as binary numbers are avoided and the entry of operands and display of results are simplified.

Although more 4-bit processors exist, more people have used 8-bit processors. This is because they are better suited to a wider range of general-purpose computing, information processing, communications, and (of most interest in our context) measurement and control applications. For these applications, they presently offer the best combination of effectiveness, cost, support devices, design aids and development equipment, existing software (portions of which can sometimes be used in new applications), availability of board- and system-level products, and general experience among designers and users. An 8-bit word size is convenient for handling ASCII characters, making 8-bit processors suitable for text processing and communications applications. The resolution of an 8-bit word is adequate for many control problems and double-precision operations are easily performed when more resolution is required.

Because several high-level languages are available for 8-bit microprocessors, they are easier to program in those cases (and there are many) where the slower speeds and greater memory requirements of high-level language programs are not impediments. It turns out that many measurement and control applications can be handled very satisfactorily when programmed for an 8-bit computer in as simple a language as BASIC.

Even so, the wider its registers, the easier a processor is to program. For this reason, there is great demand for 16-bit processors. Until recently, this demand could be met only by minicomputers and certain board- and system-level products which, although technically microcomputers, could be said to be on the borderline. Now there are several 16-bit microprocessors on the market. And, just as many 8-bit microprocessors can perform limited 16-bit arithmetic; 16-bit microprocessors can perform limited 32-bit arithmetic.

Even more important than register size is the ability of 16-bit microprocessors to address much larger memories directly. This has several advantages. One is that we can write larger programs and keep more data quickly accessible without being so concerned about running out of memory. There is a rule of thumb among programmers that, in general, there is a trade off between the memory a program requires and the time it takes the program to perform its function. A programmer can often make a program run faster if more memory is available. Thus, the ability to address larger memories directly provides opportunities for improving performance.

Another advantage is that it becomes easier for several processors to share the same memory. There is room for programs for each processor, they may share programs, and they may have access to the same data essentially simultaneously. This makes it easier for the processors to cooperate in carrying out operations. For example, one might serve as an input processor for the system, while another computes and makes control decisions, and a third serves as an output processor.

The full impact of 32-bit microprocessors remains to be seen. To the extent that they are like 16-bit processors (only "more so"), they offer better performance for the same kinds of reasons that 16-bit processors offer advantages over 8-bit processors. At first, they will be much more expensive. However, if they exhibit the same patterns of economic behavior shown by almost every other highly advanced and innovative LSI product, their costs will tumble rapidly, at a rate of 25 to 30 percent or more each year.

They are claimed, by their manufacturers, to promise performance that will exceed that of several widely used minicomputers. But their principal impact may be due to two significant design innovations.

The first of these is that they are equipped with features that make them particularly amenable to operation in multiprocessor systems. Multiprocessor systems can offer significant performance improvements over single processor systems, but their design requires consideration of certain matters relating to the simultaneous operation of cooperating and independent programs that may compete among themselves for such resources as access to portions of memory, I/O devices, and the services of other programs. Although it is possible to make effective multiprocessor systems using 8-bit microprocessors, it is easier to do with 16-bit processors because they can address larger memories and are equipped with hardware that helps programmers manage the coordination of resources. The 32-bit microprocessors take this further by offering even larger memories and more versatile mechanisms for processor and program coordination.

The second significant innovation is that they are designed to execute programs written in a high-level language in a more direct fashion and include, in their hardware, certain operating system functions that facilitate running these programs in a multiprocessor system configuration. With these devices, we begin to find mechanisms integrated into hardware that have previously been considered to belong to the "software domain."

Before closing this section, we should not forget the 12-bit microprocessor. This is a microprocessor version of the first popular minicomputer—the PDP-8 manufactured by the Digital Equipment Corporation. This processor (the 6100, manufactured by the Intersil Corporation) executes the same instruction set as the PDP-8 and is also distinguished by being one of the few CMOS microprocessors. Although one of its advantages is that it can use programs written originally for the PDP-8 (of which there are many), it never enjoyed widespread popularity—perhaps because the PDP-8 instruction set does not include some functions and addressing modes that modern programmers have come to expect and perhaps because 12 bits is sometimes an awkward word size (e.g., when one is handling ASCII characters). Nevertheless, it is an interesting branch in the evolution of microprocessors.

Packaging and Completeness

Integrated circuit packaging technology is also a vital element in microprocessor evolution. Coupled with levels of integration, it determines how "complete" a microprocessor can be. Although completeness is of first concern to microcomputer designers, it also affects users. To see the nature of the packaging and completeness issue, consider Fig. 8-10. This illustrates an important problem in microprocessor design—how to handle the many signals that are required.

First, power must be provided. In early PMOS processors, two supply levels were needed, along with the power supply return. With NMOS processors, a single supply can be used; but two pins on the

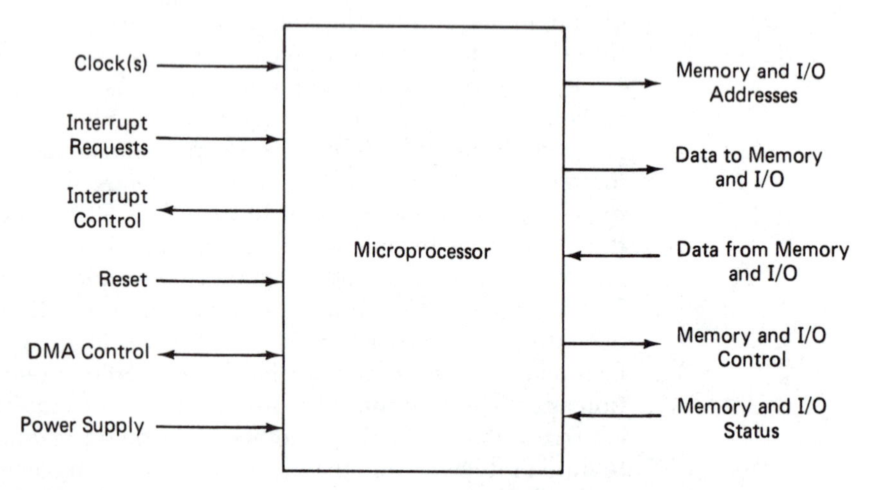

FIGURE 8-10. Many different signals are required for a small microprocessor package. How these are handled is a significant design problem.

package are still required. To ensure that the processor will start up in a known state when power is applied, and so that it can be restarted if necessary, a "reset" input will be needed—another pin.

Second, clocks are needed. Older designs used multiphase clocks (some as many as four); modern designs use single-phase clocks. If an external oscillator is used to develop the clock signal, one pin will suffice for its input to the processor, but additional external devices are needed to make up the circuit. They cost money and printed wiring board space. If an oscillator circuit can be included in the microprocessor package, two pins are needed, plus one more if the oscillator output is required to synchronize memory and input/output devices with the processor.

If external devices are to be allowed to interrupt the microprocessor (a practical necessity for most control applications), input pins will be required for these, and some means of controlling the interrupts will call for one or more additional pins. The total number depends on how sophisticated and versatile the interrupt system is to be.

When it is necessary for other devices to have access to memory or I/O devices, a means is needed to prevent the conflict that would occur if a device and the processor both attempt access at the same time. This situation arises in multiprocessor systems and also in systems that provide for *direct memory access* (DMA) input/output operations. DMA is used often for high-speed data transfers between memory and peripheral devices. With DMA, logic associated with the peripheral device is given direct access to memory so that data transfers can take place without word-by-word CPU intervention. The memory becomes a shared resource, in effect, and a means of arbitrating its use is needed. We examine DMA in detail later. For the purpose of this discussion, it means that at least two more pins are needed.

So far, we have identified a need for between nine and more than a dozen pins, depending on design features and the type of semiconductor used. We now arrive at the central requirements: interfaces with memory and I/O devices, the nature of which will be prime determinants of performance.

Sixteen signals are needed to address a 64K memory directly. Parallel word transfer requires as many more signals as there are bits in a word. Two more signals are needed to indicate whether a read or write operation is taking place. For a byte-oriented memory, this address, data, and control signal requirement amounts to another 26 pins, bringing the grand total into the neighborhood of forty. And this assumes there is some way in which all of the data bytes can be handled on the same eight lines—whether they are being read into the processor or being written by it; whether they involve memory or input/output devices. (There is a way, explained in the next section.)

An integrated circuit package with forty pins was a significant technical challenge when the first microprocessors were being developed. Now, 40-pin DIPs are common. They "just fit" the requirements of second generation microprocessors. Some "late" second generation processors go beyond, to 48- and even 64-pin packages. But there are substantial practical limits to the lengths to which the number of pins can be increased, and early processors had to use smaller packages (the Intel 8008 used an 18-pin package).

One solution used in the early days was to share the functions of certain pins and provide status output signals which identified the information that was on these pins at the moment. Control information and data were transferred by using the same pins, but at different times—in effect, it was *time-multiplexed.* This meant that the logic immediately around the processor had to be more complex. Several circuit packages were needed to form a complete working CPU. In those days, this meant a moderate number of SSI packages—MSI functions had not yet been influenced by the availability of microprocessors to the degree that they were particularly useful for this purpose, and they had not evolved as far as they have today.

The first microprocessors in 40-pin packages substantially reduced the required number of additional packages by being offered with integrated support chips designed to perform these functions and by incorporating some functions into the microprocessor package itself. Examples of these are the Intel 8080 and the Zilog Z80. Microprocessors that require external SSI and MSI devices and support chips are called *multichip microprocessors.*

These were followed by *single-chip microprocessors*, of which the Intel 8085 is an example. In its minimum system configuration, it requires only a crystal to become an operating CPU. It does this by returning to the method of sharing pins (part of the address is time-multiplexed with data) and by being offered with LSI memory and I/O chips which incorporated, in their circuit packages, the logic required to handle time-multiplexed addresses and data.

There is a branch at the next stage. On one branch, we find *single-chip microcomputers* in which memory and I/O ports have been incorporated into the package to form a complete *computer* on a single integrated circuit. Examples are the Intel 8048 and 8051. On the other branch, we find the new 16-bit and 32-bit processors, which have returned to the multichip approach and are also found to be using 48- and 64-pin packages as well. Examples are the Zilog Z8000 and the Intel 8086 and iAPX432.

With each of these approaches to packaging and varying degrees of completeness, design tradeoffs have been made which influence the effectiveness of any particular type of microprocessor for a particular application.

Interfacing to Memory and Input/Output Devices

Early minicomputers used separate busses for memory and I/O interfaces. As shown in Fig. 8-11, memory subsystems were connected to the CPU with their own address and data busses. Separate control lines were used for data and instruction word transfers. These were controlled by CPU logic for retrieving and decoding instructions and by memory-referencing instructions.

Input/output address bus, data bus, and control line activity was controlled by the programmer, using a separate set of instructions. Since the number of I/O devices is small compared to the number of memory locations, fewer address lines were required for I/O operations. I/O control lines differed from memory control lines in that they provided for such operations as skipping the next instruction in sequence if the device indicated that it was ready. DMA logic was sometimes included to provide a direct path between selected I/O devices and busses and control lines associated with memory.

Second-generation minicomputers, and microprocessors which followed them shortly thereafter, used a simpler arrangement which has several advantages and is still representative of many interfacing

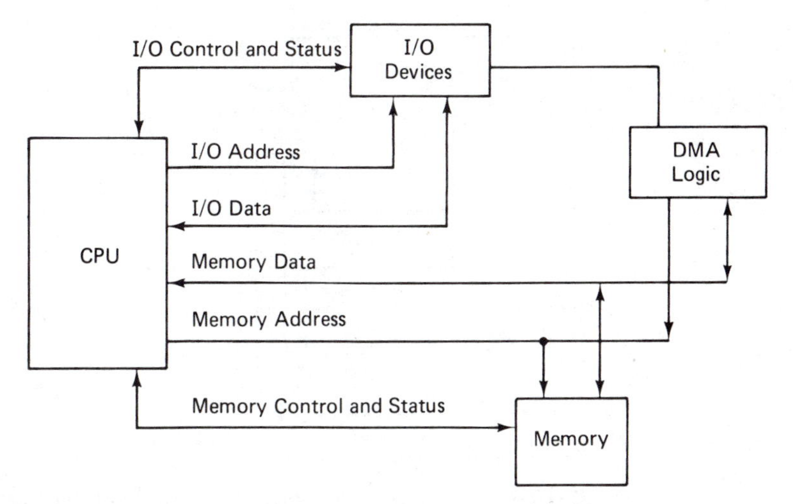

FIGURE 8-11. Use of separate busses for memory and I/O interfaces.

arrangements used today. This is illustrated in Fig. 8-12. Conceptually, it can be thought of as a *single bus structure* used for both memory and I/O devices. Address, data, and control signals are simply parts of this bus. In one variation, separate control signals are retained for memory and I/O devices, but addresses and data for both still use the same signal wires. A variation on this avoids separate memory and I/O control lines and treats I/O device ports as if they were memory locations. This is called a *unified bus structure* and the term *memory-mapped* is widely used to describe I/O addressing in such an arrangement.

The unified structure provides for using memory-referencing instructions when dealing with I/O devices. This is useful because all of the addressing modes of the processor can be brought to bear on these data transfers. It makes it possible for DMA controllers to be devices which essentially move data from one place in memory to another. DMA controllers can be dynamically reassigned to different I/O devices more easily. It also opens the door further to multiprocessor systems, since all processors on a unified bus have access to all of the I/O devices on the bus.

However, there are disadvantages. First, more complex I/O address recognition logic is required. In a unified structure, it must recognize addresses that have more bits than are actually needed to distinguish among the relatively small number of I/O ports that

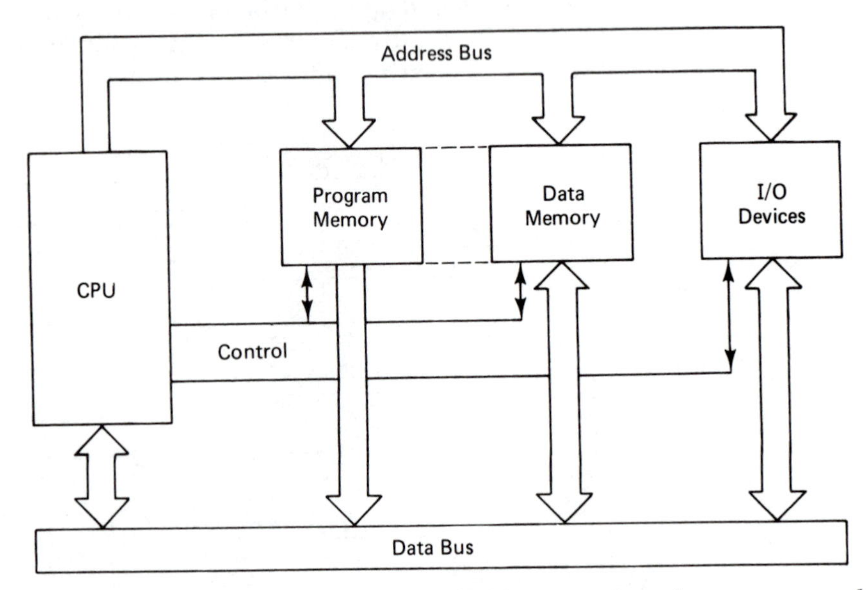

FIGURE 8-12. Use of a single unified bus structure for memory and I/O.

to a "0," it does not provide equally good response in switching from a "0" to a "1"; although it has fast *fall times*, it does not have fast *rise times*. This is because only the resistor current is available to charge the capacitance of the bus and the output transistors. Thus, wired-AND bus arrangements cannot provide the high overall speeds desired for a high-performance bus system.

The "active pull-up" arrangement of a TTL gate output stage can provide both fast fall times and fast rise times. This arrangement, which is called a *totem-pole* output stage, illustrated in simplified form in Fig. 8-15, provides a pair of output transistors, one above the other in totem-pole fashion. When the output is to be a "1," the lower transistor is switched off and the upper transistor is switched on to provide a low impedance source of current to bring the bus level up quickly.

If the currently active device (processor, memory, or I/O interface) were to drive the bus with a totem-pole output, it would have higher performance than an open-collector output. However, there is a significant problem with bussing totem-pole output stages together. Any output transistor wishing to send a "0" level must be able to "sink" the combined current from the active pull-up transistors of the other devices on the bus. In some cases, it may be unable to pull the signal low enough for receiving gates to recognize it as a "0." In other cases, the active pull-down transistor may be damaged by the high current.

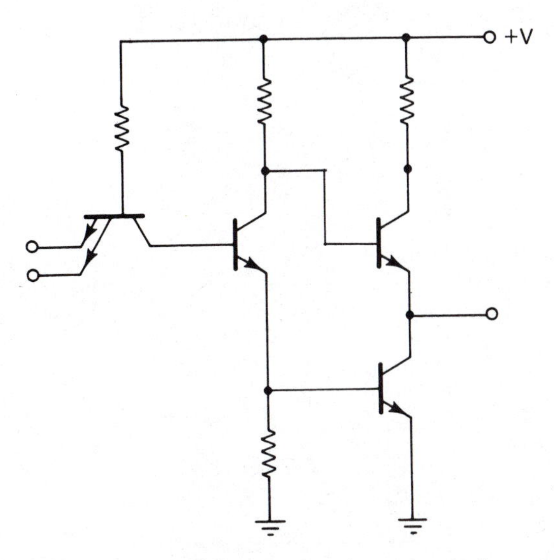

FIGURE 8-15. Simplified TTL gate with totem-pole output stage for fast signal rise and fall times.

The solution to this problem was the development of the *three-state* output stage. This concept is illustrated in logical form in Fig. 8-16. Actually, it is fairly obvious. If the pull-up transistors of inactive devices interfere, do not turn them on. In the circuit of Fig. 8-16, the added control line acts to keep both totem-pole output transistors turned off when the device is inactive (when the control line is high). When the control line is low, the remainder of the circuit acts as an otherwise ordinary NAND gate. Thus, fast rise and fall times are available to active devices, and they do not interfere with other devices on the bus when they are inactive.

The circuit of Fig. 8-16 is very simplified compared to actual devices. Typical devices include additional internal elements to improve switching speed and output drive capability and to minimize both leakage currents from disabled output stages and the loads that are presented to inputs. Some include hysteresis input circuits to help reject line noise on large bussing systems.

Most microcomputer systems utilize three-state bus drivers, even on their address busses, since these allow multiple processors and DMA arrangements. When a processor wishes to gain access to the system busses (the ones it shares with other processors), it requests access through priority and arbitration logic. When this logic grants access, the requesting processor then becomes the new *bus master*, activating its three-state drivers while the previous bus master gets out of the way by disabling its drivers. A processor may also give up access to DMA controllers by temporarily disabling its three-state bus drivers.

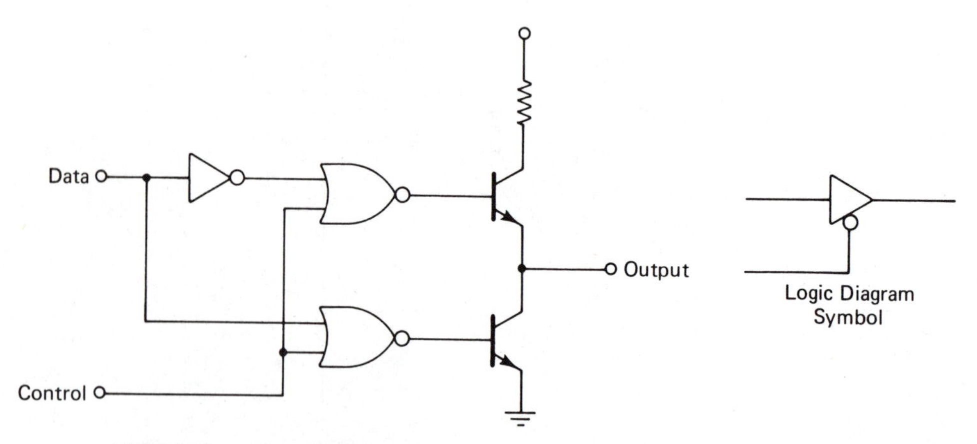

FIGURE 8-16. Simplified logic of a three-state output stage. This provides fast response yet allows devices to be bussed together.

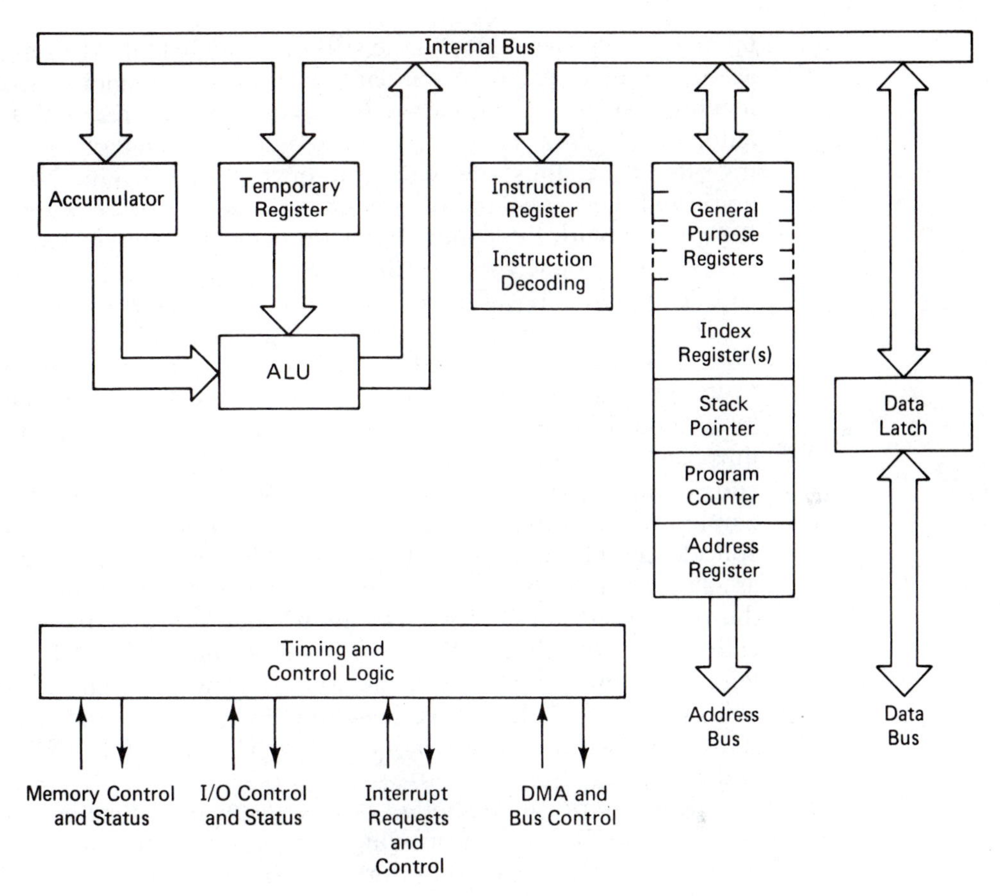

FIGURE 8-17. Internal organization of a typical 8-bit microprocessor. Note the use of an internal bus to connect the principal sections.

Internal Organization

In evaluating microprocessors with respect to internal organization, the first issue is the utility of the internal register arrangement. This is of greatest interest to programmers, because it strongly influences the ease of writing and testing programs and the speeds at which they operate. In most microprocessors, logic elements and registers are connected by an internal bus, much as illustrated in Fig. 8-17, which is typical of several popular 8-bit microprocessors. Advanced 16- and 32-bit processors are more sophisticated, but the general principles of their internal organizations are much the same.

The accumulator and a register, for temporarily holding a second operand, are associated with the arithmetic logic unit. ALU operations are performed on the accumulator and the contents of this register, leaving results in the accumulator. The temporary register is loaded automatically from a general-purpose register or memory location during one of the steps into which the instruction is divided by the operation of the logic—the programmer is not usually required to load it explicitly with a separate instruction and in many designs, in fact, has no means of accessing it directly.

The collection of general-purpose registers, stack pointer, and program counter are arranged in a *file* or *register bank* from which they are selected and moved about by the instruction execution logic. Letter mnemonics are used to denote registers. The accumulator may be termed the "ACC," the "A register," or simply the "A." General-purpose registers are termed the "B," "C," "D," or "R0," "R1," "R2," and so on, with variations if they also serve other purposes. For example, in some processors a pair of general-purpose 8-bit registers are concatenated to form a 16-bit register which the programmer can use as a limited index or memory address register. The register containing the eight most significant bits ("high-order" part) of the address is called the "H register"; the other is called the "L register" and contains the "low-order" part of the address. The stack pointer is often abbreviated "SP," the program counter "PC," and true index registers, if the processor has any, are abbreviated "IX" or "IX0," "IX1," "IX2," and so on.

In some processors, other 8-bit register pairs may be concatenated to hold 16-bit data on which simple incrementing and decrementing operations can be performed. They can sometimes serve duty as index registers or pointers to arrays of data. If they can be loaded into the PC under program control, they can be used for branching in which the destination address is computed or taken from a list of addresses stored in memory.

The number of general-purpose registers varies. Six is typical, as in the Intel 8080 and 8085 and the Zilog Z80 (which has two banks of six general-purpose registers). Some provide fewer; some provide 8 or 16. In some 16-bit processors, which tend to have more registers than 8-bit processors, some general-purpose registers also act as accumulators, with hardware automatically moving results to them from the true accumulator after the operation is complete.

In most cases, the registers are actually implemented as small read-write memories on the chip. However, some are designed to treat main memory locations as general-purpose registers. This may be done by requiring the programmer to predefine a block of locations for this purpose. Usually, the address at which this register bank starts may be controlled by the programmer, but within limits (e.g., only one bank may be active at a time; addresses start at multiples of 16).

Processors with a small number of internal general-purpose registers offer faster access to them, but programmers may find it necessary to save intermediate results more often. Registers kept in main memory cannot be accessed as quickly, but there can be more of them. The principal factor is usually the complexity of computations; the more complex they are, the more intermediate results there will be and the greater the likelihood of "running out" of general-purpose registers. This situation favors registers in memory. Simpler operations, however, will execute more quickly with internal registers that do not have to be fetched from main memory each time they are used.

Latching registers and drivers are connected to the address and data bus pins of the IC package. Although these provide enough drive capability in most microprocessors to handle a small number of external devices, additional buffering is generally needed in most general-purpose microcomputer designs. The remaining internal elements consist of logic that controls timing and sequencing, controls signals to memory and I/O devices, and monitors and responds to interrupt requests. Instruction decoding and execution steering logic often uses logic arrays similar to those mentioned earlier.

There are two other common variations on internal organization. In the first, additional general-purpose register banks are included, with instructions provided to switch between them quickly. Thus, one register set may be reserved for use by a fast response interrupt routine or a frequently used subroutine. By switching to another set of registers, these routines can avoid having to save the contents of the main registers they use. This can improve response and execution times considerably. The Zilog Z80 and the Intel 8051 are examples of this internal organization.

In the other variation, the low-order part of the address register is multiplexed onto the data bus early in the instruction execution cycle. External devices receive and latch this part of the address while decoding the high-order part, which is presented directly from the IC package pins. Actual data transfer is then done during a subsequent part of the cycle. This relieves eight pins on the package for other uses, as noted earlier. The Intel 8085 and 8051 and the National Semiconductor NSC800 are examples of microprocessors which use this variation.

Addressing Modes

Most microprocessors provide several different modes of addressing memory locations. The basic general types (direct, indirect, and relative addressing) were explained in Chapter 7. Most of the general-purpose 8-bit processors commonly employed in industrial applications offer these modes. In this section, some variations are discussed.

The number of locations that can be addressed directly by an instruction depends on the number of bits allocated to its address field. As we saw in the last chapter, variable length instructions can provide added direct address bits, but at the cost of a more complex design. Processors with fixed-length instructions can provide for more locations by dividing their memory spaces into *pages*, the size of which is determined by the number of address field bits. Within a page, locations can be addressed directly in a natural manner. The particular page is selected by the contents of another register (a *page register*), which is loaded with the *page number* by a separate instruction. Thus, the program can work freely in a particular region of memory and can switch to another region by loading the page register with its page number. This *fixed page addressing* is found in the Intersil 6100. Paged addressing works reasonably well when programs can be arranged so that related data is stored in close proximity. It provides for greater total memory capacity at less hardware cost. However, the boundaries of these pages are at fixed absolute addresses in the memory.

A variation of fixed-page addressing is *floating page addressing* or *relative addressing*. This provides better memory utilization and convenience, but can lead to errors if it is not used carefully. Instead of fixing page boundaries at particular addresses, the page "floats" along, with its center located at the instruction currently being executed (defined by the contents of the program counter) or at an address contained in an index or relative address register. *Offsets* (or *displacements*) from this center address are taken from the address field of the instruction and added to a *base address* taken from the PC, index, or relative address register. Although this requires extra hardware, the programmer can access memory locations 100 bytes, say, ahead of the current instruction, four bytes behind it, et cetera. In a typical processor with this type of addressing, an 8-bit offset field is used to provide a range of 256 bytes, up to 127 ahead or 128 behind the base address. Usually, relatively addressed branch instructions are also included so that the program can jump forward or backward within this range.

It must be used with care, because subsequent changes in the program can invalidate previously used offsets by changing the actual number of bytes between a location and instructions that refer to it. An important benefit, however, is that it makes it possible to write small inherently *relocatable* routines that can be placed anywhere in memory. Thus, the program can be divided into parts, written and tested separately, which are then more easily combined to form a larger program. Because it is so useful, most modern microprocessors are equipped with relative addressing.

If the base value of the relative address is taken from a register, then the mode is a mixture of indirect or indexed addressing and relative addressing. Terminology for addressing modes varies among manufacturers, so it is wise to avoid taking the terms at face value. It is better to study their user manuals carefully to be sure you understand them. Properly used, they can help you write very effective programs; improperly used, they can result in arrangements of program instructions which are error-prone or difficult for others to understand.

The size of addressable elements is also a consideration. In most 8-bit processors, only bytes can be addressed, with some extensions for loading and saving the contents of concatenated register pairs. Sixteen-bit processors generally offer two levels of addressing. In one, 16-bit words are addressed directly; in the other, bytes are addressed. Thus, a single byte or a consecutive pair of bytes may be transferred. Usually, pairs must begin at even-numbered addresses. This helps to simplify hardware and speed up data access by allowing the least significant bit of the second byte's address to be generated with a simple OR instead of a formal increment operation. Some 16-bit processors also have "long word" addressing in which four consecutive bytes are transferred as a single 32-bit quantity.

Several microprocessors can also address to the level of individual bits. This is useful in discrete control applications where most variables are single-bit inputs or outputs and logical combinations of these. Be cautioned, however, that this is not always so versatile as it seems. This is because the address of the bit within a memory word or I/O port usually must be a fixed constant in the address field of the instruction. It cannot be varied or computed; this can be a significant impediment to its effective use. In processors that do not address to the bit level, it is necessary to use AND operations in which one operand has only that bit set which corresponds to the bit to be accessed. Such operands are usually called *masks*, since the bits that are not set "mask" off unwanted bits from the remainder of the byte.

Instruction Sets

If every variation on every instruction were counted, the typical microprocessor would seem to have a very large number of instructions indeed; yet many would be of marginal value to the programmer. Among those counted would be all variations on register-to-register move operations; for example, instructions which move the contents of the A register to the A register. These do nothing but take up time (which, it must be admitted, is of value when it is necessary to insert a short delay into the program). Sheer numbers of instructions are not adequate measures of microprocessor capability.

It is usually best to divide the instruction set into classes of instructions and critically examine the processor's capabilities for each class. For example, instructions are often grouped into the following classes:

Data transfer—Instructions that move data from one place to another (register-to-register, register-to/from-memory, I/O).

Arithmetic—Instructions that perform and support addition, subtraction, increment, decrement, multiply, and divide operations.

Logical—Instructions that perform and support Boolean logic and the handling of masks and data fields that are smaller than the microprocessor word size.

Control transfer—Instructions used to branch and call and return from subroutines and interrupt routines, either unconditionally or depending on the results of previous operations and data values.

Miscellaneous—Instructions that do not fit any of the above categories, such as interrupt enable/disable and priority control, delay and timing operations, and register bank switching.

Other groupings and finer divisions are possible, of course. The idea is to evaluate the instruction set for its effectiveness for the application one has in mind. It is, therefore, best to form groupings that relate well to the application. The grouping above is reasonable for an application in which no particular type of operation in overwhelmingly important, such as general-purpose industrial applications, which tend to be rather balanced in their need for all types of instructions. Applications requiring a great deal of computation place greater emphasis on arithmetic and data transfer operations. Applications requiring a great deal of communication place greater emphasis on data transfer and treat input/output and interrupt handling operations as separate categories.

It is a good idea to select several important application program functions and code abbreviated models of portions of them. They need not be strictly rigorous in terms of language and details of operation—just enough for you to get the feel of the instruction set and an idea of how much memory and time they will require in the final program. The user's manuals will tell you how many bytes each instruction uses and how much time they take to execute. By adding them up, you obtain estimates that help to compare one processor with another.

These are called *benchmarks* and are often used to compare the capabilities of processors. If you have the opportunity to run and time these benchmarks on working systems, so much the better. Usually this is inconvenient or unnecessary for simpler processors, since it is easier to figure out the times by hand, using data from the processor's manuals. It may be difficult to do this for complex systems, however. Pipelining in 16- and 32-bit processors complicates the figuring.

Some manufacturers provide benchmark data. This can be helpful, but requires careful scrutiny of the conditions they assume. For some reason, benchmark data supplied by manufacturers always seem to indicate that their processors are the fastest and use the least memory. It is difficult to believe that they are all the best. Some independent writers provide benchmark reports in trade journals. Usually, these indicate that one processor is best for one class of function and another is best for something different; somewhat more credible results. Do not try to extrapolate these estimates into total operating time and memory requirements for any but simple programs without adding liberal factors to accommodate general "overhead" and things you have not thought of yet. Most programs end up needing functions you did not imagine when you began.

Using the general classes listed above, we now examine some variations found in processors commonly used for industrial applications.

Data Transfer. Register-to-register operations almost always provide complete "mobility"—the contents of any register can be transferred to any other register in the same register bank with a single instruction.

Register-to/from-memory operations usually take a little more trouble. In these, a relative address or index register must be loaded with the memory address before the transfer is executed, requiring two instructions to access an arbitrary location. In some processors, an exception is made for the accumulator in that a 3-byte instruction is provided with a full 16-bit address field so that it may be loaded or stored by a single instruction.

Some processors provide memory-to-memory transfer operations. The programmer sets up a concatenated register pair with the address of the first byte to be moved (the source address) and another pair with the address at which it is to be stored (the destination address). Another register is loaded with the number of bytes to be moved. At this point, execution of the multibyte transfer instruction causes the complete block of data bytes to be fetched and stored in its new locations, one byte at a time. Variations on this are found in block I/O transfers which refer repeatedly to the same source or destination addresses in the input/output device address space.

A variation on such single instruction/multiple operation instructions is the simpler one of automatically incrementing or decrementing an index or relative address register each time it is used. The term *auto-index* is used to refer to this type of operation. In some processors, particular registers are set aside as auto-index registers and are incremented or decremented each time they are used. In other processors, whether a register is treated as an auto-index register depends on the particular instruction used to refer to it.

Arithmetic. Virtually every microprocessor provides add and subtract instructions that operate on the accumulator and one other register or memory location, leaving the result in the accumulator. Some include variations that incorporate carries or borrows from previous adds and subtracts. These make multiple precision arithmetic easier. For arithmetic operations with data from the memory, an index or relative address register must usually be set up beforehand.

Increment and decrement instructions usually apply to all of the general-purpose registers as well as the accumulator, and sometimes to register pairs and to index, stack pointer, and relative address registers. In some processors, individual bytes in memory can be incremented or decremented.

Few 8-bit processors provide hardware multiply/divide instructions. These operations can be performed by software routines, but take a substantial amount of time, especially if they must deal with signed operands. For applications that require substantial computation in a limited time, it is a good idea to look for a microcomputer equipped with an arithmetic support chip like the ones described in the next section.

Virtually every application requires testing and comparing values (to see if they are zero or nonzero, positive or negative numbers, greater than, less than, or equal to other numbers). These characteristics are derived from the results of ALU operations and recorded in carry, zero, and sign flip-flops which are parts of a program status word (PSW) which is accessible to the programmer as a register. Individual bits in the PSW (called *flags*) can be tested and used to cause conditional program branches by members of the Control Transfer class of instructions.

In order to avoid altering the data being tested, most processors are equipped with a *compare* instruction. This subtracts the contents of a register or memory location from the contents of the accumulator, but refrains from loading the result back into the accumulator as it would in an ordinary subtract operation. Relevant PSW flags are updated, however, and may then be tested by conditional branch instructions.

Instructions that shift and rotate bits within registers and between registers may be classed as arithmetic or as logical operations, depending on your point of view—they are equally necessary to both. Just as adding a zero to the right side of a decimal number (in effect, shifting it left one digit) has the effect of multiplying by 10, shifting a binary number left one bit has the effect of multiplying it by 2. Likewise, shifting right can be interpreted as division by 2. Shift instructions are used for these and also for maintaining the location of the imaginary "binary point" in arithmetic operations on binary mixed numbers. The most commonly used processors provide only for single-bit shifts per instruction. To shift two or more positions, instructions must be repeated. However, some provide for swapping the four bits in each half of an 8-bit accumulator—in effect, a four-position rotate operation.

In shift and rotate operations, each bit is copied into the position to its left or right, depending on the direction selected by the instruction. Several types of instructions might be offered which differ in the way in which they dispose of bits on the end of the register. In some, the end bit is copied into the carry flag. In others, it is simply dropped out of sight. In rotate instructions, it is copied into the other end of the register. Similarly, there are variations that use the carry flag as the source of an incoming bit at the opposite end of the register or that bring in selectable values of "0" or "1." Some shift right instructions copy the most significant bit to the right. This is useful for a form of fixed-point binary number representation known as "2's complement," which is the most common binary number system used in microprocessors.

Generally, shift and rotate operations apply only to the accumulator. Bits shifted out of either end of the accumulator may be copied into the carry flag and held there while the accumulator is exchanged with one of the general-purpose registers. The bit in the carry flag can then be shifted into the new data in the accumulator. In this way, longer words can be shifted.

Logical. Among the logical operations, one generally finds all the basic operations: AND, OR, and NOT (which is generally called complement in an instruction description). Exclusive-OR is often included, sometimes called "add without carries," although this term has gone out of general use because of possible confusion with addition instructions that include the carry flag as an aid to multiple precision arithmetic. Compare instructions are sometimes placed in the category of logical instructions. Shift and rotate instructions may also be so classified, as noted earlier.

In typical applications, logical instructions are used to isolate and set, reset, or test individual bits or groups of bits within a word. A status bit (e.g., "ready," "busy") from an I/O interface port can be tested using these instructions by loading the port's contents into the accumulator, masking off the other bits by an AND with a byte in which only the corresponding bit is set, and then examining the zero flag from the PSW by means of a conditional branch instruction. If the result of the AND is nonzero, it means the bit was set at the time the port was loaded into the accumulator.

Generally, all logical instructions operate on the accumulator. In processors that provide addressing to the bit level and accompanying bit-level logical instructions, the carry flag in the PSW is used as a kind of "Boolean accumulator." The Intel 8051 is an example.

Control Transfer. This term has its origins in the idea that the program "controls" the computer. These instructions "transfer" this control from one part of the program to another. It is not a very good term, but no one seems to have come up with a better one, so just about everyone uses it. Instructions in this class are called *branches* or *jumps* and the ones that transfer control to subroutines are generally called *calls*. Instructions that "return control" from a subroutine to the routine that called it are, of course, called *return* instructions.

Note, however, that some instructions can have branching effects but do not have any of these terms associated with the name of the instruction. An example is the "PCHL" instruction in the Intel 8080 and 8085 and the Zilog Z80. It loads the program counter with the contents of the concatenated H and L registers; in effect, an indirect branch instruction.

Virtually every processor provides a reasonable set of unconditional branch and subroutine call and return instructions. The most important distinctions are in how they handle the *conditional* forms of these instructions. Although every processor offers several conditional control transfer instructions, in some it is possible to jump, call, *and* return conditionally on either the true or false conditions of different PSW flags. Thus, one can call a subroutine if the parity of the accumulator is even or odd, if the contents of the accumulator are zero or nonzero, positive or negative, or if the carry flag is set or reset. The subroutine may also base its return on the same set of conditions and, of course, branches may do the same. These can be extremely useful. For industrial control applications, the utility of control transfer instructions and especially conditional instructions should be weighted heavily in choosing a processor.

Miscellaneous. It is in this class that interrupt control instructions are found. These are derived from the features of the interrupt system with which the processor is equipped and always include instructions for enabling and disabling interrupts.

Some processors offer several *interrupt modes*. In one mode, interrupts trap programs to one of several reserved memory locations, each associated with a specific priority level. In another mode, the device that caused the interrupt may be required to supply a *vector*—an address to which the program is to be trapped. Although this requires more complex control hardware associated with the device, it is very flexible and can permit the processor to handle interrupts from a larger number of devices more quickly. Instructions that control which mode is to be used and, in some processors, the priority levels of interrupts that will be accepted are in this class.

Most microprocessors include a NOP (which stands for "no operation") instruction. This is usually used to form short delays and to adjust the amount of time a part of the program takes. This is acceptable practice when the delay need not be precise or is relatively short, but it should not be used if the delay must be critically matched to an external event. Examples of acceptable use of such "software timers" are in flashing lamps for operator display or waiting for peripheral devices to warm up. Some support chips may take one or two microseconds to respond to a command. In that case, it would be all right to use one or two NOPs in order to delay before giving it the next command, since it would be ready by the time a more elaborate and precise timing action could be set up.

Uses to be avoided, if possible, are those in which several milliseconds might elapse, it is important that the program not miss the event, and the time for the event might vary from one system to another, or over time. An example would be one in which the program commanded a solenoid to one position, waited for it to get there, then commanded a related switch to close. Actuation times for solenoids vary from one to another and as the solenoid ages. If a closely set software timer is used, someday the system may fail to operate properly, and figuring out why could be considerable trouble. It would be better to use some "positive feedback," such as a limit switch that indicates that the solenoid has completed its travel.

Support Chips

Support chips is a general term for a collection of LSI devices that perform certain processing and I/O functions in conjunction with microprocessors. For the most part, these are special designs for particular functions, but some are actually microprocessors in their internal organization.

Devices that serve as aids in "completing" the microcomputer are included among support chips. Examples are address latch devices for processors with multiplexed busses, clock generators, and "system controllers," which are devices that convert status and synchronization outputs from the processor into more useful memory and I/O control signals. These are used only with the processors for which they were designed. Although necessary, they are of little general interest to the microcomputer user.

Of great interest, however, are the more general-purpose support chips: devices for parallel and serial I/O, computation, timing and counting, and communication system support. Although these are most easily interfaced to microprocessors made by their manufacturers, with a little extra trouble (and often no trouble) they can be interfaced to almost any microprocessor. In this section, we review some characteristics of commonly used support chips.

Parallel I/O. Parallel I/O support chips usually include control logic and support circuitry for two or more parallel I/O ports of eight bits each. One of their most important characteristics is that they are programmable in the sense that the directionality, formats, and functions of their ports can be changed by means of mode control output bytes transferred to them from the microcomputer with which they are used.

Each port can be so programmed to be an input or an output port. For some devices, a port can be programmed to be bidirectional. This is useful for interfacing with microcomputer peripheral devices that have adopted bidirectional interface formats. In some devices, the 8-bit ports can be divided into a pair of 4-bit ports, each with different directionality. This is useful for situations in which inputs and outputs do not fit well into strict byte divisions; e.g., 12 input bits and 4 output bits.

When programmed for output, these devices latch the output data. When a device programmed for input is addressed, data are gated through the device and onto the data bus interfacing it with the microprocessor. Data thus transferred are in the state they were in during the transfer. Some programmable parallel I/O support chips add input latching so that data can be written into the chip from an external device and read by the microprocessor later. Some are also designed to generate interrupt requests when such events occur.

The output drive capability of typical parallel I/O support chips is somewhat limited, perhaps enough to handle one or two standard TTL device inputs or several low-power or LSTTL inputs. In some cases, they can drive transistor amplifier bases directly. Usual practice, however, is to provide buffer circuits between the chips and the devices they drive.

Direct memory access support chips are also used to support parallel I/O operations. These DMA controllers are equipped with internal address registers and counters—the number varies from two to four such sets and determines the number of DMA channels that it can control concurrently. These are initialized by the program. The DMA controller is also equipped with "request" and "acknowledge" signal pins, one set for each channel. These are used as handshaking signals between I/O devices and the DMA controller.

When a device has a byte to be transferred to memory, it requests the transfer by activating the "request" input to the DMA controller. The DMA controller obtains control of the system busses, fetches the byte from the device's port in the memory or I/O address space, acknowledges receipt of the byte with its "acknowledge" signal, and writes it to the memory location specified by the address register for that channel. The address register is then incremented and the corresponding counter is decremented. This process is repeated until the counter has been decremented to zero, at which point the specified number of bytes will have been transferred and the DMA controller may notify the program by interrupting it. DMA output operations work in roughly the opposite way, the "request" now indicating that the output device is prepared to accept a byte and the "acknowledge" indicating that the byte has been moved to its port by the DMA controller.

Direct memory access devices are used most often to support interfaces with fast external data and program storage devices, such as magnetic disks and tapes, and with high-speed communication channels. In industrial applications, they are used in recording and playback of process data and in distributed control systems where large amounts of command and process data must be exhanged between elements of the control network.

Serial I/O and Communication. The first widely used commercially available LSI input/output device was actually a support chip developed by the General Instruments Corporation. This was introduced shortly before the first microprocessor and was used as a serial interface for minicomputers. It was called a "Universal Asynchronous Receiver/Transmitter." This was abbreviated *UART*, a term still used for what is very likely the most common support chip. Virtally every general-purpose microcomputer includes a serial interface based on a UART. There are now many different types, and some have been extended to include capabilities for synchronous, as well as asynchronous, communication.

The UARTs contain serial-to-parallel and parallel-to-serial shift registers and timing and control logic. These form two essentially independent data channels, one for output from the computer and one for input to it. They are used to support communication interfaces among computers and for interfaces to *interactive terminals* (input/output devices that include a keyboard for operator input to the computer and a cathode ray tube (CRT) for displaying computer output to the operator). They are also used often in interfaces to small tape cassette and cartridge data recorders.

In serial communication, data are transmitted and received one bit at a time. Transmitted characters and *packets* of characters are preceded and followed by bits and characters that serve the functions of delimiting them, formatting them, and providing for detection (and, in some cases, correction) of errors that may be caused by noise during their transmission. The UARTs accept bytes to be transmitted from the microcomputer, surround them with the required formatting and error-detecting bits and transmit the resulting character serially. Similarly, received serial data are accumulated by the UART so that the data byte may be read by the microcomputer in parallel. Formatting bits are removed from the received data and limited error checking is performed by the UART.

Devices that are capable of supporting both synchronous and asynchronous communciation are called *USARTs*. Some are more elaborate than others. They may include the ability to dynamically change the rates at which data are transmitted and received, to automatically detect synchronization characters, to control handshaking signals used with devices that provide the electrical interface to the communication medium over which data are transmitted (e.g., the telephone system), and so on. As they become more elaborate, they are sometimes called "communications controllers." This level of these devices might include very highly integrated devices to provide the processor with substantial support in handling complex communications protocols.

The principal benefit of the UART is that it relieves the microcomputer from the tasks of continually monitoring the serial input signal for data and having to dole out each bit when transmitting. These operations must be precisely timed, and this is difficult to do if the microcomputer must handle other functions at the same time (which is usually the case).

Computation. Because few 8-bit processors include hardware for multiply and divide operations and because the dynamic ranges of numbers used in many applications necessitate the use of multiple precision arithmetic, computation support chips are frequently used to extend the microcomputer's capability in the realm of "number crunching"—a term used to characterize applications that require a great deal of computation.

These devices (variously called "arithmetic chips," "arithmetic processing units," "APUs," and "numeric data processors") can perform arithmetic operations on fixed and floating point number formats that exceed the resolution and dynamic range of the microprocessor's internal registers and can do so at higher speeds. Some of these devices can also evaluate trigonometric, logarithmic, exponential, involution, and square root functions, as well.

In typical configurations, these devices occupy two or three locations in the memory or input/output address space. Numbers are passed to them at these addresses one byte at a time, two to four such transfers being required, depending on the resolution and formats of the numbers being handled. These are placed on a push-down stack, implemented as a small read-write memory in the device. When the stack has been loaded, command bytes are transferred to the device, and it begins its computation work.

Commands include such operations as "Add the two numbers in the first and second positions of the stack, leaving the result at the top of the stack," "Raise the number at the top of the stack to the power in the second position of the stack," "Compute the square root of the number at the top of the stack," and so on. The time required for a computation may vary from several microseconds for simple addition of fixed-point numbers to several milliseconds for a logarithmic or exponentiation computation. When the computation is finished, results may be transferred back to the microcomputer or left on the stack to be used in further computations.

These can be thought of as special-purpose microprocessors that can operate in parallel with a microcomputer to carry out computing functions that would take much longer were they to be implemented in software on the microcomputer. The arithmetic chip can usually interrupt the microcomputer when it is finished, or the microcomputer can poll its status. Even if the microprocessor simply waits, considerable improvements in the computing power of the system are obtained. However, the most powerful configurations are those in which commands and data are transferred by means of a DMA arrangement. This provides very high-speed operation and can yield a computer that rivals minicomputers in "number crunching" capability.

Timing and Counting. In applications requiring counting and precise timing, LSI "counter/timer" support chips are useful. Generally abbreviated "CTCs," these include several 16-bit registers that can be incremented on the basis of one of several selected inputs. As was the case with parallel I/O chips and some UARTs, the operating

TABLE 8-1
Characteristics of Typical Microprocessors

Manufacturer	*Processor*	*Technology*	*Size*	*No. Instr.*	*Instr. Time Range (μsec)*	*Gen. Purp. Registers*	*Address Range*	*Package Pins*	*Remarks*
AMD	2900	Bipolar	4	16	See Note 1	16		40	Microprogrammable bit-slice computer component
Intel	4004	PMOS	4	46	10.8–21.6	16	4K	16	"Founding father"
Intel	4040	PMOS	4	60	10.8–21.6	24	8K	24	Improved 4004
Intel	8008	PMOS	8	48	12.5–37.5	6	16K	18	First 8-bit general-purpose microprocessor
Intel	8080A	NMOS	8	78	1.5–3.75	6	64K	40	First "2nd generation" device
Intel	8085	NMOS	8	80	0.8–5.2	6	64K	40	Improved 8080
Intel	8048	NMOS	8	96	2.5–5	8	4K	40	Single-chip microcomputer
Intel	8051	NMOS	8	111	1–4	128	128K	40	Single-chip microcomputer
Intel	iAPX86/10	NMOS	16	97	See Note 2	8	1M	40	

Intel	iAPX432	NMOS	32		See Note 2				Multipackage device, executes high-level language
Intersil	6100	CMOS	12	81	2.5–5.5	0	4K	40	Same instruction set as DEC PDP-8
Mostek	MK3870	NMOS	8	70	1–6.5	64	4K	40	Single-chip microcomputer
Motorola	M6800	NMOS	8	72	1–2.5	0	64K	40	
Motorola	MC68000	NMOS	16	61	See Note 2	16	16M	64	
National	NSC800	CMOS	8	150	0.5–2.88	14	64K	40	Z80 instruction set, 8085 interface
National	NS16016	NMOS	16	78	See Note 2	8	64K	40	
RCA	1802	CMOS	8	91	3.2–4.8	16	64K	40	First widely used CMOS microprocessor
Texas Instruments	TMS9995	NMOS	16	72	See Note 2	256	32K	40	
Western Digital	WD-16	NMOS	16	116	See Note 2	6	64K	40	
Zilog	Z80B	NMOS	8	150	0.7–4.2	14	64K	40	
Zilog	Z8002	NMOS	16	110	See Note 2	16	354K	40	

Note 1: This is a component of a microprogrammed computer. Instruction times and other characteristics depend on the way the computer in which it is used is designed.

Note 2: These processors include very complex instruction sets. Benchmark programs are the only fair ways of comparing them.

mode of a CTC is programmable, in the sense that it can be changed under program control.

The CTC inputs may be signals from limit switches, pulse trains from tachometers or position transducers, signals from flow transducers, or any other essentially digital time-based waveform. When programmed as counters, CTCs then accumulate the number of pulses in these signals. When programmed as timers, they can be used to measure the elapsed time between transitions and, therefore, can be used to measure such signal parameters as frequency.

Most CTCs can operate in modes in which initial timer values are retained and automatically reloaded into the internal register each time the register overflows. By incrementing the register from a separate signal source, such as the microprocessor clock, the CTC output produced at each overflow becomes a signal with a selection of possible frequencies. This can be used to control certain actuators (e.g., the stepping rate of a stepping motor) or as a source of timing for serial communications devices such as UARTs.

Summary. Support chips are variations on one way to make more capable computing systems—spread out the work to be done among several devices and let them work in parallel. This way differs from more general multiprocessor arrangements in its scale and in the somewhat specialized functions performed by these devices, but the basic principle is much the same. The development of support chips has now reached the stage in which microprocessors, designed not to "stand alone" but to serve input/output support roles, are available so that users can custom-tailor their own support chips for interface functions. Basically, these are single-chip microcomputers with on-chip data and program memories and on-chip I/O ports. They are called "universal peripheral interfaces." Although this term overstates their capabilities somewhat, they are nevertheless very useful, because they can take up much of the "central" microcomputer's work load and leave it free to concentrate on the loftier matters of directing the general operation of the system and making more complex control decisions.

CONCLUSION

We have examined the early history of microprocessors, the ways in which they are made, and their general characteristics. There are many types of microprocessors. New ones are announced so frequently that it is virtually impossible to be up to date on them all. Nevertheless, some stand out over the others and are widely used for various reasons. To provide a capsule list of the names and characteristics of those that seem to be most often used in the kinds of appliations in which we are most interested, Table 8-1 is offered. This table also lists several 16-bit microprocessors and one 32-bit microprocessor which will appear increasingly in future applications. Several industrial control systems are using 16-bit processors, and the reader can expect to encounter them in many situations.

Because it is so difficult to keep up with the rapid pace of microprocessor technology and the development of related hardware and software techniques, it is fortunate that the electronics industry is blessed with the attentions of a well-informed, competent, and responsible trade press. This trade press is a valuable source of information on recent developments, as well as informative tutorial articles on particular subject areas within electronics and computer science. Their regular study is highly recommended to the active student of microcomputing. Several of the more prominent ones are listed in the bibliography for Part II.

The manufacturers of microprocessors and associated componentry have also gone to great lengths to supply the users of their products with helpful manuals and application notes. These are also worthy of study. In fact, the student may find that he or she is faced with almost too much in the way of printed information about these devices and the uses to which they can be put.

CHAPTER 9

Memories

"'It's a poor sort of memory that only works backwards,' the Queen remarked."

Lewis Carroll
Through the Looking-Glass

Memory system technology and architecture have been as important to the automation of industrial processes and machinery as the microprocessor itself. The effectiveness of a computer system owes as much to its memory as to any other aspect. Beyond simply being a place to keep programs and data, memory affects the overall design of the system and its application: processors, programs, input/output devices, interfaces with primary elements and actuators. There are many ways to organize and use memories. Modern developments in computer architecture, system software design, and LSI semiconductor technology are often strongly based on memory design developments, and vice versa. There is a strong mutual dependence among all of these things.

In this chapter, we discuss some general memory considerations, how they are organized, and the component technologies on which they are based, with emphasis on the types commonly found in industrial control applications.

GENERAL CONSIDERATIONS

In a way, memory is like money. You should have enough and be able to get at it reasonably quickly. Having much more than you need is nice, but spoiling in a way that is hard to explain. Having too little can make life miserable. Memory is a necessary resource and the first line of defense when a program is in trouble. One of the worst things you can do is use it all up and not be able to arrange for any more. Squeezing programs into a limited memory is enormously costly. Yet, so long as there is no immediate danger of running beyond easily increased amounts, there is little value in trying to conserve, because the cost of a byte of memory is small compared to the programming work needed to avoid its use. Spending memory wisely is a valuable talent in a programmer.

Most microcomputer applications fit comfortably within the directly addressable memory space of a typical 8-bit microprocessor, 64K bytes. A great deal of effective control and information processing software and data can be kept in a memory of this dimension, especially when the programs are written in assembly language, where there is an essentially one-to-one correspondence between program statements and computer instructions. When one is using high-level languages, the conditions change substantially, and the need for more memory increases for several reasons.

One reason is that even the simplest high-level language programs require a set of routines (a *run-time package*) that supports the program in various ways—routines that almost certainly contain some code that you do not need for your particular application, but which would be troublesome to excise. As a system becomes more sophisticated and powerful, providing additional preprogrammed services and functions, this system software overhead burden becomes greater and its efficiency, in terms of memory utilization for a particular application, may be reduced in ways that depend on the application. Though such systems require more memory than less powerful systems, they return the greater investment in memory in other ways, so the net effect is better overall. But they still require more memory.

Some memory-intensive software systems treat application programs as though they were data. Application program statements are fetched from memory by the software system and interpreted by it, to carry out more powerful individual functions than could be executed by a "directly" programmed system. Although this requires more money (and execution time), they are much easier to program and to modify as design errors are found and corrected and as requirements are refined and changed. But it is not uncommon to find that such operating systems require more than half (sometimes as much as three-quarters) of the directly addressable memory space.

Even microcomputer systems programmed in assembly language benefit from services provided by operating systems. Operating systems support the development of larger programs by making it easier to combine the work of several programmers into larger programs.

Division of labor among people—to get more work done in less time—carries over into many programs combined to form a larger program. Programmers, being human, make mistakes and suffer misunderstandings. In large systems, it is more likely for an error in one program to alter part of another program so distantly related that finding the cause of the error can be exceedingly troublesome. Since a common cause of such problems is an intrusion into the victim's allocated memory area, means of protecting areas of memory are highly useful features in such systems.

Although it is nearly impossible to provide total protection from programming mistakes (which can go undetected for incredibly long periods of time), it is possible to make the system "robust"—that is, tolerant of errors to the extent that it retains enough control to maintain a minimal degree of function or, at the very least, to fail in what is sometimes called a "graceful" fashion. Much of this robustness can be had by taking care to ensure that the operating system is well protected. Protection of sensitive operating system and application program areas can be provided by the way in which the memory system is designed.

Large computer systems must often go through elaborate procedures when they are turned on or off. These procedures are usually aided by a human operator, who is responsible for setting up peripheral devices and readying them for being shut down. But the situation is very different for most industrial control microcomputers. They are turned on and off with the machines they control, often abruptly. They must put up with both intended and unintended interruptions of power. They must bring themselves into full operation, without direct operator intervention, moments after power is applied, In order to do this, their programs must be in memory and ready to go or be able to be loaded automatically.

These requirements, among others, led to widespread use of read-only memory in microcomputer systems. Read-only memories are ready to go when power is applied, do not lose their contents when power is removed, and provide a substantial degree of protection from accidental alteration by program errors and by electrical incidents.

Because there are degrees to which one may nevertheless wish to change the contents of even a "read-only" memory, there are several types which can be used. The choice depends on the number of identically programmed systems to be produced, the degree to which the programs are believed to be free from errors, the frequency with which changes might be necessary, and the possible need for limited on-line changes.

Memory that can be written freely is generally called *read-write* memory or, especially in the context of microcomputers and modern minicomputers, *RAM* (Random Access Memory). Prior to RAM, the most commonly used memories were based on storing information by temporarily magnetizing small iron cores. Because core memory cannot be integrated to the same degree as RAM, it has been largely supplanted by it. But common semiconductor RAM is "volatile"—it does not retain its contents when power is interrupted.

The origin of the term "RAM" is interesting. Prior to the development of core memories, many computers used rotating drums on which information was stored by magnetizing small regions of an iron oxide coating. As the drum rotated beneath read/write heads (similar to the types used in audio tape equipment), information was read or written one bit at a time and assembled in shift registers to form complete words. Access to a particular datum required waiting until it came under the read/write heads. It was not "random"—the program did not have free and immediate access to every location. With core memories, any location could be accessed in random order, unconstrained by the order in which it had been written. Hence the term "random access memory" came to be used at that time.

The term "core" had become a common shorthand term for the main memory of a computer when the first semiconductor memories came into use. During that time, writers were often careful to use the term "semiconductor RAM," in order to distinguish it from "core RAM," even though the simpler term "core" was more commonly used. Core memories were never widely used with microcomputers. In the early days of microcomputing, one could identify programmers who did not have much experience with microcomputers because they tended to call the main memory "core," even though it was actually semiconductor RAM or read-only memory. Gradually, the terms "RAM" and "ROM" came to be commonly used to refer to any read-write and any read-only memory, respectively.

Because read-only and read-write memory components are not universally interchangeable, individual memory subsystems are often designed for one or the other; a different printed wiring board is used for each type, or different areas of circuitry on the same board are used for each type. Thus, modern semiconductor memories are not "homogeneous" in the same way that core memories were. Now programmers are privileged to have the opportunity to run out of more than one kind of memory.

As memories have become larger and more variegated, memory reliability has become more important. Although semiconductor memories enjoy excellent intrinsic reliability, as they grow larger simple statistics increase the probability of a malfunction. Because systems are often applied to critical operations where malfunctions can be very costly, reliability becomes increasingly important.

Semiconductor memory reliability has suffered from time to time from its very newness, its reliance on properties and materials that were not completely understood. An example is the tendency of certain types of programmable read-only memories to lose data due to a phenomenon by which selectively blown fuses in the device actually grow back, resuming their original conducting state. Another is the recent discovery that highly integrated RAM memories are subject to temporary loss of data (so-called "soft-errors") caused by an interaction between memory cells and alpha particles, subatomic particles emitted by minute traces of radioactive elements in integrated circuit packaging materials.

Thus much attention is devoted to elements of memory system design that enhance reliability. Among the techniques used are lateral and longitudinal parity schemes and elaborate methods of encoding stored information redundantly, which permit automatic detection of most errors and even correction of certain soft errors. Though costly, these methods can yield memory systems with extraordinarily high levels of reliability.

Modern microcomputer systems often contain a variety of memory components. Almost without exception, their main directly addressable memory is partitioned into a mixture of read-only and read-write memory. Beyond this are extensions that provide the effect of increased main memory and extensions that rely on bulk storage of programs and data in peripheral devices. The characteristics of these memory subsystems are generally described in the following terms:

Density—The amount of memory that can be contained within a certain volume.

Speed—The rapidity with which data can be accessed and written.

Size—The number of words that can be held by the subsystem.

Volatility—The ability to retain information when not powered.

Power requirements—Both current consumption and the number of supply levels may vary.

Reliability—The likelihood of soft and hard (permanent) malfunctions.

In the next section, we examine some ways in which memory subsystems are organized internally and interfaced to microprocessors.

GENERAL ORGANIZATION OF MEMORIES

So they can be used in the following discussion, we briefly define terms commonly used for the types of memory elements and components frequently found in microcomputer systems. More detailed descriptions are given later.

Memory Terms

Under the general heading of read-only memories, there are four principal types. The contents of purely read-only memories (*ROM*s) are determined when they are manufactured. The data they contain are determined by metal patterns laid down on the ROM chips during their final manufacturing stages. They cannot be altered further in any way. Although the term *ROM* is often commonly, and loosely, applied to all memories that have restricted alterability, strictly speaking it should be used only for this type.

The abbreviation *PROM* stands for "programmable read-only memory." This term is likewise applied loosely to all restricted alterability memories which can be "programmed" by their users. More strictly, it is used to identify devices that can be programmed only once and that cannot be erased and reprogrammed.

The abbreviation *EPROM* stands for "eraseable programmable read-only memory." EPROMS can be programmed, erased, and reprogrammed with different information many times. They can be changed and reused.

The abbreviation *EEPROM* (sometimes written *E²PROM*) stands for "electrically erasable programmable read-only memory." These are relatively new devices. EPROMs are erased by exposure to a sufficient dosage of ultraviolet light of a particular wavelength. Ordinarily, this means EPROMs must be removed from their printed wiring boards for erasure. EEPROMs, on the other hand, can be erased in place by the application of proper input signals and voltages, and reprogrammed on-line.

In the context of these memories, the term "programming" is used to describe the act of writing data into them. It may represent programs, fixed constant data, and parameters. Such programming is performed by using special equipment designed for this purpose. In the case of some recent types, the programming circuitry needed is sufficiently simple that it is practical to include it on an otherwise normal memory subsystem printed wiring board.

Under the general heading of on-line read-write main memories—for which RAM is a practical synonym in the context of microcomputers—there are two principal types. *Dynamic RAMs* operate on a stored charge principle. Because this charge must be refreshed from time to time, dynamic RAMs require support circuitry which may represent a burdensome overhead for a small memory. For larger memories, it is bearable because dynamic RAMs can be more densely integrated and are less costly, bit for bit, than the second principal type, which is *static RAM*. Static RAMs store data in integrated flip-flop circuit arrays. Because they do not require refresh circuitry, they are better suited for relatively small memories.

The term *bulk storage* is generally applied to memory subsystems used to store much larger amounts of data. Access to data in a bulk storage device is not as direct as it is with main memory devices. When needed, it is transferred from the bulk storage device into buffer areas in the main memory. After it has been generated or updated, it is written back into bulk storage. Bulk storage devices include large semiconductor RAM-based memory subsystems and peripheral devices as well.

Peripheral memory devices are usually represented by magnetic tape and disk devices. (It is common in the jargon to spell "disc" with a "k" when we refer to these devices.) With these devices, information is recorded in magnetized iron oxide deposited on a medium which is moved past a read/write head. Movement of the medium may be directed by the computer or the peripheral memory controller (as in the case of tape-based systems, where the forward and reverse motion of the tape is directly commanded) or the medium may be in continuous motion (in the case of disks, where the controller waits for the desired location to come under the read/write heads before starting the transfer of information).

Other types of bulk storage include devices based on magnetic "bubble" and charge-coupled device (CCD) semiconductors. These are relatively new technologies. There are analogies between their operation and that of rotating memories in that data are continually shifted between their storage elements while they are operating, periodically appearing at their read/write elements, just as data on a disk can be thought of as rotating to appear periodically beneath the read/write heads.

Charge-coupled device memories are volatile; bubble memories are not. The CCD memories shift stored charges from one element to the next. Bubble memories shift small magnetized "domains" ("bubbles"). The commercial future of these devices is in some doubt, since several manufacturers have discontinued them as product lines because they have not fulfilled their original economic promise.

Cache memories are very fast memories of limited size. In the typical case, the cache memory may be loaded with a portion of the program transferred from a larger main memory. Because of its high speed (due to the use of very fast bipolar RAM components), programs executed from this memory operate more quickly. When a different portion of the program is needed, it is read in from the main memory. Detecting the need for a new portion of the program, and the resulting transfer, are handled automatically by logic associated with the CPU. Cache memories are rare in microcomputer systems, but they will come to be used more as processor speeds improve to the point at which they provide a reasonable return on the cost of the additional logic required.

Timing

Memory is characterized by its *cycle time*, the time that must elapse between the initiation of two successive memory-referencing operations. Cycle time is composed of three parts:

Latency—The time required to initiate the operation, decode the address of the location referred to, and prepare to make the data transfer.

Transfer—The time required to make the actual physical transfer from the bus into the memory elements in which the data is stored or from the memory elements onto the bus.

Recovery—The time required to prepare for the initiation of the succeeding memory reference operation.

The combination of latency and transfer time is often called *access time* with regard to general memories.

Although the magnitude of each component of cycle time depends on the design of the subsystem and the kinds of memory elements it uses, there are many similarities and analogies between them for different types of memories.

Contributors to latency in a main memory subsystem include time for it to recognize its subsystem address in the most significant subfield of the address, time for it to use lower-order subfields to select individual chips on its printed wiring board, and time for the selected chips to decode the least significant address field in order to select individual bits.

Transfer time in reading from the memory is made up of the time required to transfer information from an individual memory cell to internal busses within the chip and thence to its output buffers, which must also be switched from their high impedance state to their totem-pole configuration in order to transfer the data onto the driven bus.

In writing to the memory, data must be gated through the chip's input buffers onto its internal busses and into the selected memory cells in which they are to be stored. When the speed of the complete system is figured, these times must also account for processor response times—the memory and the processor must be matched, as it were. If the processor can accept data faster than the memory can deliver them, the processor must wait for the memory.

In a typical semiconductor memory, recovery time includes the time required for output buffers to return to their high impedance state. In some forms of memory, the readout of information from the memory cell is destructive. The data are altered by reading and must be rewritten into the cell before succeeding cycles can be initiated. In memories equipped with error detection and correction circuitry, time for these functions must be included.

In peripheral memories, latency is due to the time required for motion of the medium to bring the desired location under the read/write heads. In a tape-based system, it may be necessary to go from one end of the tape to the other—a process that can take anywhere from seconds to minutes, depending on the type of tape system. In a rotating memory, the time required is always limited by the rotation speed of the medium; it has a fixed maximum on the order of milliseconds. Data transfer time for peripheral memories is composed of the time required to assemble the serially recorded word and transfer it through the peripheral interface to the microcomputer's bus when reading, and to perform the reverse of this process when writing.

Analogies to recovery time are less direct. For example, some devices perform "read-after-write" checks to ensure that data were recorded correctly. This is done by using a read head which is offset from the write head. The check must wait for data to arrive at the read head, to be read back, and to be compared to a copy of the data held in a register in the peripheral device controller. Until this has been done, new data cannot be transferred.

Because blocks of data are usually transferred and recorded in a contiguous region of the medium, cycle time is not uniform on a per byte basis. In reaching the area of recording, most of the latency time is spent. Once there, the latency time for cycles for succeeding bytes is very much less.

Logical and Physical Memory Arrangements

Prior to the use of semiconductor memories, most computers had main memory arrangements which could be thought of as essentially homogeneous arrays, typically core memories. Because core memories could hold information when unpowered, programs could be left in memory. In a dedicated system, this could be the application program. More generally, a small "bootstrap" loading program was left in memory. This was used to read in larger, more capable loading programs which could then carry out the loading of more complex program and data files from peripheral memories. The bootstrap loader was kept small so that it could be keyed in with control panel switches if it happened to have been overwritten due to a program error or lost due to a malfunction.

Typical microcomputer systems still use addressing schemes that treat main memory as if it were homogeneous. That is, the microprocessor itself does not distinguish between, say, program and data memory. It does not attribute any structure to the memory beyond that implied by paged addressing modes adopted by its instruction set. In terms of relations with the microprocessor, any given location could be read-only or read-write in nature. (Some single-chip microcomputers, the Intel 8051 for example, have departed from this scheme in that they do distinguish between program and data memory, but they are presently in the minority.)

Thus, we typically have the situation depicted in Fig. 9-1(a): a contiguous array of memory words. In the case of a typical 8-bit microprocessor, which uses 16-bit addresses, it can contain up to 65536 (64K) bytes, numbered from 0 to 65535. Pictures or diagrams that show memory arrangements and the allocated purposes of regions of memory are called *memory maps*. The terms "upper" or "higher" and "lower" are generally used to describe the location of a region with respect to the range of address numbers. Locations nearer to word 0 are "lower" than those nearer to word 65535. Some writers show higher addresses at the top of memory maps, to preserve the sense of being "higher" in their diagrams. Others feel that it is more natural to preserve the sense of the contents of memory in its relation to a "list"—the first item in a picture or diagram of a list is usually placed at the top. Both forms are commonly used in literature and hardware manuals and will not be confusing if you are aware of the distinction.

There may be some internal logical structure imposed on the memory by paged addressing modes, as illustrated by Figs. 9-1(b) and 9-1(c). But the fixed paging of Fig. 9-1(b) and the floating paging of Fig. 9-1(c) are embodied in the way the processor forms addresses from its instructions and have no relation to the physical organization of the memory subsystem or to the logic it uses to decode addresses and select elements within itself.

Physically, regions of memory may be read-only or read-write. Which is which is determined by the memory hardware design. Programs are not prevented from attempting to write to a read-only memory element. Nothing will happen; the location will not be changed.

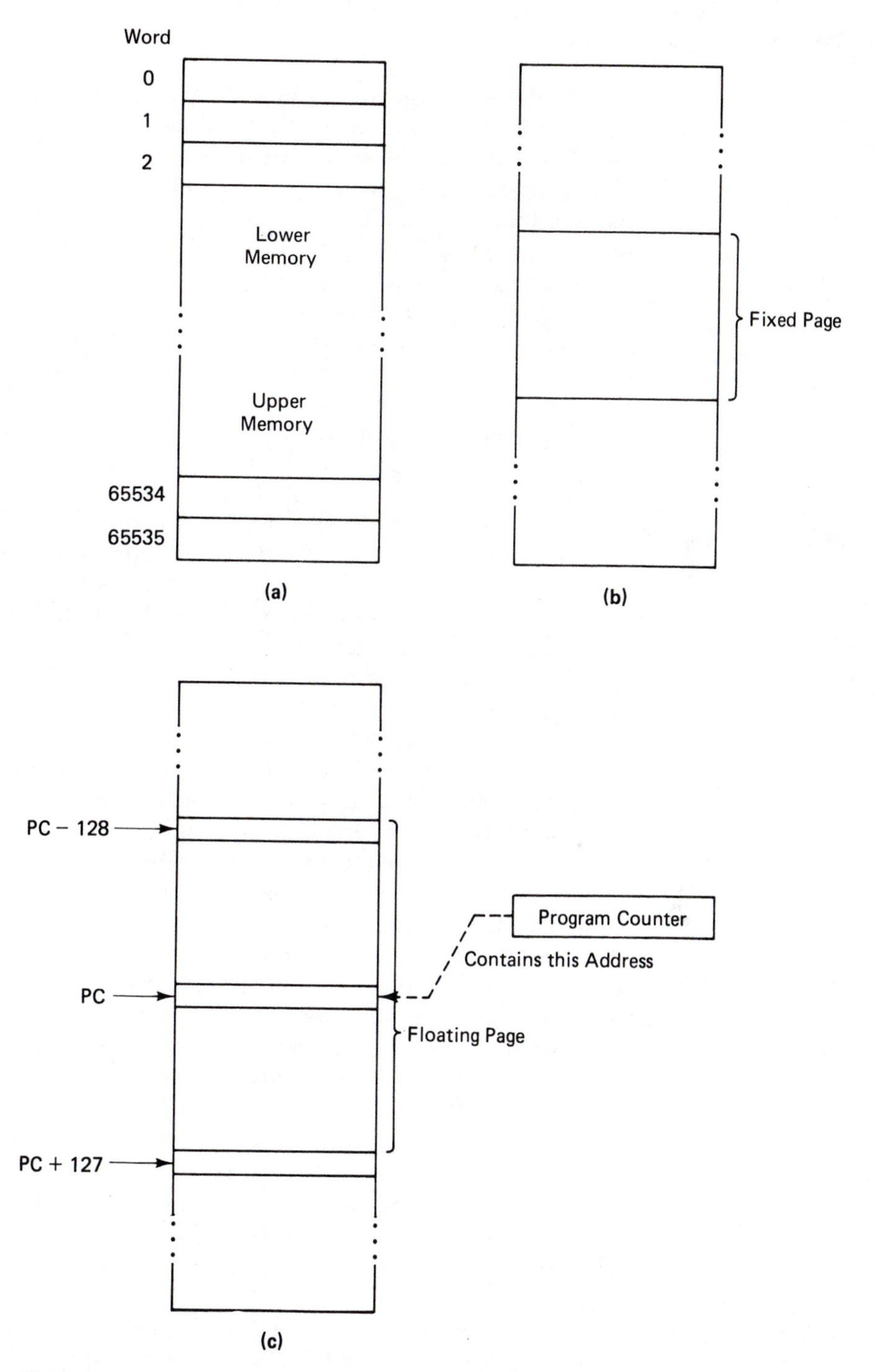

FIGURE 9-1. Memory addressing methods: (a) directly addressable contiguous array, (b) fixed pages, (c) floating pages (address range centered on instruction address in the program counter).

Different applications require different amounts of read-only and read-write memory. The amounts of each actually available depend on the way in which the memory is packaged. With board- and system-level products, one may choose boards that contain only RAM, boards that contain only read-only memory, and boards that contain a combination of both. These may be combined up to the maximum size of the memory addressable by the processor.

Figure 9-2 illustrates the internal organization of a typical board-level memory subsystem. This example is organized around an array of eight integrated memory circuits, each containing 2K bytes, for a total of 16K bytes. The input and output pins of these chips corresponding to the same bit within a byte are tied together on a bus which is internal to the memory subsystem and connected to the system data bus through bidirectional three-state bus drivers.

This subsystem divides the address into three fields. The most significant 2-bit field is recognized by logic on the subsystem board and used to select the subsystem board being addressed. Because there are four possible codes in this field, there may be up to four such boards in the system, for a total of 64K bytes. This decoding is gated by memory control signals from the system bus. Recognition of its address causes logic on this board to take two further actions: (1) enabling and setting direction controls of the bidirectional bus drivers according to whether the operation is a read or a write, and (2) enabling chip select logic. The chip select logic decodes the next most significant 3-bit address field to yield a single chip select signal which enables one of the eight integrated memory circuits. The circuit thus enabled then decodes the remaining eleven bits in order to select the specifically addressed byte from the total memory space.

Details of the internal arrangements of memory subsystems such as this depend on the types of memory chips they are designed to use. Memory chips are available in many configurations. The EPROM and EEPROM devices usually provide an 8-bit readout from the chip. The RAM chips generally offer more variety by offering readouts of 1 bit, 4 bits, or 8 bits.

Terms commonly used in the literature and the encoding of part numbers for memory devices refer most often to the number of bits the device contains. Further description then goes on to specify its organization in the form "(words) × (word size)." For example, the 2716, 2732, and 2764 are related EPROM devices. The "27" identifies the general type (EPROM) in this part numbering scheme; the last two digits are the number of bits in the device in units of K. They all read out 8-bit bytes. Thus, the 2716 is a 16K (bits) EPROM organized as a 2K × 8 memory (like the devices used in Fig. 9-2); the 2732 is a 32K (bits) EPROM organized as a 4K × 8 memory, and the 2764 is a

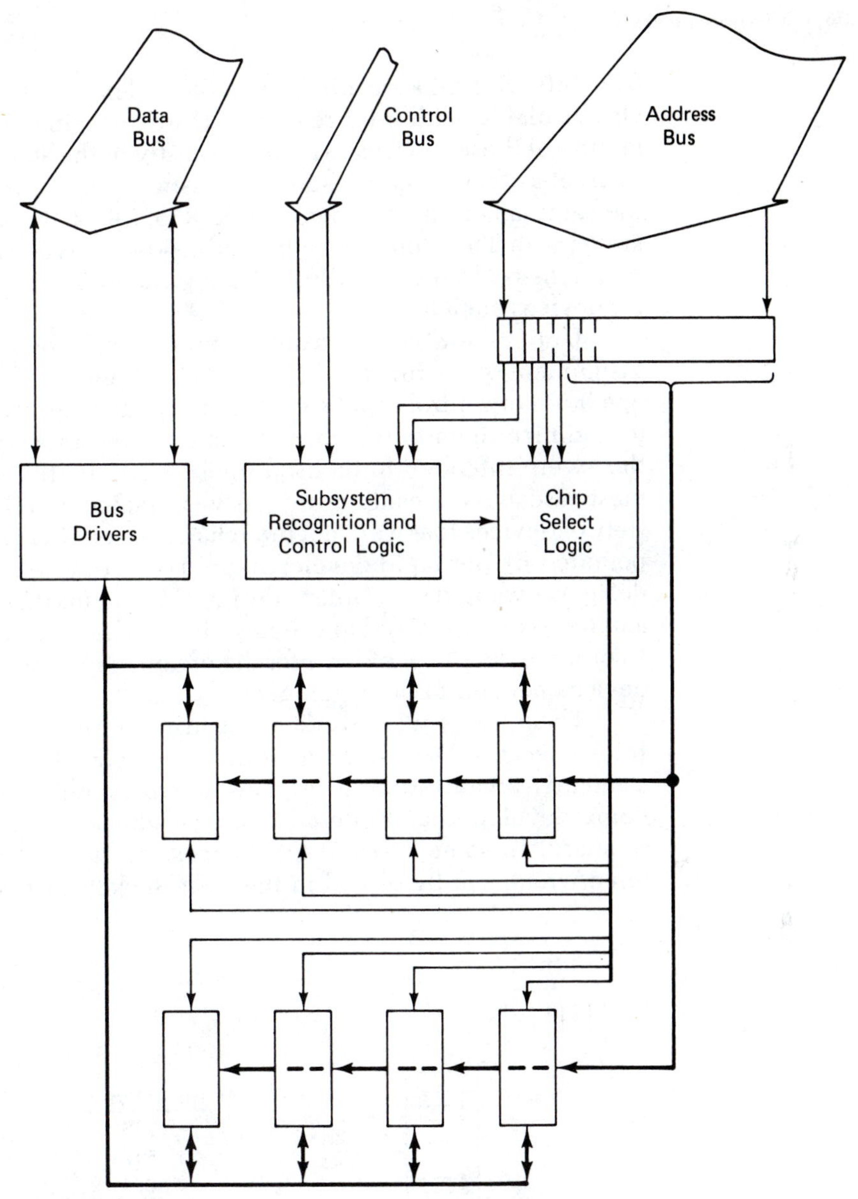

FIGURE 9-2. Internal organization of a typical board-level memory subsystem. Note the division of the address into three fields for board, chip, and byte selection.

64K (bits) EPROM organized as an 8K × 8 memory. RAM devices are also available with 8-bit readouts. However, since the level of integration of these devices (which are mostly of the static type) is not yet as great as that of their read-only cousins, they are not available with the same amounts of memory—1K × 8, 2K × 8, and 4K × 8 types are typical. They tend to lag behind read-only types by factors of 2 to 4 in density. Figure 9-3 illustrates typical address, data, and controls for devices such as these.

Until recently, each manufacturer used its own approach to the assignment of pin functions ("pinouts" or "pinning") and rules for the operation of control input signals to these devices, usually attempting to optimize them to the characteristics of the processors with which they were intended to be used (those made by the manufacturer, in most cases). As a result, designers were faced with a bewildering variety of devices that were not interchangeable. This problem was compounded by the rapid development of new devices. Memory system designers wanted to be able to design subsystems that could anticipate and be easily upgraded to use new devices as they became available. In order to do this, standards for the organization and pinouts of future devices needed to be developed.

These rules were worked out in the form of a standard by the Joint Electron Device Engineering Council (JEDEC), an industry standards organization. Most new microcomputer memory devices being developed and offered today are designed to conform to this standard. In some cases, RAM devices are being developed and offered which can be placed in the same sockets as read-only devices.

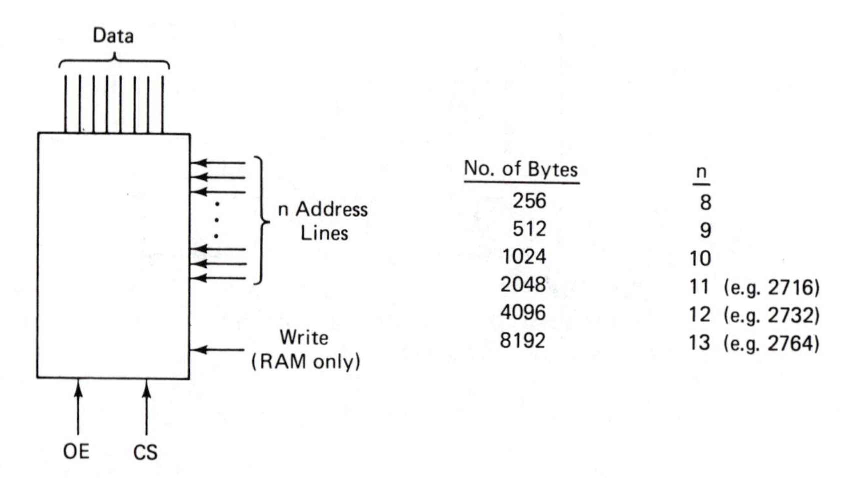

FIGURE 9-3. Address, data, and control signals for typical memory devices.

This makes it possible to configure memories with a greater variety of read-write and read-only arrangements using the same memory subsystem printed wiring boards.

Bus contention is a general term for situations in which more than one device is attempting to drive a bus at the same time, even if only momentarily. Bus contention can cause several problems, ranging from actual device damage (unusual) to noise generated by the currents involved (rather common), all deleterious to system performance.

Early designs used approaches in which a single chip-select signal enabled the selected memory device. This signal was based purely on address decoding and enabled both the access to the addressed memory element *and* the output drivers of the memory chips. Because device timing varies from one device to another, and with temperature, and because the time required to decode one address differs from that required for other addresses (due to different logic paths in the decoders), it was necessary to design systems with very generous timing margins in order to ensure quiet and reliable operation.

More modern designs now use two control signals, as shown in Fig. 9-3. The chip-select signal (CS in the figure) affects only internal access to memory cells within the device and does not enable the output buffers. A separate output enable signal (OE in the figure) controls the output buffers of the selected device. That is, outputs are enabled only when both its CS and its OE signals are enabled. This allows address decoding for one cycle to begin while devices addressed on the previous cycle are still in the process of being deselected ("getting out of the way"). Thus, faster memory performance is available without incurring the problems of bus contention.

Four-bit and 1-bit readouts are common among semiconductor RAM devices and ROMs and PROMs. Combinations of these can be used to form memories of various widths in order to match the word sizes of a variety of processors and also to accommodate error detection and correction codes and parity bits, which are often used to provide protection against soft errors. The ROMs and PROMs in these configurations are also well adapted to use as logic elements in special purpose designs.

Figure 9-4 shows how a pair of 4K-bit 1K × 4 static RAM devices may be combined to form a 1K × 8 memory section. Figure 9-5 shows how 4K-bit 4K × 1 RAMs may be combined to form a 4K × 8 memory section with a parity bit for each byte. As data are written to this, parity generating logic computes input byte parity and generates a ninth bit, making overall parity of the 9-bit word even or odd, to be stored at the same address. When data are read, parity checking logic examines the parity of the word. If it is not correct, the memory subsystem sig-

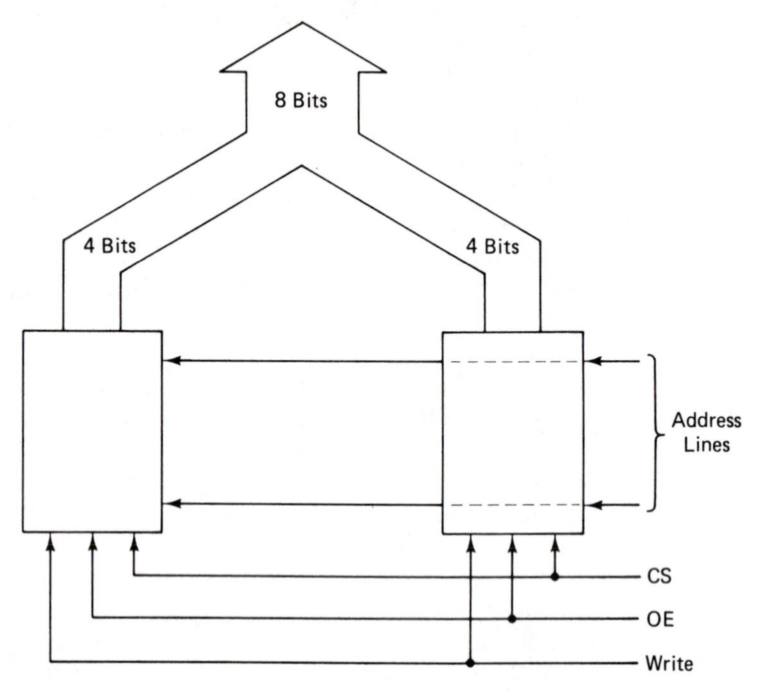

FIGURE 9-4. A pair of 1K × 4 RAM devices arranged to form a 1K × 8 memory.

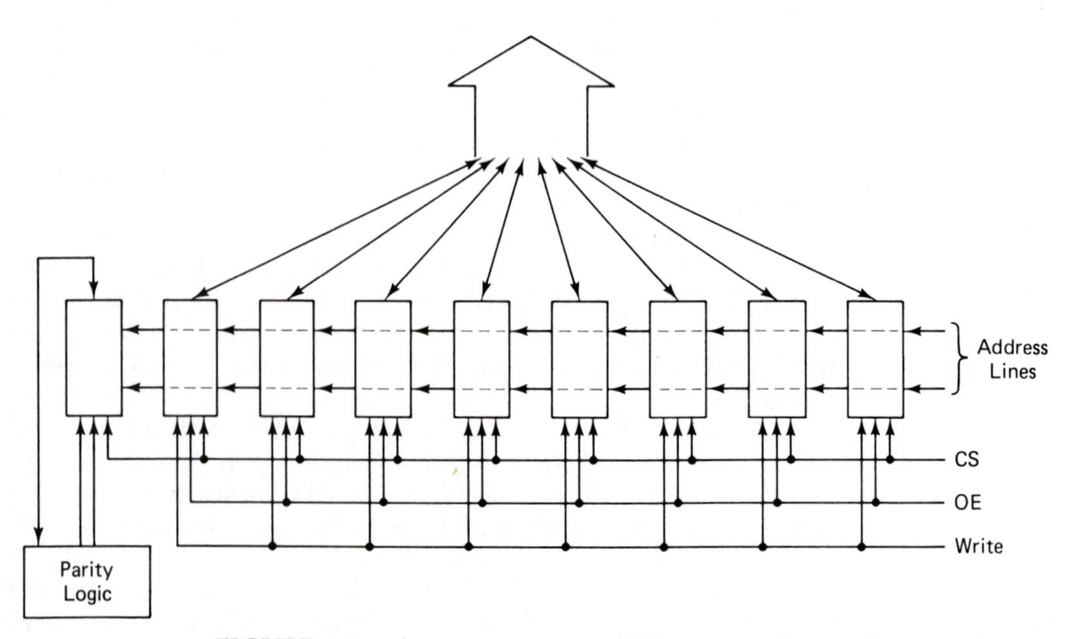

FIGURE 9-5. An arrangement of 4K × 1 RAMs, combined to form a 4K × 8 memory with parity.

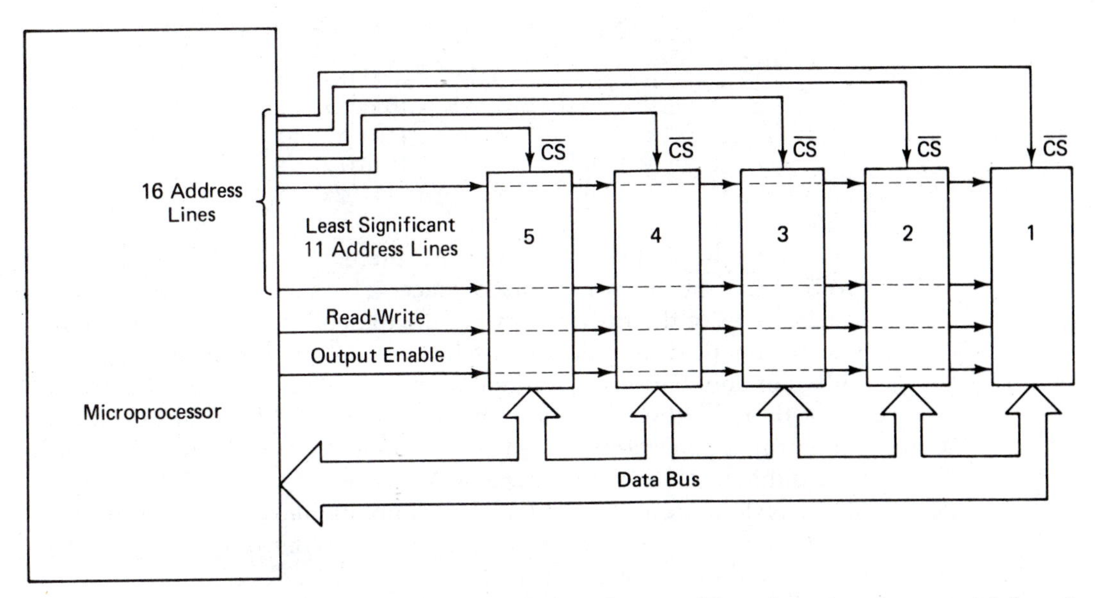

FIGURE 9-6. Use of address bits as chip select signals to avoid decoding in a small memory system.

nals the processor (usually by generating an interrupt request) so that it can take whatever action may be called for by the application. In more sophisticated memory subsystems, additional devices can be used to store the additional bits needed to make up complete codes that provide improved detection of errors (parity schemes can detect only single errors—multibit codes can detect some multiple bit error conditions) and, in some cases, automatic correction of certain types of errors.

Microcomputer systems that need only small amounts of memory can be designed to virtually eliminate decoding and address recognition logic external to the memory devices themselves. Figure 9-6 shows one way to do this. In this arrangement, the most significant five address bits are used *individually* as chip select signals for five separate memory chips. Each chip could be a device containing up to 2K bytes for a total capacity of 10K bytes. Chip select signals for memory devices are usually active-low signals. Thus, a particular chip will be selected if the address bit wired to its chip select pin is a zero. The program must be careful not to refer to any address in which more than one of the five most significant address bits is zero, since this would enable more than one device at the same time.

Note also that this memory is *not* contiguous in the address space. This can be seen by listing the correspondence between valid combinations of address bits and selected devices:

Device	Address
1	01111xx ... x
2	10111xx ... x
3	11011xx ... x
4	11101xx ... x
5	11110xx ... x

Thus, the first device represents memory in the last 2K bytes of the first half of the 64K address space. The second device is in the last 2K bytes of the third quarter of the total address space, and so on. This means programmers must take care not to allow the program to spill over and "fall through the cracks" into memory space that really is not there. Although an arrangement such as this would be inconvenient in a general-purpose microcomputer system, it may be worth the trouble in return for hardware cost savings if many identical copies of the system are to be made. Arrangements like this can be used in small microcomputer-based instruments and controllers, such as data loggers and digital versions of PID controllers.

Extending Memory Limits

We have talked about memories that fit within the normal directly addressable space of the processor and about ways to minimize logic for memories that can be much smaller than this. What does one do if the amount of memory needed is more than the directly addressable space?

One way, which has been used for many years, is to employ *overlay* techniques. The feasibility of overlays is based on the assumption that not all parts of the program or its data need to be in the memory at the same time. Only some of the routines which compose the complete program are used so often, or are shared by a sufficient number of software functions, as to demand that they be continuously resident in memory. These *resident programs* and their *resident data* areas (which include sections of memory used for communication between programs of both resident and nonresident varieties) are kept in memory permanently. Resident programs may be stored in read-only memories. The remaining programs and their data are then loaded into read-write memory areas as they are needed, "overlaying" programs and data that were previously in these areas, but which are temporarily unnecessary.

This scheme is illustrated in Fig. 9-7. Overlay programs and data are kept in a bulk storage memory device, such as a disk memory, until they are needed. Overlay programs need only be loaded, but old overlay data must be written back to the bulk storage device before

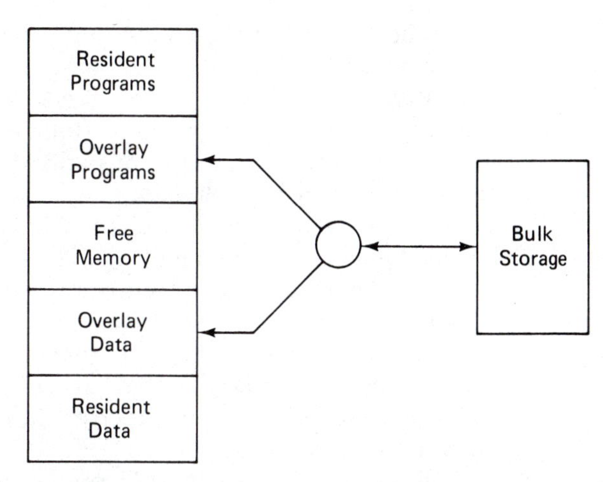

FIGURE 9-7. Use of overlays. Programs and data are read from peripheral memory when needed.

new overlay data are read in. By placing resident programs at one end of memory and resident data areas at the other end, some additional flexibility is obtained in that overlay programs and overlay data may grow toward one another into a free area located between them. If overlays are relocatable, more flexibility is obtained in terms of the system's being able to hold a wider variety of combinations of programs and data files, as they are required.

Overlays can be very cumbersome and complex. Operating system software is needed to manage free memory space allocation and to handle transfers between the memory and the bulk storage device. The operating system design must consider the possibility that program states may exist in which more overlay space is needed than can be supplied from the free memory. The time required to release an area of memory and load a new overlay must be considered in determining whether the overall system response time will be adequate for the application. The system may have to include provision for a priority system in order to make the necessary choices when two different overlays are called for at the same time and there is insufficient room for both of them.

Overlays may be practical in situations where the system has little or no real-time response requirements or all of the routines responsible for real-time functions can be resident programs which can be invoked quickly. Examples of industrial applications meeting these criteria are data analysis, as in a laboratory or computing center, for reduction of data collected by separate real-time computing systems, and the off-shift production of reports. Systems that are principally real-time in most of their functions must generally look to other approaches if they require more main memory.

Figure 9-8 illustrates the *bank switching* method. The main memory is divided into sections (four in the figure). Other divisions are possible, of course. How it is divided depends on the hardware design, but 16K-byte divisions are common. One section (denoted "0/0" in the figure) is always present in the main memory address space. Memory management elements of the operating system software are kept here, along with the real-time routines that have the shortest response time requirements. "Behind" each of the other sections (figuratively speaking), lies one or more additional 16K-byte sections (called *banks* and denoted "1/1," "1/2," etc., in the figure). These may be switched into the directly addressable memory space under software control.

Banks can be switched in several ways. For example, the logic that controls which banks are currently "active" may be directed by assigning it addresses in the input/output address space. Each bank would correspond to one memory subsystem board. Each board would be equipped with its own enable/disable logic, addressed as if it were an I/0 device. Commands to this logic are thus independent both of the instruction set of the processor and of the memory configuration active at the moment. A "read" of its address might serve to enable a bank, a "write" to disable it.

In the form shown in the figure, a particular memory bank can appear only in the corresponding section of main memory. For example, section 1/1 cannot be switched to appear in the last main memory section, where 3/0 appears. Note also that a routine in bank 2/1, say, cannot have access to routines or data in banks 2/0 or 2/2. If banks

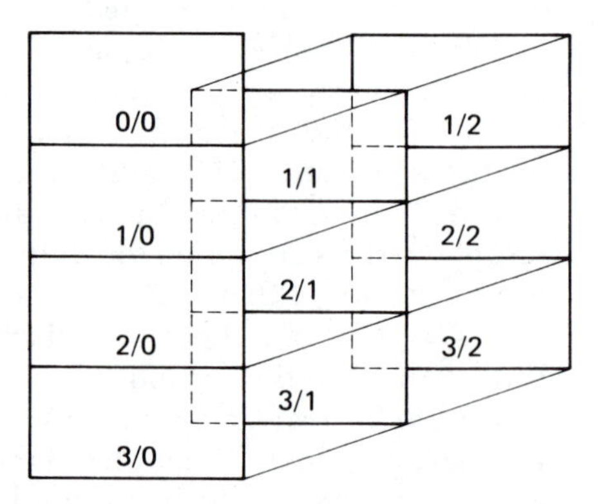

FIGURE 9-8. In bank switching, sections of memory are electronically switched into active memory as they are needed.

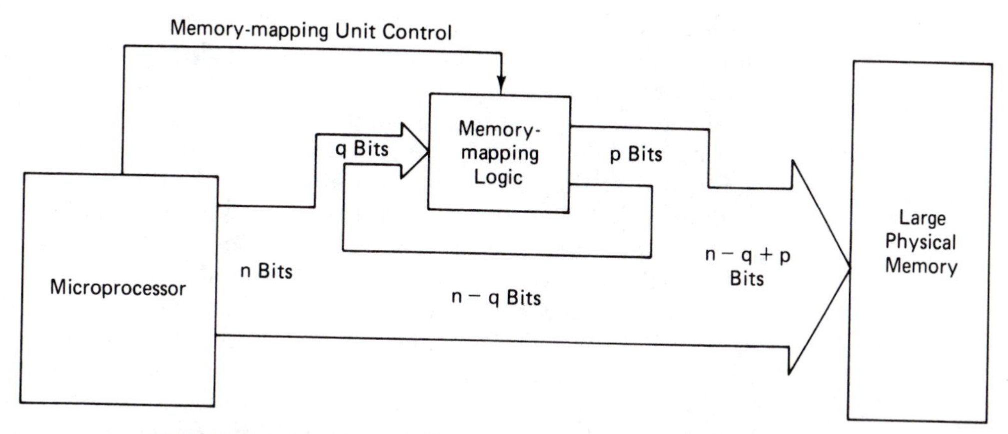

FIGURE 9-9. Use of memory-mapping hardware to extend the number of bits available for addressing a large physical memory. An n-bit logical address is extended to an (n − q + p)-bit address by words stored in the memory-mapping logic.

could be assigned to any section, it would be possible to have freer communication among routines at the cost of greater complexity. Subsystem enable/disable logic could be adapted to accept addresses of main memory sections to which they would respond in place of simple enable/disable commands. This amounts to address relocation of 16K segments.

However, software becomes more complex. Programs would have to be able to operate from a variety of initial addresses, and this might be difficult to accomplish. Routines accessing other routines and data would need to know where they had been placed by the memory management software and would have to adapt themselves to new locations or would have to direct assignments and provide for waiting if the requested assignment were unavailable because of use by other routines.

For these reasons, the methods shown in Fig. 9-8 are not widely used as general solutions to memory expansion problems. It is rather the case that they are used to provide extra room for large data files that are accessed infrequently and by as few programs as possible (say, by a single service routine which could be called by others in order to read or write this file). Before we end our discussion of this method, it should be noted that this approach is sometimes used in system-level products to hold loading and initialization routines which are simply switched out of the way when they are no longer needed.

Figure 9-9 illustrates a more versatile way of expanding memory that is much easier to use. Here we introduce the concepts of physical and logical addresses. Put most simply, the *physical address* of a location is the address decoded by a memory subsystem to gain access to that location. The *logical address* is the address used in the program to refer to the location. In the approaches we have discussed so far, physical and logical addresses have been the same—numerically identical. But they need not be numerically identical. If the correspondence between physical and logical addresses can be changed under cooperative software and hardware control, a much larger and more easily accessed memory can be obtained.

In the arrangement shown in Fig. 9-9, *memory-mapping* hardware is introduced between the processor and its memory system. This hardware contains a small memory of its own (usually a bipolar RAM, for speed). As indicated in the figure, its words are p bits wide and there are 2^q of them. The n-bit direct address from the processor is divided into two fields. The most significant q bits of this address are brought to the memory-mapping hardware, the remaining $n - q$ bits are taken directly to the main memory system (which, of course, may be composed of several subsystems).

The most significant q bits are used to retrieve a word from the small memory in the memory-mapping hardware. This word becomes a p-bit supplement to the $n - q$ bits of the original direct address, forming a physical address that is $p + n - q$ bits wide. Since $p > q$, this address contains more bits than the original n-bit address and, therefore, can access a much larger memory. The small memory in the memory-mapping hardware can be loaded by independent methods, for example, via the input/output address space. This function is part of the processor architecture in some 16-bit microprocessors. The effect is to provide simple addressing of a large physical memory that is divided into *segments* of 2^{n-q} locations. There are 2^p such segments, 2^q of which may be active at any time. Sections of the small memory in the memory-mapping hardware may be allocated to portions of the program so that they can control the mapping of their own segments independently.

The memory expansion methods we have discussed can be used, in principle, with any of the 8-bit microprocessors commonly used in industrial applications. However, such cases are relatively rare, for, as we have noted, most processors provide a direct address capability that is adequate for the majority of applications. The more modern 16-bit and 32-bit microprocessors provide greater memory capabilities both by the use of wider direct addresses and by providing instruction set elements and LSI support chips that support the management of large segmented physical memories handled by memory-mapping methods similar to the one we have just discussed.

MEMORY COMPONENT TECHNOLOGY

In this section, the technology of individual types of memory components are examined. In the figures and descriptions in this section, we concentrate on methods used to store information in semiconductor circuits. Only a few elements and their relations with their immediate neighbors are treated.

There are many ways in which these elements are organized in actual integrated memory devices, but they all have the common char-

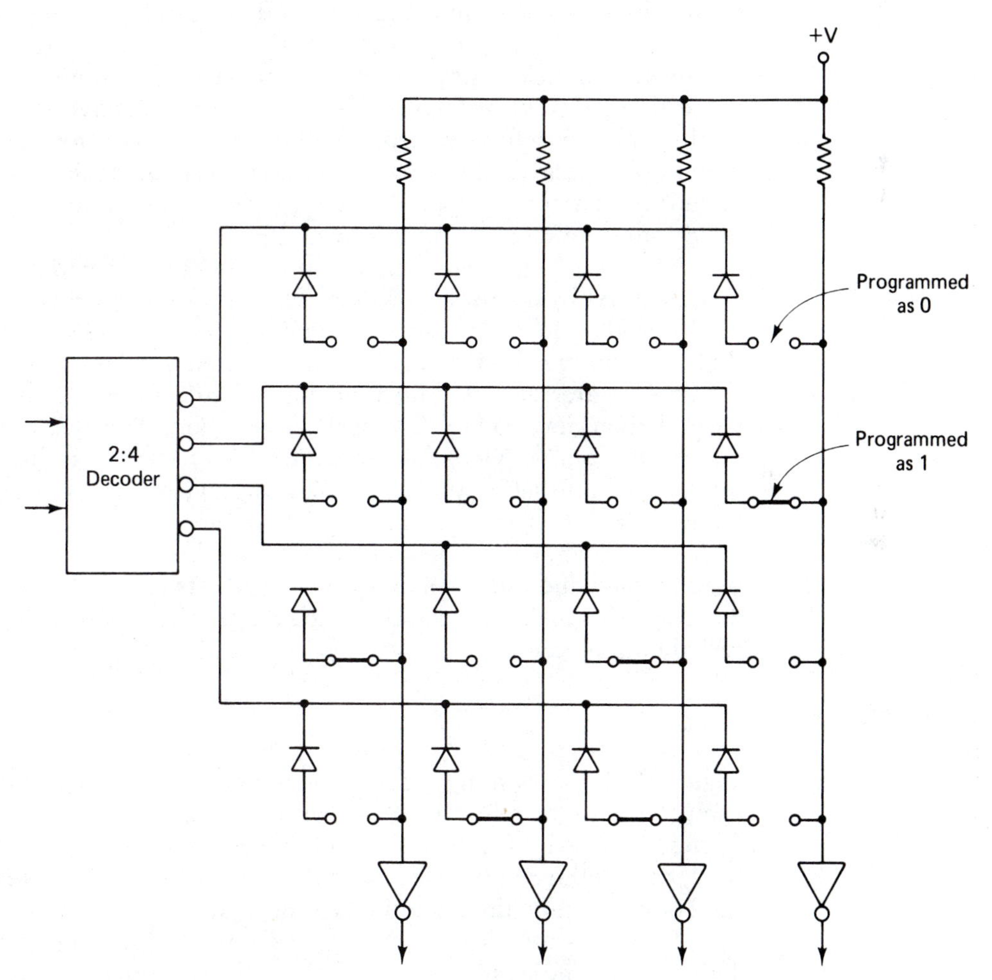

FIGURE 9-10. A 4 × 4 read-only memory (ROM) array. Diodes are selectively connected by masking operations when the device is made.

acteristic of being arranged in arrays in which individual elements are selected by the "intersection" of the outputs of pairs of decoders—the array decoding methods that were described in Chapter 7. As they become larger, arrays of arrays are used, the designer's purpose always being to keep the organization as regular and simple as possible.

ROM

Figure 9-10 depicts a 4 × 4 (4 words of 4 bits each) ROM array. The active decoder output selects one of the four rows of diodes. Connected diodes in the selected row pull their column lines to a low voltage, which is inverted to a high ("1") output by the output drivers. Connected diodes in other rows do not affect the column lines, because they are reverse-biased. The diodes are selectively connected to the column wires in order to determine the information programmed into the device. These connections are made during the final metalization stages of manufacturing, before the device is encapsulated into the integrated circuit package.

The connections are based on programming data supplied to the manufacturer by the user. The manufacturer uses these data to prepare masks used in the metalization/programming of the ROMs (which are, therefore, often called *masked ROMs*). Although the memory devices themselves are relatively inexpensive, manufacturers usually charge several thousand dollars for mask preparation. The devices are, in effect, "custom" devices for the user. This charge must then be amortized over a relatively large number of devices for them to be economical. For this reason, masked ROMs are usually used only for microcomputer systems of which a great number are to be made. Once programmed and encapsulated, masked ROMs cannot be changed in any way. This makes it critical that the programs and data programmed into them be correct.

PROM

Figure 9-11 shows four adjacent elements of a bipolar PROM array, each consisting of a fuse in series with a transistor emitter. If the fuse is intact, active-high selection of the transistor (decoder signal into its base) causes the column line to be driven to a "1." If the fuse is open, the column line is unaffected (remains a "0").

Elements are programmed by selectively blowing their fuses. This is done one word at a time. Column lines corresponding to fuses to be opened are held low, the word is selected by the decoder, and current pulses are applied to the +V lines. When a sufficient number

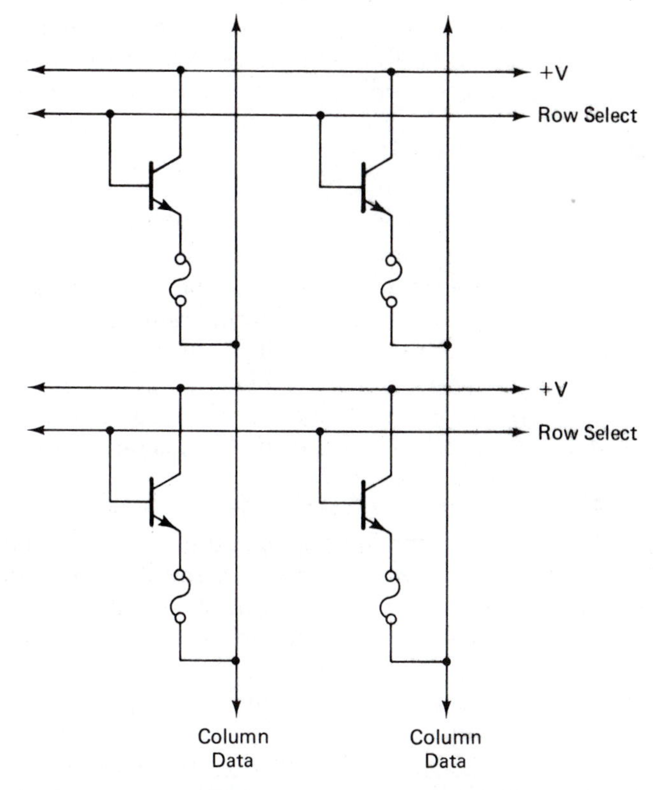

FIGURE 9-11. Elements of a bipolar programmable read-only memory (PROM). Data is programmed into elements by selectively blowing small polysilicon fuses.

of pulses of sufficient current have been applied, the fuse opens, leaving the element programmed to a "0." Because this can be done with relatively simple equipment, it can be done by the user. It is also possible to make limited corrections later. Although, once blown, a fuse cannot be replaced, it is possible to come back later and change "1s" to "0s". Thus, PROMs can be used economically for smaller numbers of systems which need not be strictly identical.

The MOS devices cannot readily handle the high programming current levels. For this reason, PROMs are usually bipolar devices. Since bipolar devices are not as highly integrable as MOS devices, PROMs usually have fewer bits than other forms of read-only memory devices. Two different types of fuse materials have been used. Early devices used nichrome (a nickel-chrome alloy), but it was discovered that remnants of blown fuse material exhibited a tendency to grow back in some devices, reconnecting the fuse after the device had been in service for some time. Newer devices use polycrystalline silicon ("polysilicon") as fuse material in order to avoid this problem.

EPROM

Eraseable programmable read-only memory is widely used as program memory in industrial microcomputer-based systems. It combines the important advantage of reliable program retention with the ability to be easily erased and reprogrammed many times. This makes it desirable for use with limited numbers of systems. Not only is it relatively inexpensive, but changes can be made easily, even after the systems have been installed—so that program errors can be corrected and improvements can be installed later, as they evolve and are enhanced with new features.

The EPROMs were introduced in the form of 256 × 8 devices, at about the same time as the first microprocessors. They have evolved to the extent that 4K × 8 and 8K × 8 devices are now common and 16K × 8 devices are appearing. For many applications, one such memory chip is large enough to hold the entire program. Most of these devices are based on the floating gate avalanche-injection MOS (FAMOS) charge storage device, which was developed by Intel Corporation.

Figure 9-12 shows a cross section of a FAMOS device. It is like an MOS transistor, but with an unconnected gate of silicon buried in the silicon dioxide layer between the drain and source regions. Application of a high junction voltage to the device results in the injection of high-energy electrons into the floating gate. Once the high junction

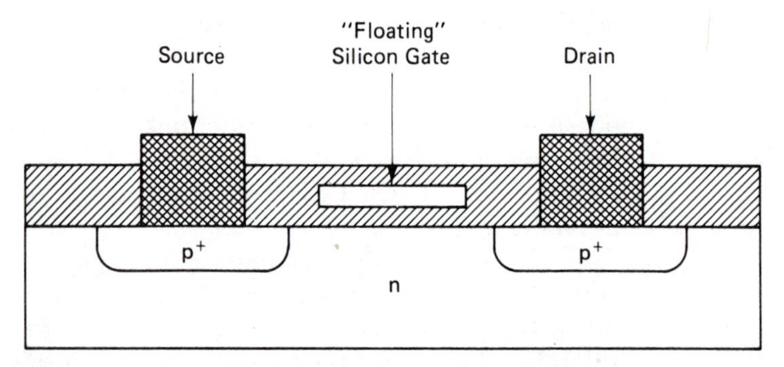

FIGURE 9-12. Cross-section of an erasable programmable read-only memory (EPROM) floating gate avalanche-injection MOS (FAMOS) charge storage device. Data is represented by charges trapped in the "floating" silicon gate.

voltage is removed, these electrons are essentially trapped in the gate by the surrounding silicon dioxide insulator. The electric field due to the trapped charges affects conductance between the drain and the source. Other circuits in the integrated memory then measure whether a charge is present by measuring this conductance.

The trapped charge eventually dissipates, but the time required for enough dissipation to change the measured state is well in excess of ten years, even at relatively high device temperatures—long enough for the life of the devices to be comparable to the useful life of the remainder of the system in the great majority of applications. The trapped charge can be dissipated rapidly by exposing the device to a sufficient dosage of ultraviolet illumination. Photocurrent induced by this illumination can erase a FAMOS memory device to its initial unprogrammed condition in a matter of minutes. After erasure, the device may then be reprogrammed. So that erasing illumination can reach the device, integrated circuit packages used for FAMOS memories have a quartz window positioned directly above the chip.

Generally, FAMOS memories must be removed from their sockets on printed wiring boards to be erased and reprogrammed. Some recent types can be programmed "on-board" because their programming circuits are sufficiently simple that room for them can be provided if the board designer so chooses. However, these are rare situations. Usually, erasing and reprogramming is done by using ultraviolet lamps and EPROM programming devices that are separate from the application system itself.

EPROM programming systems are commercially available for operation as "stand-alone" equipment, with data supplied to them by keyboard, tape input, or serial interface to microcomputer development systems. Others can make many copies of a master EPROM simultaneously. Others are parts of development systems, integrated with the hardware and software used for inputting, assembling, running, and debugging programs.

EEPROM

A recent development in EPROM technology is the electrically erasable programmable read-only memory (EEPROM). Many of these devices are based on a floating-gate tunnel oxide ("Flotox") mechanism (also developed by Intel Corporation), which is similar in some respects to the FAMOS mechanism. This mechanism is too complex to explain properly in the space available here. Suffice it to say that it provides significant advantages for industrial microcomputer applications. These derive from its ability to erase and reprogram individual locations within the memory, using simple external circuits (electric currents are used for erasing instead of UV illumination) that can be incorporated easily in memory subsystem designs.

Unlike EPROMs, it is not necessary to erase the complete device. This means that individual sets of data or programs within the chip may be erased and reprogrammed under the control of the software, on-line. It can be argued that a device that has evolved to this level of on-line alterability hardly deserves the appellation "read-only." Indeed, its only true claim to a "read-only" component in its name is based on the fact that very particular combinations and durations of applied signals are required to alter its contents—combinations so particular that accidental alteration is very unlikely.

The EEPROMs can, of course, be used in the same way as EPROMs for storing programs and constant data. They have the same characteristics of nonvolatility and "hardiness." But they also make it possible for parameters and data to be retained by a system for long periods and to be updated by the system itself, and they make it possible for the system to change its program or to accept changes and corrections to its programs that may be transmitted to the system via serial communication links from remote sources. Thus, EEPROMs are important developments that further our ability to make remotely located, adaptive control systems more versatile.

Dynamic RAM

Dynamic RAMs are used as components of medium-to-large main read-write memories. Figure 9-13 shows two adjacent cells of an MOS dynamic RAM that uses a single capacitor/single transistor storage mechanism. This structure evolved from earlier designs which used

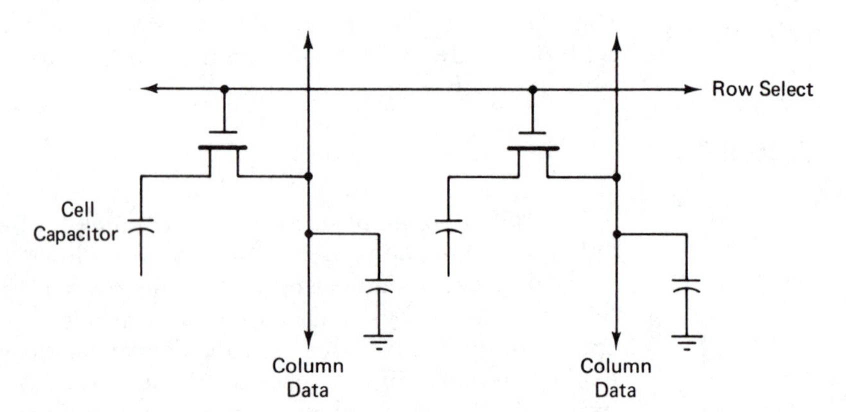

FIGURE 9-13. Two cells of an MOS dynamic RAM. Data is stored as charges on the cell capacitors, which must be refreshed periodically.

multitransistor circuits and relied on storing charge in transistor gate capacitance and in parasitic capacitances in those regions.

In reading, the row select signal transfers charge, stored in the cell capacitor, via the transistor, to the column data line capacitor. If the cell capacitor is not charged, prior charge on the column data line capacitor will be reduced. Sensitive latching sense amplifiers on the column data lines measure the small voltage changes that result and produce a suitable readout of the data. Since reading an element discharges its capacitor, information originally stored there must be written back into the element. This is done automatically by circuits that recharge the cell capacitors to the state that was read out. A similar operation is used to write new data into the cells.

If left alone for a few milliseconds, charges stored in the capacitors will dissipate and result in loss of the data. Reading a cell, as just noted, causes the charges to be *refreshed.* In fact, the devices are designed so that all cells on the selected row are refreshed. This provides the basis for continual refreshing of the contents of dynamic RAMs. Dynamic RAM subsystem printed wiring boards are equipped with refresh circuitry that consists of a counter, having enough bits to count through a complete set of row addresses, and some additional circuitry to control the timing and operation of this counter. When the subsystem is not being accessed, this circuitry performs a series of row accesses to all of the memory chips on the board in order to ensure recharging of all the storage elements.

Although this additional circuitry overhead is easily borne by medium-to-large read-write memory subsystems, it is less economical for smaller memories. For smaller memories, static RAMs are generally the preferred choice.

Static RAM

Figure 9-14 depicts a typical static MOS RAM cell. Six transistors are used to form a flip-flop, in which the datum is stored, and gating elements to provide access to the cell. Transistors Q1 and Q2 serve as drain loads for the transistors Q5 and Q6, which are cross-coupled to form the bistable circuit element. Q3 and Q4 are controlled by the select line to gate the datum from the flip-flop to the data lines in its true and complement form and to gate new data into the flip-flop.

Because so many more active devices are required for a static RAM cell, static RAMs cannot be as highly integrated as dynamic RAMs. However, they have the advantage of being nondestructive read-out devices and of being able to hold data indefinitely, so long as they are powered. As a result, they require no refresh circuitry and are much easier to use.

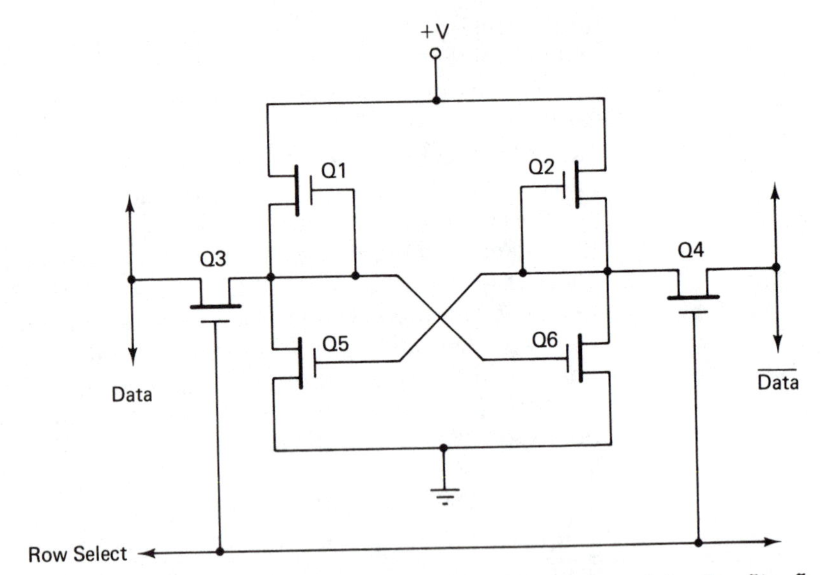

FIGURE 9-14. A static MOS RAM cell. Data is stored in the flip-flop formed by the cross-coupled transistors Q5 and Q6.

Most static RAMs have 1024 × 1, 256 × 4, and 1024 × 4 organizations. An interesting recent development in static RAMs is the availability of devices with 1024 × 8 and 2048 × 8 arrangements, with pinning that matches that of the new JEDEC standard EPROMs. This makes it possible to have a single memory subsystem in which both EPROMs and static RAMs can be intermixed freely.

Magnetic Tape

The principal types of magnetic tape-based bulk storage devices are *cartridges, cassettes,* and *reel-to-reel* systems. Figure 9-15 illustrates the components that are common to all types. Variations between types have to do with mechanical designs of tape-handling mechanisms and particular methods of encoding recorded data on the tape.

Data may be exchanged between the computer and the tape drive controller serially or one byte at a time—both methods are common. Data are recorded by magnetizing iron oxide deposited on mylar tape wound on a pair of reels. The control logic in the device directs motor-driven rollers which move the tape past read and write heads that sense the fields generated by the magnetized oxide and generate fields to alter its magnetization, respectively. In some devices, the use of physically separated heads provides the ability to read back data immediately after they are written so that they can be checked.

In most devices typically used with microcomputers, data are recorded serially on one or two tracks on the tape, with formatting marks to separate data into words, records, and files. Data encoding methods involve variations in the polarity or phase of the recorded signals. There are several popular methods, each having various relative advantages of reliability, noise immunity, and ability to record data densely.

Tape cartridges include both reels, mounted on a metal plate and covered with a plastic shell, and incorporate a single drive roller within the cartridge. The tape is moved past an opening in the shell by means of a single roller in the transport that acts through the drive roller in the cartridge. Cassettes contain both reels, but they are separately driven. In general appearance, these are similar to the cassettes used in home high-fidelity sound systems but are manufactured to more exacting standards. Reel-to-reel systems use separate removable reels. Tape must be unwound from the supply reel once it has been mounted on the drive and threaded through the read/write head channels.

Cartridges have several mechanical advantages over cassettes and reel-to-reel systems. These include the use of a single-point drive, which helps to ensure uniform tape motion independently of the amount of tape remaining on the reel, and a simpler, more rugged mechanical design. For these reasons, they are preferred for industrial applications in which the tape may receive less than gentle handling.

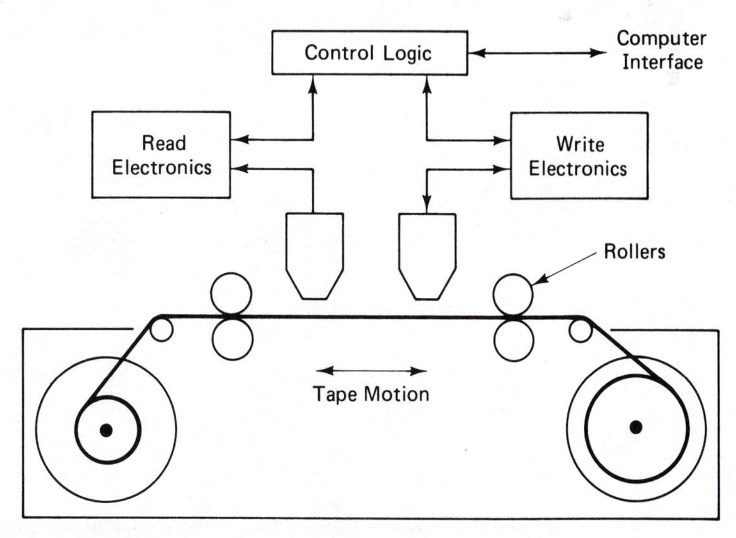

FIGURE 9-15. Components common to magnetic tape-based bulk storage devices. Data is recorded on magnetized iron oxide deposited on mylar tape moved past read and write heads.

Disks

Figure 9-16 shows the basic components of a disk system. Again, a medium with deposited iron oxide is used. In this case, the medium rotates under the read/write heads. As with tape, data may be exchanged between the computer and the disk controller in either serial or byte-parallel form. Data are recorded on the disk medium in *tracks*. Each track is divided into *sectors* (128 or 256 bytes per sector being common). The number of tracks and the number of sectors per track vary. More advanced designs have 100 or more tracks and in some cases record information on both sides of the disk.

The two principal variations on disks are *moving-head* and *fixed-head* types. In moving-head disks, the controller positions the read/write heads over the track to be accessed. Stepping motor-driven lead screw mechanisms are commonly used to accomplish this. The controller watches the data read out for timing marks, which identify the starting points of tracks and sectors. When it comes to the sector to be read or written, data are transferred serially to or from the disk. Fixed-head disks provide a read/write head for each track. They offer faster access since head positioning is not required, but they are more costly because of the extra heads required.

Disk drives are further distinguished by whether or not the medium is removable. In nonremovable media drives, stable metal disks are used, and the complete mechanism is sealed at a positive pressure

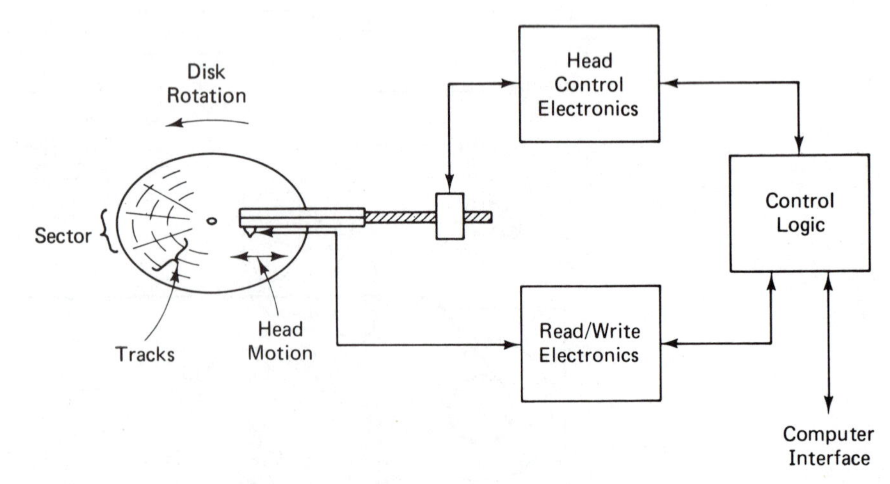

FIGURE 9-16. Basic components of a disk system. Data is organized in circular tracks divided into sectors. Continuous rotation of the disk moves data under the read/write heads, which are positioned over the selected track by the head control electronics.

to keep dust particles and other contaminants out. The heads ride very close to the disk, so closely that a particle of dust can cause serious problems, even damage. However, because of the proximity of the heads, data can be recorded very densely, so these devices can record much greater amounts of information.

In removable media drives, mylar disks are commonly used, a protective coating is placed over the oxide, and the read/write heads contact the medium. Attainable recording density is thus reduced, but the removability of the media provides access to larger amounts of data, since operators can insert new disks as they are required. Although these are less susceptible to dust and contaminants than fixed-head disks, they wear out more quickly. The principal example of this type of device that is commonly used with microcomputer systems is the so-called "floppy disk."

CONCLUSION

We have reviewed the characteristics of memories at a level that should enable you to understand most of their general principles and considerations and to evaluate their use in industrial microcomputer-based control systems.

Developments in on-line and peripheral memory technology are driven by needs for larger, denser, more reliable, and less expensive devices. Because they receive so much research and development attention from semiconductor, computer, and peripheral device manufacturers, newer and better devices become available very rapidly. Close attention to announcements in the trade journals is therefore highly recommended. Several prominent trade journals to watch are listed in the bibliography for Part II.

CHAPTER 10

Interfaces

> "Dust as we are, the immortal spirit grows
> Like harmony in music; there is a dark
> Inscrutable workmanship that reconciles
> Discordant elements, makes them cling together
> In one society."
>
> William Wordsworth

The purpose of microcomputer interfaces is to reconcile discordant elements, to make the microcomputer and the things it monitors and controls cling together in one harmonious society. A great deal of dark inscrutable workmanship is sometimes found in computer interfaces. There are few strong interface standards. Those which exist are subject to certain latitudes of practice, perhaps because of the variety of elements to be mated, each with its own set of rules and requirements.

It is in interfacing that the industrial controls engineer often finds the greatest challenges and opportunities for original work. Even when working with board-level and system-level microcomputer products, it is likely that some aspect of the project will require solving an interfacing problem.

GENERAL CONSIDERATIONS AND ISSUES

General Aspects and Organization

We can classify aspects of microcomputer interfaces in three categories: electrical, mechanical, and logical.

Electrical aspects have to do with such things as drive capability, circuit input loading, noise immunity, reliability, circuit speed, and response time. *Mechanical aspects* deal with physical volume required by interface elements, types of connectors, terminals, wires, and cables required for interconnections, dissipation of thermal energy, and ways in which the microcomputer, interfaces, and monitored and controlled devices are located with respect to one another. *Logical aspects* have to do with timing, data path mismatches, addressing, operating sequences, and ways of representing information.

It is important to note that many of these aspects may affect software profoundly, yet the issues of software effects are sometimes ignored. This is unfortunate because proper consideration of software effects and ways in which software can help with interfacing problems can go a long way toward improving the cost and performance of a system.

It is sometimes useful to imagine an interface as a series of buffers, as shown in Fig. 10-1. "Buffer" is used in the sense of its dictionary definition—something that lessens the impact of things on one another. It is often used as a synonym for electronic devices with strong driving current. It is also commonly used by programmers as a term for areas of memory in which data can be held temporarily. Here, we extend it to devices that resolve mechanical, logical, and electrical incompatibilities.

Several electrical and mechanical aspects of interfacing (such as reliability, connectors, terminals, wires, cables, thermal considerations) are very broad in scope and applicable to many other areas of system design. In this chapter, we concentrate on logical aspects and the most important electrical aspects. These are interspersed in discussions of the operation of several types of interfaces.

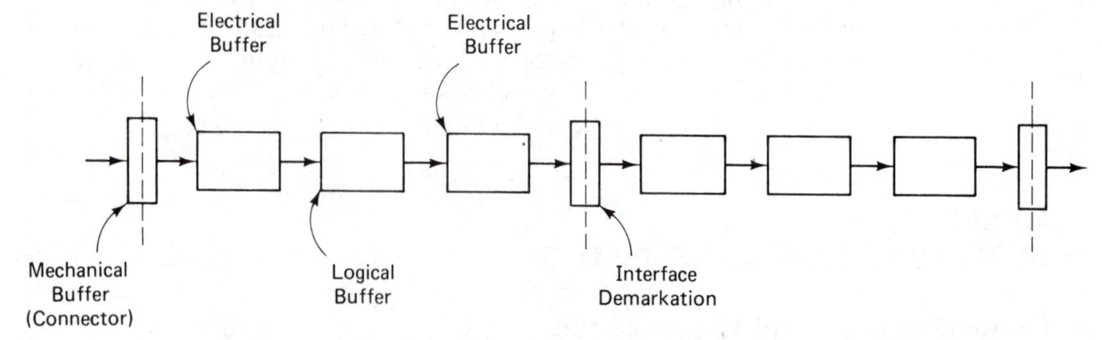

FIGURE 10-1. Interfaces can be thought of as a series of buffers—mechanical, electrical, and logical—that lessen the impact of one system element on another by resolving incompatibilities.

Information

The most important discordances among elements of computer-based systems arise from different ways of representing the same information. Many interface aspects are related to the veracity with which information passes through the interface from one element to another.

Information is represented by amplitude, phase, and frequency of electrical, optical, and mechanical signals. Interfaces can be viewed as transducers which alter the way this information is represented. They convert amplitudes, phases, and frequencies into digital numbers (and vice versa); they accumulate pulses as counts; they rearrange parallel data into serial form and serial data into parallel form. They hold information until its intended receiver is ready to accept it. They warn sources of information that receivers are in need of information.

In the ideal case, resolution of incompatibilities should not alter the meaning of information as it passes through these buffers. This is difficult to realize in practice, especially when the interface involves conversion of the form of information from analog to digital or vice versa.

Digital and Analog Interfaces

There are two main categories of interfaces: digital and analog. The microcomputer is the ultimate sink of incoming and source of outgoing information and is essentially digital. Elements on the other side of the microcomputer interface may be either digital or analog. Digital devices have, for practical purposes, a finite number of discrete states. Analog devices have an essentially infinite number of states—they handle continuous or piecewise continuous signals.

The greatest difficulties are encountered at boundaries between digital and analog elements. Here, profound changes take place in the form of information—the conversion of an information-bearing analog signal to digital form, the conversion of a digital word to an equivalent analog signal bearing the intended information.

Interfaces between digital elements generally confine their reforming of information to regrouping. Individual bits may be combined into a word or distributed from a word. The combination may take the form of collecting serially received bits into a larger word for parallel transfer to the computer (as in communications or pulse counter interfaces). The reverse of this process is used for transmitting data or controlling pulsed actuators, such as stepping valves and stepping motors.

Although serial interfaces are used for many purposes, they are usually used for communication among computers and terminals. They are a sufficiently important subcategory of digital interfaces to deserve special treatment.

Timing

The computer and the elements that compose its environment each have their own "agendas." Each operates in its own fashion, with differing sequences and speeds, having little to do with the meaning of information but much to do with the extent to which it can be captured and transferred. Failure to capture or transfer information properly can alter its meaning and cause subtle errors in operation.

Although no physical action can take place in "zero" time, it can sometimes seem to have done so. If the receiver of information does not look quickly or often enough, its perception of the received signal will be altered in ways that misrepresent reality. An example of this effect is seen in films of turning spoked wheels. Because the projected image on the screen is really a series of images taken at a fixed rate unrelated to their rotational speed, the wheels of a stagecoach in a western movie often seem to be turning backwards or at rates that are incompatible with the speed of the stagecoach itself. This phenomenon has bothered several generations of children at Saturday matinees and has disturbed the operation of some systems designed by those who grew up to become computer engineers and programmers.

Issues of rates of change of information and bandwidths of sources and receivers of information are well-known to many engineers, especially those involved in digital signal processing. For the purposes of this chapter we must simply keep in mind that an important event (such as a change in an input signal) may be missed or misinterpreted if the interface design does not consider it.

Transient information must be captured and delivered to its destination (the computer programs that process or respond to it) soon enough to be used properly. The programs must respond with their outputs soon enough to have the desired effects. This is the essence of *real-time* computer systems. In a sense, *all* computer systems are "real-time." Even general ledger and payroll programs (not usually thought of as being real-time) operate under time constraints that differ only in their scale.

Many mechanisms are available for resolving timing discordances. Most are logical in nature. Interrupts are an example. They force a synchronism between the microcomputer and external events. First In/First Out (FIFO) buffers are another. They collect bursts of incoming data and hold it until the microcomputer can accept it. These mechanisms are not restricted to implementation in hardware—many can be implemented in software, extending the concept of interfaces into the "interior" of the microcomputer.

Some interface operations must be made to seem to be single actions when they actually consist of a sequence of steps. One example is the output of digital values to a digital-to-analog converter. If the word size of the computer is less than that of the converter, more than one data transfer is required. The converter must not be allowed to change its output until it has received the complete, multiword digital value. Another example is updating a counter driven by input pulses. Carries in such a counter may take place while the computer is reading it. If this happens, the value read in may be incorrect. These two examples are also related to issues of data path mismatches, addressing, and operating sequences.

Noise Immunity

Noise can be admitted into the system (and even created) by the interface. Remembering that noise is *any* unwanted signal component helps to focus this issue.

We might, for example, design a digital switch interface that produces an excellent reproduction of the voltage across the switch, forgetting that this reproduction will include effects caused by bouncing of contacts as the switch is opened and closed. This is a case in which the interface should reject some information. But if our purpose is to *test* switches, we *want* this information to be included in the resulting input.

If an input signal represents a process variable known to be capable of changing only very slowly, we can assume that rapid variations represent noise. This can be rejected by filters or components with slow response times, but this interface cannot then be used for signals that contain rapidly changing *wanted* information. When it comes to noise immunity, we cannot have our cake and eat it too.

Potentially damaging voltages and currents are also unwanted signal components. This makes the design of "rugged" highly accurate wide bandwidth analog interfaces for industrial applications a very challenging task. Although there are many ways of isolating a system from potentially damaging inputs, each has its own effects on the information content of the signal finally transferred from the interface into the microcomputer.

Addressing and Digital Data Transfer

We have discussed addressing schemes which place I/0 devices in either the main address space or their own address space. These give programs access to ports between the microcomputer and interface subsystems. We discussed how data are transferred between the computer and ports, with handshaking logic to indicate new data presence at the port and receiving device readiness to accept new data. We discussed the use of unidirectional and bidirectional busses as pathways for data transfers. All these are found in interfaces between the computer and I/0 devices at the boundaries closest to the computer.

As we move away from the microcomputer and closer to the interfaced device, we still find interface elements that must be monitored and controlled by the microcomputer programs. Examples are multiplexers which select analog signals for conversion, modems which constitute the electrical interface between a communication interface and the medium used to transmit data, and digital logic used to control the collection and distribution of signals associated with discrete sensors and actuators. Because programs must have access to these, methods of communicating with them must be provided. However, because each interface has its own rules and requirements, we find a greater variety of mechanisms here than at the boundary of the microcomputer itself.

In some complex interfaces, we may find organizations that rival the CPU and its memory systems in architectural complexity. For example, it is increasingly common to find interfaces organized around another small microcomputer system. This is especially so with peripheral devices, such as printers and interactive terminals. In industrial interfaces, it is sometimes useful to have a small microcomputer embedded in the interface subsystem and entrusted with responsibility for watching inputs for changes. This strategy helps widen the bandwidth of the system without overloading the "central CPU." It also provides the opportunity to have programmable interfaces that are more easily adapted to different missions without a need for hardware changes.

In a system that must handle many different types of signals, complexity grows rapidly. This is due more to type variety effects than to the complexity of handling any one kind of signal. The latter grows linearly, while the combined effect of different types grows more than quadratically. These effects are felt most strongly in software, where they sometimes result in programming errors that go undetected for incredibly long periods of time before the system finally encounters that particular combination of circumstances that brings them to light. For these reasons, an important interface design principle is to keep it as simple and as "regular" as it can be made. This should not be taken to mean that *every* opportunity to reduce software workload should be taken. There are many cases where software can help simplify interfaces and make them more effective. Interface designers should be alert to these and take advantage of them wherever possible.

DIGITAL INTERFACES

We begin with interfaces of the digital type, since they are usually simpler. Interface subsystems tend to be similar at their boundaries with microcomputer data, address, and control busses, so we can also use digital interfaces to illustrate organizational features of the other types, as well.

General Organization

Figure 10-2 is a generalized illustration of a digital interface. It is composed of three major areas: (1) a *microcomputer boundary area,* which serves as the interface between the subsystem and the microcomputer busses; (2) an *internal logic area,* which embodies the logical operation of the subsystem; and (3) a *signal conditioning and isolation area,* which acts as the interface between the internal logic and the discrete sensors and actuators served by the subsystem. The

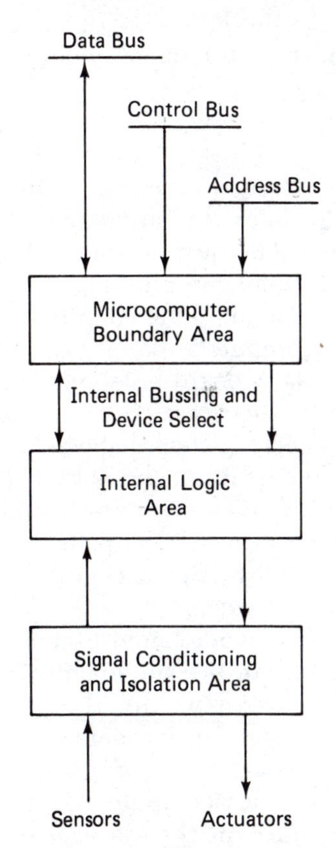

FIGURE 10-2. Generalized form of a digital interface.

microcomputer boundary and internal logic areas are interconnected by an internal bussing arrangement supplemented by device and function select signals generated by decoders in the microcomputer boundary area.

Microcomputer boundary areas are usually similar in operating principles. Differences lie in the number of address bits decoded by device select logic, which must decode a sufficiently large address subfield to make as many device and function selections as required by the interface. Signal conditioning/isolation areas are tailored to the interface type and differ greatly from one to another. In a communications interface, it includes circuitry to perform signal level conversions needed for the communications media used. In an analog interface, it includes filters, gain amplifiers, and isolation devices. Analog interfaces are more likely to be isolated by ferromagnetic or flying capacitor techniques, digital interfaces by optical techniques. The internal logic area of an analog interface board contains analog-to-digital and digital-to-analog converters. In a communications interface, it would contain one or more UARTs or USARTs.

In the following detailed examinations, we begin with internal logic and work our way out to each side.

Internal Logic Area

The principal elements of the internal logic area are gates and registers. Gates route inputs from signal conditioning areas to the internal busses. The desired states of output signals are held in registers that drive signal conditioning and isolation circuits. In some designs, registers also latch the condition of input signals for later reading by the microcomputer.

Many board-level interface products use LSI parallel I/0 support chips in this area. These can be thought of as sets of "programmable" gates and registers that can be configured (within limits) for different numbers of input and output signals. This is done by command words from the microcomputer which are latched and decoded within the device to set up data paths between its input/output pins, which are connected to signal conditioning/isolation area circuits, and the internal bus area.

This work can also be accomplished by a small microcomputer (e.g., one of the single-chip variety) on the interface board itself. In such a design, we find the small microcomputer embodying the gates and registers in its memory and I/0 ports. The advantage of such a design is enhanced programmability and off-loading of work from the "main" microcomputer. To the extent that the main microcomputer can control the the interface microcomputer's operation, it may assign it the tasks of monitoring inputs for special conditions and taking control actions when certain events occur.

Some designs utilize MSI components in this area. These are satisfactory for most discrete interfaces in which the relative number of input and output lines is known beforehand. It is also possible to provide programmability with MSI components by arranging sockets and printed wiring board traces so that the specific number of inputs and outputs can be configured (again within certain limits) by substituting different ICs in the sockets. This approach is inexpensive, but it does not utilize printed wiring board area as efficiently as with LSI support chips.

The internal bussing arrangement depends on how the internal logic area is implemented. In some designs, separate bussing is provided from gates to the data bus interface and from the data bus interface to registers. This is common with MSI implementations. With LSI support chip implementations, the internal bus is usually organized as a bidirectional bus that is similar to a microcomputer data bus.

Figure 10-3 shows two ways in which internal bussing is often arranged when MSI components are used. Part (a) shows a bidirectional arrangement; part (b) shows a unidirectional pair, one for input signals and one for output signals. One bit of the internal bus is shown in the figures, with two output signals driven from it and two input signals received by it. In both cases, MSI input gates can be implemented with three-state devices to multiplex inputs onto the internal data bus. Unidirectional arrangements can use MSI multiplexers in place of three-state devices. If equipped with three-state outputs, MSI multiplexers can also be used with bidirectional arrangements. Multiplexers have fallen from favor in recent years because simple and versatile three-state gate devices are now commonly available in octal packages.

The internal data bus is joined to the microcomputer data bus by bus drivers. Since the microcomputer data bus is usually bidirectional, these are arranged for bidirectionality on that side of the boundary. On the interface side of the boundary, they must, of course, be arranged to match the internal bus structure.

In a very small system, boundary bus drivers may not be needed. If only a few devices are on the microcomputer data bus, the interface registers and gates may present only a small load that can be connected directly to it. Care must be taken, however, because interfaces have recovery time characteristics analogous to those of memories and may interfere with one another and with system memories unless proper timing margins are provided by address recognition and device selection logic.

Figure 10-4(a) illustrates the use of a pair of parallel I/0 support

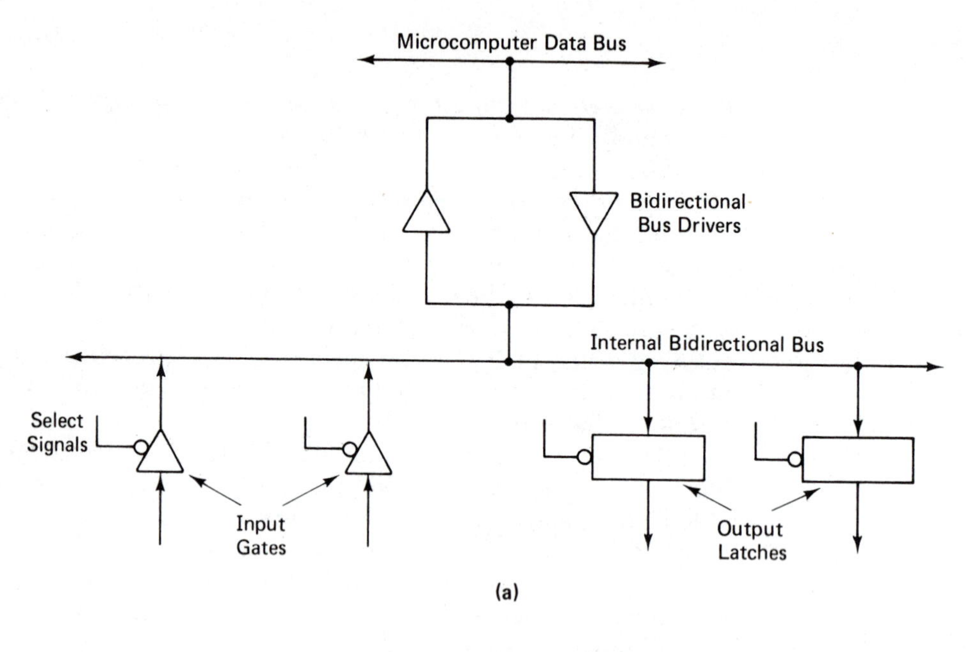

(a)

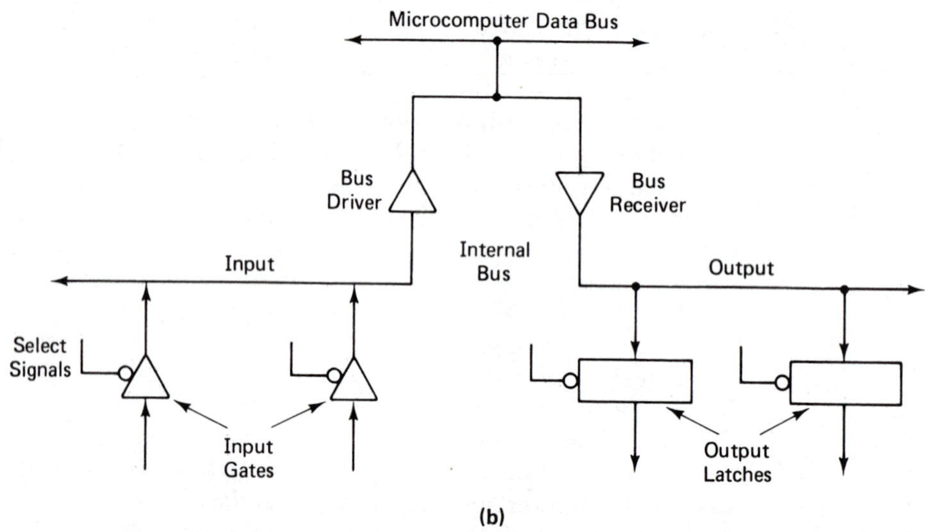

(b)

FIGURE 10-3. Two internal bussing arrangements for use with MSI interface components: (a) a bidirectional bus that carries both input and output signals, (b) a unidirectional bus with one section for input and one for output.

chips and their relation to the internal bus, signal conditioning/isolation area, chip select signals, and address lines. Each chip handles three ports. For typical chips of this type, port directionality is set by command words from the microcomputer. The figure suggests one way in which the ports may have been set up by such a command. Two address bus lines, brought through simple buffering if needed to minimize loading, are decoded within the chip. One of the four codes available might select the command word; the other three might select the port to be accessed.

Figure 10-4(b) shows how I/0 line programmability can be achieved with MSI components. One signal line is shown, with a representative latch or gate from an MSI integrated circuit package. The signal from the signal conditioning/isolation area is brought to both the gate input and the latch output. The select signal is brought to the gate enable input and to the latch strobe input. If only the latch device is installed, the circuit will serve as an output. If only the gate device is installed, the circuit will serve as an input. In the figure, the internal bus is bidirectional. If it were unidirectional, it would be possible to include the input gate with the latch. This makes it possible for the microcomputer to read the contents of the latch as well as write to it. This is useful for self-testing, since diagnostic programs could then transfer information to the interface, read it back, and compare it to see that the data lines and select signals were operating properly.

Microcomputer Boundary Area

The organization of the microcomputer bus boundary area is shown in Fig. 10-5. Note that it is similar to the memory subsystem illustrated in Fig. 9-2. As with a memory subsystem, the microcomputer uses its data, control, and address busses to get access to interface subsystem elements. The data bus may be dedicated to input/output or shared with memory, depending on the organization of the microcomputer busses. If it is of the memory-mapped type, the control bus signals will be the same ones used for memory control, and the address bus will be the same full direct address bus (most commonly 16 bits wide).

In this case, the subdivision of the address bus differs from that of a memory subsystem, but the device select logic serves almost the same function as memory chip select logic. This is usually the deepest level of addressing. That is, no further decoding is done within the interface. Thus, if the interface contains, say, 16 logical elements which must be accessed individually, the device select logic decodes the least significant four address bits, and the output of this decoding enables one of the 16 elements.

Digital interfaces with highly integrated components may call

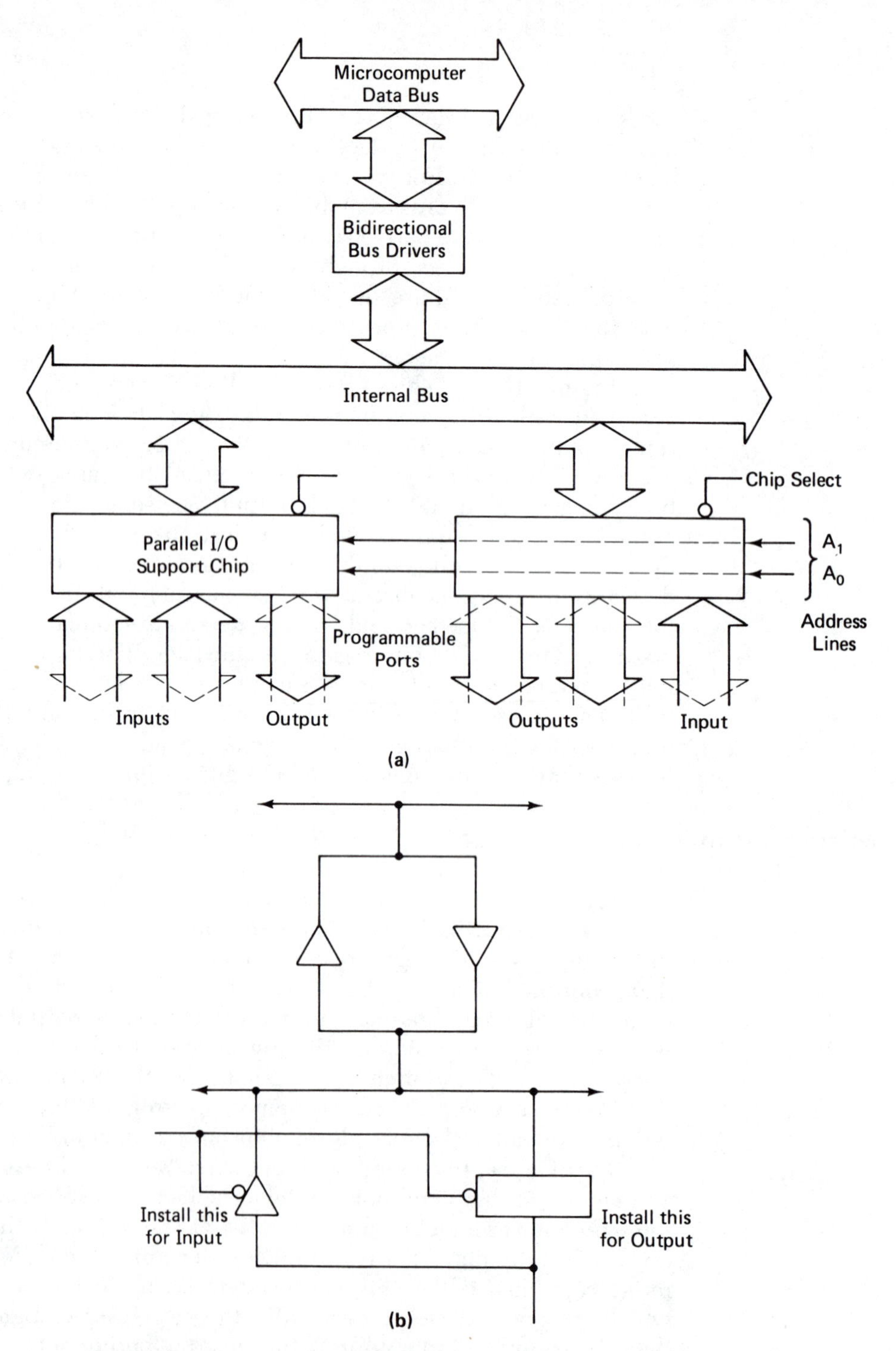

FIGURE 10-4. Programmable I/0 port directionality: (a) use of parallel I/0 support chips, in which directionality is set by software commands, (b) MSI-based interface programmability by selective device installation.

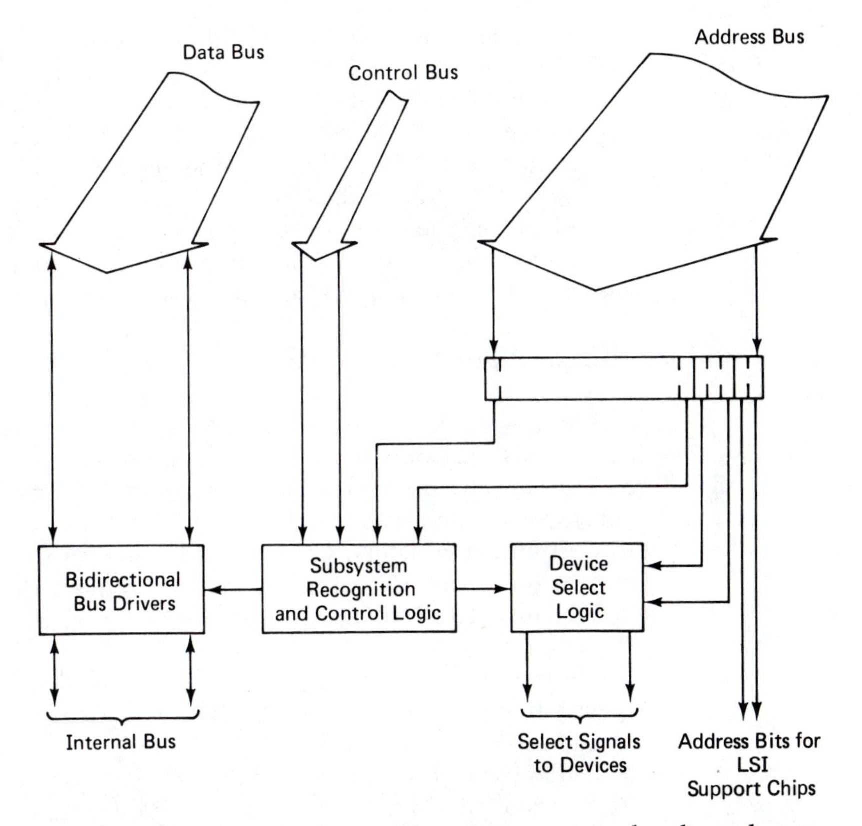

FIGURE 10-5. Organization of the microcomputer bus boundary area. The principal difference from a memory subsystem is the subdivision of the address bus.

for another level, using 1 or 2 bits to select functions within the LSI component. In this case, the field decoded by the device select logic is moved 1 or 2 positions toward the more significant bits, and the least significant address bits go to the LSI component. Note that the LSI components are accessed in a way which is similar to the way individual memory chips are accessed. The principal difference is the number of elements addressed within the LSI component and, therefore, the number of address bits required.

Subsystem recognition logic must recognize its own address on the bus. This is often done by comparing an address bus subfield to a set of data switches mounted on the board. The number of bits it must recognize depends on how the address space is partitioned in a memory-mapped I/0 organization. To avoid wasting addresses, it must recognize *all* of the remaining bits. Less complex recognition logic may be realized with fewer bits, but this causes a *de facto* partitioning of the address space that sets aside more locations than are usually needed for interface subsystems, denying them to memory subsystems. If the system is not memory-mapped, the recognition logic need only handle the remaining bits of the I/0 address bus, which is 8 bits wide in the most commonly used microcomputers.

Upon recognizing its address, the control logic enables the device select logic and the data bus interface. The data bus interface also requires that the direction of transfer be specified. This is derived from control bus signals in the same way as for memory subsystems.

Signal Conditioning and Isolation Area

If the subsystem is an interface to devices that are not compatible with TTL signals, its boundary with them must be populated with signal conditioning circuits to provide voltage and current level translations. Inputs from switches sensed by alternating currents, for example, require dropping resistors, bridges, and filters to provide a dc input. Outputs to solenoids and motor starters require drivers to provide enough current and voltage to operate electromechanical or solid-state relays.

In typical industrial applications, serious electrical noise is sufficiently likely that some form of isolation is virtually mandatory. Digital interfaces often use optical isolation techniques. Optical couplers are available for this purpose in integrated packages, each containing 1 or 2 sets of LED/phototransistor circuits which, when properly applied, provide 1000 to 2000 volts of isolation. Solid-state relays with integral optical couplers are also available. These have generally replaced custom circuits and transformer isolation techniques used in the past.

Digital interface subsystem cards are usually installed in the same card frame as the microcomputer, memory, and other interface boards. The physical space required for conditioning and isolation devices, and concern about electrical interference from nearby high voltage and current wiring, have motivated most designs to be arranged so that signal conditioning for discrete sensors and actuators can be mounted away from the system, with connectors on the edge of the card used for cabling to separately mounted signal conditioning and isolation circuits.

For small board-level systems, where there may be more latitude in card arrangements, interface cards that include small relays or isolated inputs are sometimes used. For custom-designed systems, where the designer can control the placement of components and the types of signals they handle, it may be possible to include them on the same cards with the microcomputer, memory, and other logic. This should be done only under carefully controlled conditions and only after careful consideration because such arrangements may suffer from noise problems that are difficult to resolve.

Digital Interface Application Notes

In this section, several ways of applying digital interfaces are discussed.

It is often imagined that there is a one-to-one correspondence between bits handled by a digital interface and discrete sensors and actuators to which the interface is connected. If 32 switches are to be monitored, 32 bits of input are thought to be required. If 16 BCD digits are to be displayed, 16 × 4, or 64, bits of latched output are thought to be required. In fact, several principles of logical organization that we have seen in previous chapters can be applied to extend the capability of a digital interface so that it can handle more information than might be immediately apparent.

Figure 10-6 shows a portion of a TTL digital interface. Here, we

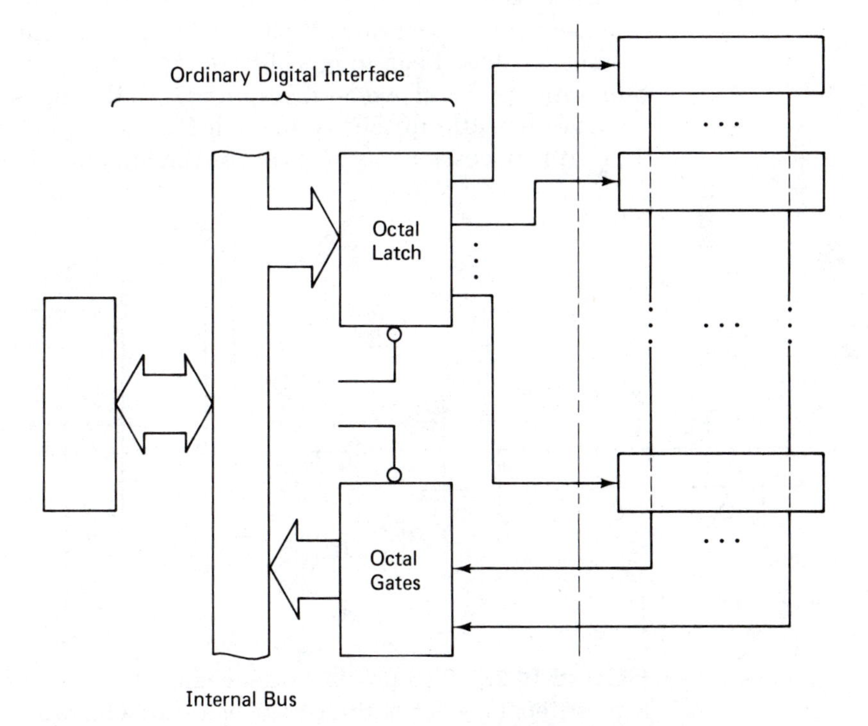

FIGURE 10-6. Basic arrangement of an interface for inputting more bits than are provided for directly by the interface ports.

are working with one input port and one output port. To the right of the figure, there are eight "octal" circuits, the details of which we address in a moment. The outputs of these circuits, generally formed from components not mounted on the interface board, are bussed together to the digital interface input gates. Each output port bit is wired to one of the external circuits.

Note the similarity between this arrangement and that of a read-only memory. The interface board input gates clearly can be used to transfer information from the bus between these external circuits to the microcomputer. It remains only to select among the external circuits. This is the purpose of the interface board output latch. Assuming that each external circuit can be enabled by an active-low signal, we see that the microcomputer can select one circuit from among the eight external circuits by loading the latch with a byte in which all but one bit is a "1." The single bit containing a "0" enables the circuit to which it is wired.

The external circuits can take many forms. Two possibilities are shown in Fig. 10-7. Part (a) is an octal arrangement of three-state gates. Part (b) is a set of eight switches with backing diodes that is essentially the same as a ROM. In general, any "bussable" device can be used to make up these circuits. What is achieved by this is an input subsystem capable of handling 64 bits of input, but formed from only one 8-bit input port and one 8-bit output port. If the input signals are TTL-compatible, little further is needed. If they are not compatible with TTL, it is necessary to put level conversion (and, possibly, isolation)

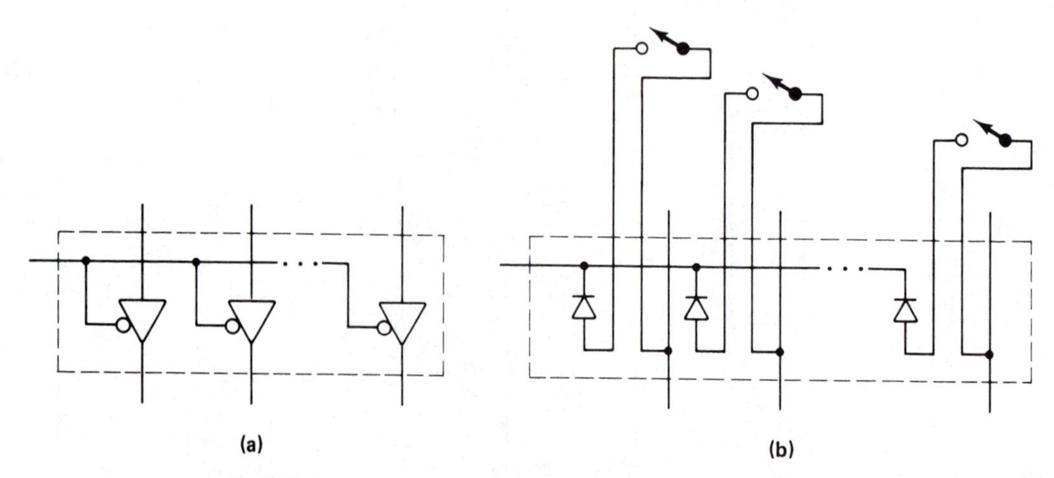

FIGURE 10-7. Two possible octal circuits for use with the interface of Figure 10-6: (a) a set of three-state gates for TTL inputs, (b) an 8-switch circuit. Used with the circuit in Figure 10-6, these provide for up to 64 bits of input via a two-port interface.

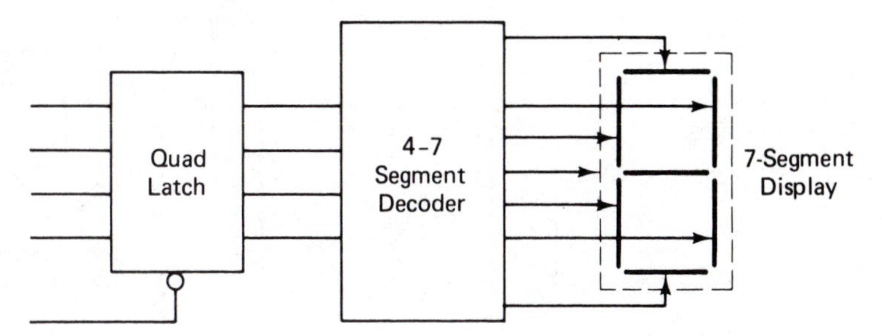

FIGURE 10-8. Organization of logic for the display of numeric data using 7-segment display devices.

circuits on the lines between the ports and the external circuits. This arrangement can be used to read values set into BCD thumbwheel on an operator control panel. It can be used to monitor a substantial number of limit switches and other discrete machine sensors.

Control panels often require numeric displays, implemented with seven-segment display devices. These may be based on LED, liquid crystal, or other types of electroluminescent devices. The lighted elements are arranged as shown in Fig. 10-8. Four-to-seven segment decoders convert 4-bit BCD digits to the proper combinations of segment illuminations to form the digits from "0" to "9." In some devices, additional elements are added to provide decimal points and plus and minus signs. Although these devices are found in several forms, the easiest to use are those in which a 4-bit latch, 4-to-7-segment decoder, and 7-segment display device are all integrated within a single circuit package. Figure 10-9 shows how they can be combined to form a 16-digit display controlled from a pair of digital output ports. The structure of this arrangement is similar to that of Fig. 10-6.

Such digital interface extensions save on hardware costs, but at the expense of a slight software complication. Rather than addressing the I/0 ports directly, programs must take some extra steps to accomplish the desired operations. Usually, these are not onerous burdens and can bring about substantially reduced hardware costs. In the case of the numeric display, the steps the software must take are as follows:

1. Each successive pair of BCD digits to be displayed must be combined into one byte.
2. This byte must be transferred to the second output latch on the interface board.
3. A byte must be transferred to the first output latch, in which only the bit corresponding to this pair of digits is a "0."
4. A byte containing all "1s" must then be transferred to the first output latch.

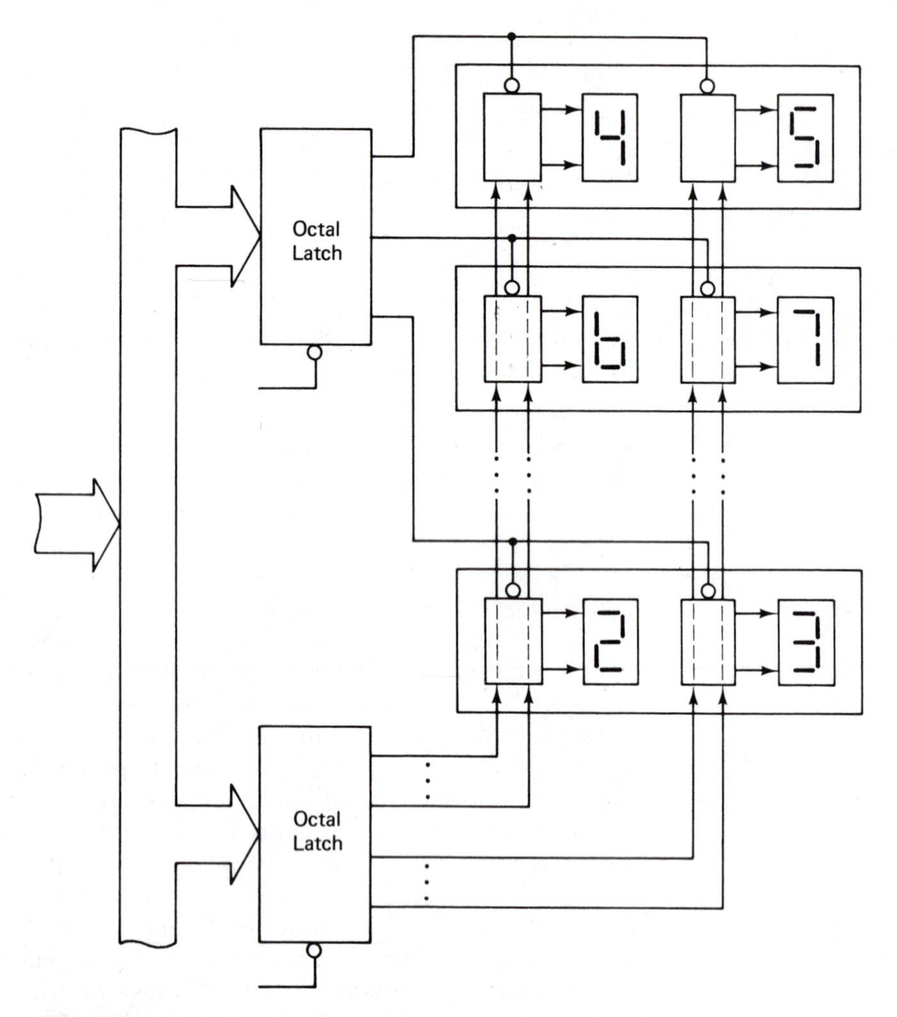

FIGURE 10-9. Combination of integrated display devices to form a 16-digit numeric display drive by only two digital output ports.

The last two steps create a low-going pulse on the strobe inputs to the display devices. The duration of this pulse is several instruction times in length but is, nevertheless, a pulse. This same technique can be used as a software pulse generator for control of other discrete stepped devices.

ANALOG INTERFACES

The microcomputer boundary area of an analog interface is often virtually identical to that of other interface types, but its internal logic and signal conditioning areas are quite different. They are complicated by their principal mission (conversion between analog and digital representations of information), by their need to coexist with noisy digital logic, and by sometimes conflicting requirements of rejecting noise while still providing accurate digital reproductions of input signals.

In this section, we examine the general operation of analog-to-digital (A/D) and digital-to-analog (D/A) interface subsystems. Although A/D and D/A interfaces are discussed separately, both types are often found together on board-level interface products. It is common, for example, to find a 16- or 32-input A/D and a 2- or 4-output D/A on the same printed wiring board, sharing only "real estate," power, and microcomputer boundary area logic.

Analog-to-digital Interfaces

Function Configurations. The principal functions of an A/D interface are signal conditioning, multiplexing, sampling, and conversion. There are several ways to arrange the hardware that performs these functions. Figure 10-10 shows three possibilities; the one shown in part (a) is the most common.

The figure shows signal conditioning on each input channel. This can range from nothing at all to complex, well-isolated instrumentation amplifiers. Some board-level product A/D interfaces provide signal conditioning in the form of unfilled component positions for resistors to convert 4–20 mA signals to voltages and, in some cases, also for current limiting resistors or clamping diodes to protect input circuits. It is common practice to leave the installation of these components up to users. If further signal conditioning is required, it is mounted on a different board or external to the system.

Though not shown in Fig. 10-10, signal conditioning may be shared among channels by preceding the signal conditioning circuits with analog multiplexers. However, two compromise issues are raised by such an arrangement.

The first is the degree to which interchannel isolation can be maintained. Bringing several unconditioned channels into the same multiplexing circuit increases chances for noise or potentially damaging high currents and voltages to interfere with operation. Flying capacitor multiplexing is sometimes used in such a situation, but is less than ideal because of electromechanical reliability issues. Its relays can generate switching noise, and their contacts may be unsuitable for precision measurements, since they form junctions that may have thermocouple and parasitic capacitance effects.

The second issue is response time. Shared signal conditioning must respond to each signal each time it is switched. After switching,

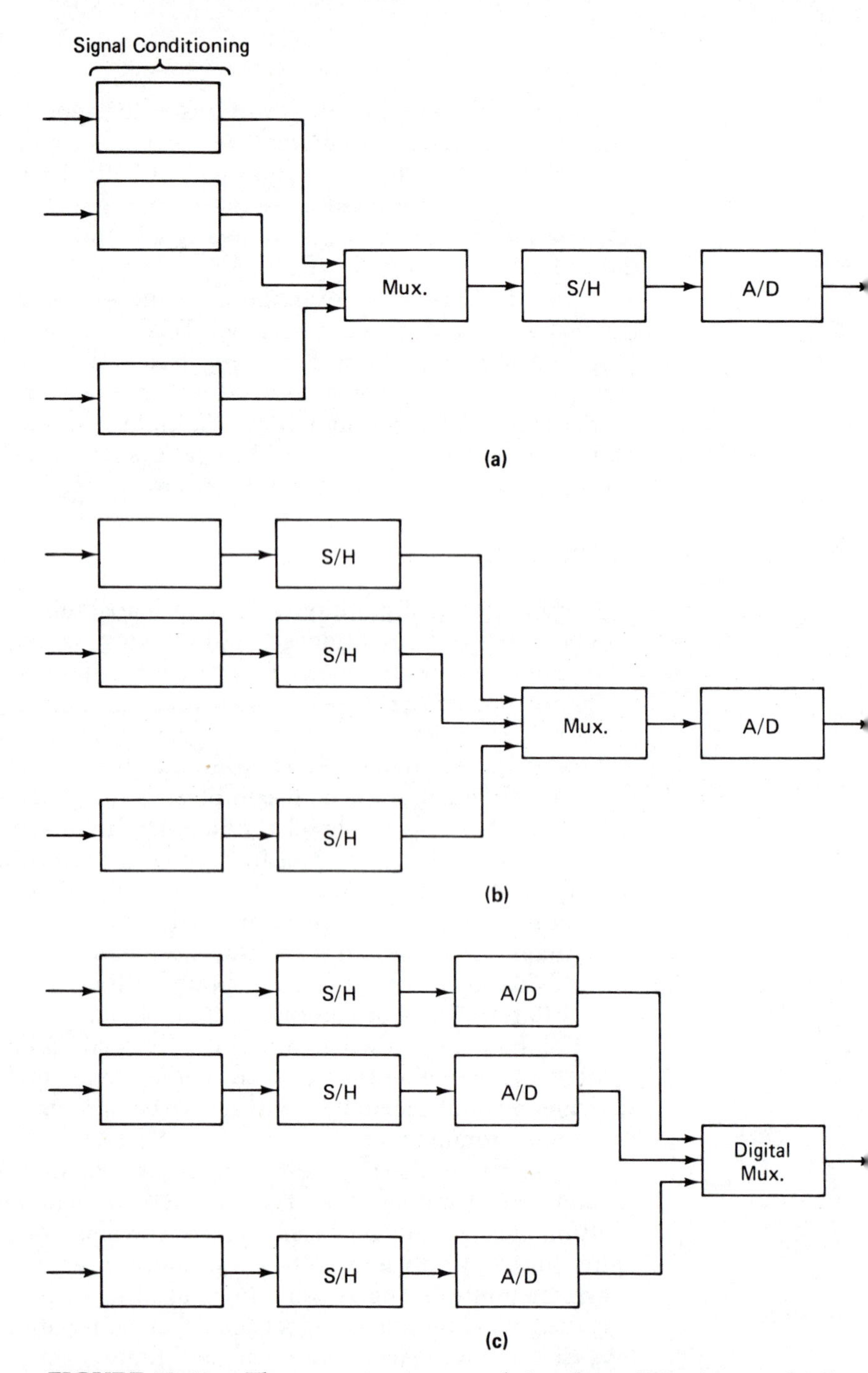

FIGURE 10-10. Three arrangements of signal conditioning, multiplexing, sampling, and conversion for analog input interfaces: (a) is the most common, (b) and (c) provide simultaneous sampling, and (c) is the most accurate simultaneous sampler.

the system must wait for conditioning circuits to settle before taking a reading and moving on to the next input channel. In highly isolated and filtered systems, settling time can be rather long and can reduce data acquisition rates to only a few samples per second.

For these reasons, it is common to equip each channel with its own signal conditioning and multiplex only when minimum cost considerations dictate it and the application can tolerate slow sampling rates.

The *sample-and-hold* (S/H) function is included in virtually every A/D interface subsystem. Its purpose is to "capture" the amplitude of the input signal at a particular point in time and hold this value long enough for it to be converted by the A/D converter. For certain combinations of slowly varying signals and fast A/D converters, the S/H may not be required, but these are unusual circumstances.

The configuration in part (a) of Fig. 10-10 samples different signals at different points in time, since there is only one S/H. In most applications, the time between samples is either negligibly small or sufficiently regular that values of different signals at the same point in time can be determined to sufficient accuracy by interpolation or extrapolation. However, some applications require simultaneous readings. These would use the configurations shown in parts (b) or (c) of the figure. Some high-speed digital signal processing systems for radar and communications require this approach.

The S/H operates by storing input signal voltage as charge on a small, high-quality capacitor. Because no capacitor is completely free from leakage effects, the sampled signal slowly deteriorates. For greatest accuracy, it must be converted as quickly as possible. Measurement accuracy deteriorates progressively from the first S/H to the last in configuration (b), as the converter works its way through them, one by one. In some cases, converter speeds and reading rates are fast enough so that adequate accuracy is obtained. If the converter and system are not fast enough, it may be necessary to equip each channel with its own converter, as in part (c) of the figure. Of all the configurations, this is the most costly.

Because the configuration of part (a) represents the best compromise between performance and cost for the majority of applications, it is the one most often used in industrial microcomputer applications.

A Typical A/D Interface. Figure 10-11 is a block diagram of a typical board-level product A/D interface subsystem. It includes most of the elements found in general-purpose A/D interfaces and is, therefore, useful in discussing their organization and operation.

The general flow of data is from one edge of the printed wiring board to the opposite edge. As indicated in the board outline inset in

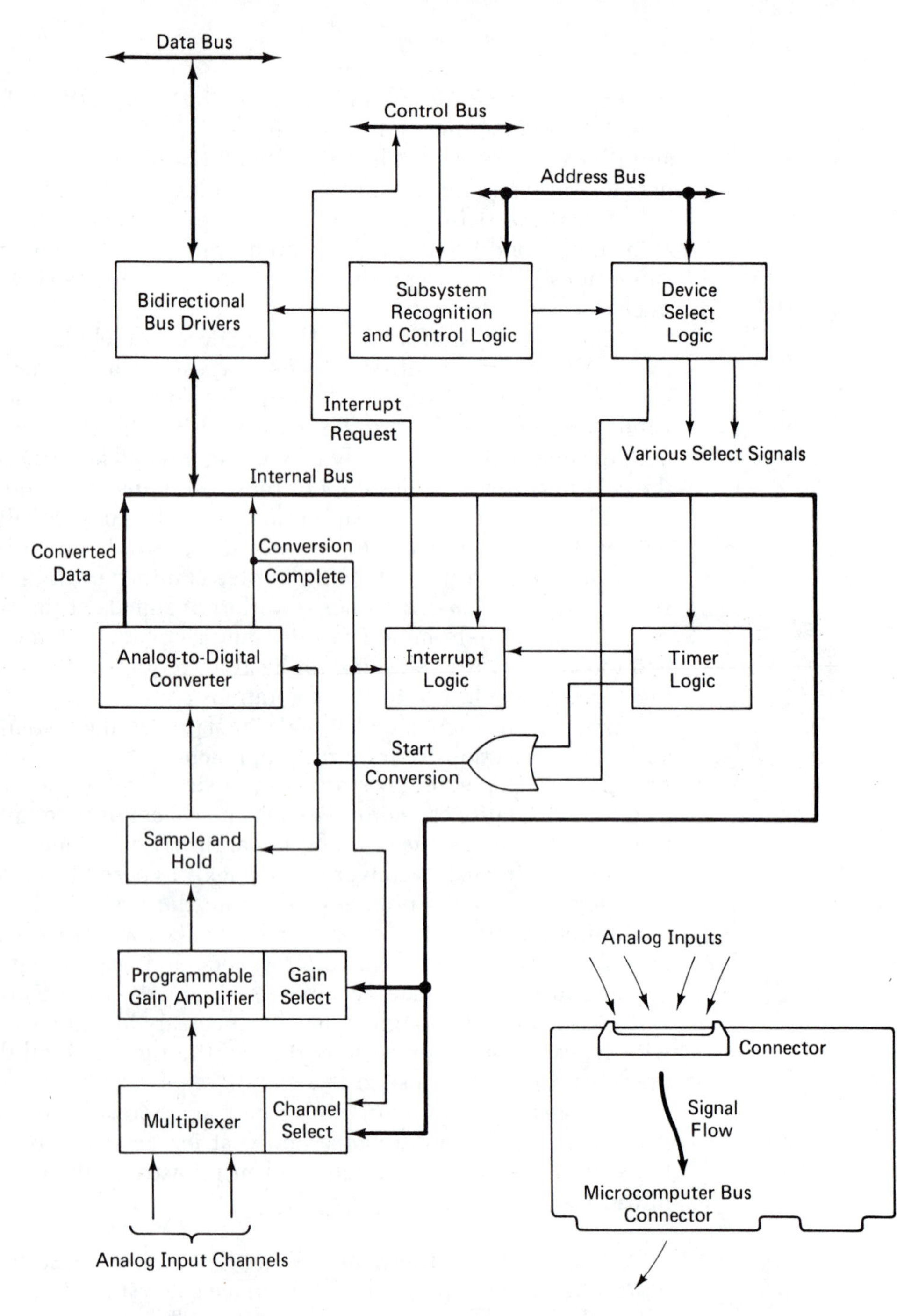

FIGURE 10-11. Block diagram of a typical board-level A/D interface subsystem. The inset board outline shows general signal flow through the board.

the figure, analog input wiring is at the top edge, where it is easily accessible in most card frame installations. Usual practice is to use a plug-in connector that is compatible with flat or braided wire cable. This connector mates with the board by means of a receptacle or card edge connector on the board. If space is provided for current loop or input protection components, they are accommodated by mounting holes or solder terminals in the vicinity of the analog input connector. Connection to the microcomputer bus is at the opposite edge of the board. In between are the elements shown in the block diagram.

Next, we encounter the multiplexer and channel select logic associated with it. In most designs, the multiplexer is implemented with CMOS FET switches arranged to handle single-ended (SE) or differential (DI) signals. If the interface is designed for *n* single-ended inputs, then it handles *n*/2 differential inputs. Jumper wires or user-set switches configure it for one of these two input choices. Figure 10-12 shows one way in which a pair of quad switch packages can be arranged, with configuration jumper wires for single-ended or differential operation. This provides for eight SE or four DI inputs, common numbers of inputs for small or medium-sized interfaces. Choices of 16 or 32 SE inputs are often found on larger boards.

The switches are controlled from the channel select logic. The microcomputer provides the number of the channel to be selected, usually in a control byte field that may also contain other function control data. The decoded channel number selects a single switch if the interface is configured for SE inputs or the proper pair of switches for DI inputs. Some designs provide for incrementing the channel number automatically after each A/D conversion. This is useful when

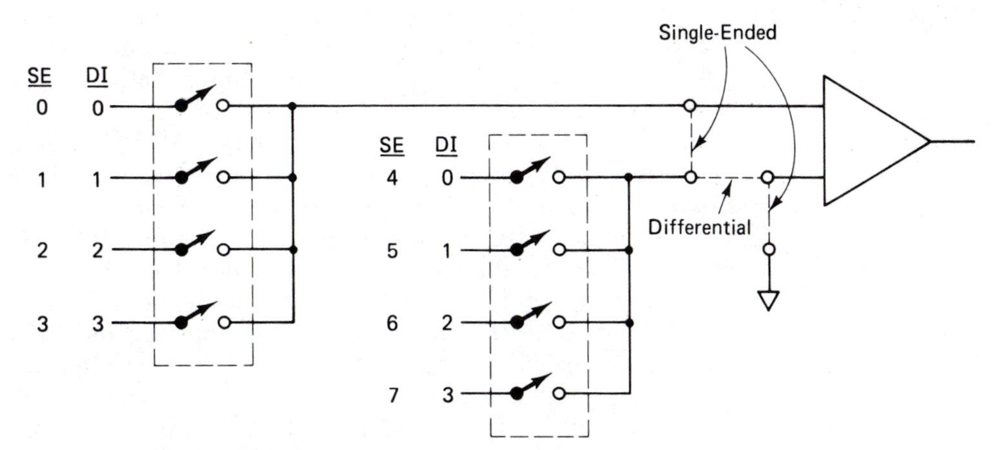

FIGURE 10-12. Arrangement of quad CMOS FET switches to handle both single-ended and differential input signals.

input signals are to be scanned in regular order, since the software need not be concerned with stepping from one channel to the next.

The output from the multiplexer goes to an input amplifier. On many boards, this amplifier provides programmable gain settings which may be specified by hardware reconfiguration, software control, or a combination of both. Gains are changed by switching to different resistors in the amplifier feedback network. Users may install different resistors, or the software may select the resistor to be used from one of a set by controlling another arrangement of CMOS FET switches similar to those used in the multiplexer.

Gain factor combinations of 1, 2, 4, and 8 are commonly found on interfaces with software controllable gain. It is difficult to provide a broader set of choices from a single amplifier stage. One reason is that resistance ratios are easily disturbed by resistances inherent in the FET switches. Though individually small, they are of the same order of magnitude as the gain resistors themselves at the upper end of the gain range. When a broader set of choices is needed, an additional amplifier stage is sometimes used. For example, two series stages, each with gains of 1, 2, 4, and 8, provide total gains of all integral powers of 2 up to a gain of 64.

Gains of 1, 2, 4, and 8 are satisfactory for most signals with spans of 2.5 to 10 volts or so and for inputs from 4–20 mA current loops across appropriately sized input resistors. But many industrial transducers have substantially smaller output signals. For example, these gains are insufficient for thermocouples, RTDs, and strain gages. Board-level products with special gain resistors can be used, but the range of choices may have to be given up. If all of the inputs to the board can be handled with a single special gain value, this will work. If several gain values are required, the next best method is to provide amplification in the signal conditioning circuitry. This method is common in industrial systems based on board-level and system-level microcomputer products.

The amplifier output is then applied to the S/H circuit and the S/H output to the converter itself. There are several methods of performing analog-to-digital conversion. Because an understanding of digital-to-analog conversion is useful in explaining A/D conversion, we hold off examining the operation of the A/D until later in this chapter and treat it as a "black box" for the moment. Figure 10-13 is a more detailed block diagram of the converter and its surroundings. This section provides digitized readings. It is controlled by a "start conversion" command, which may be an output from the device select logic, activated by an output instruction having its address. In the simplest designs, only this method of starting conversions is provided. In more sophisticated designs, the start command may also be pro-

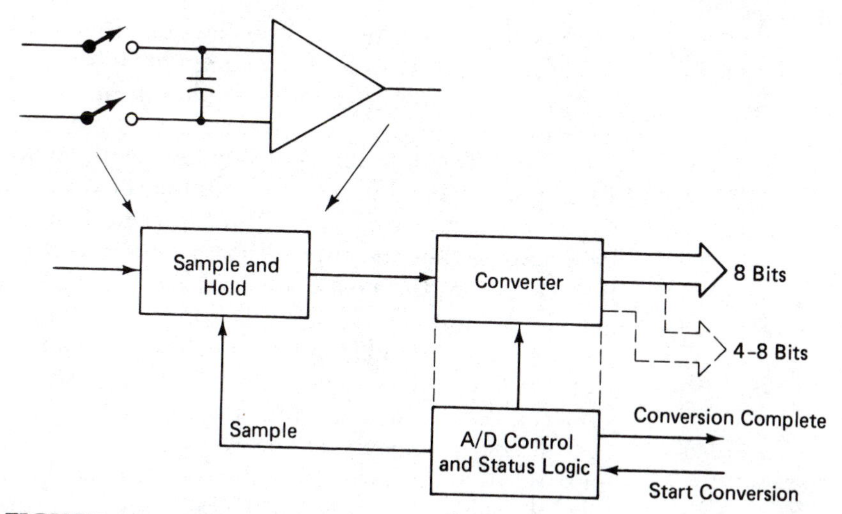

FIGURE 10-13. Relationship of the A/D converter, sample and hold, and control/status logic.

vided from a timer, as indicated in Fig. 10-11. More on this in a moment.

The start conversion command first triggers the S/H function, which begins to capture the analog input voltage value at that moment. The time required for the S/H to settle on this value is called its *aperture time.* Once it has settled to this value, the A/D control logic initiates the actual conversion. In some cases, the S/H is allowed to track input signals until time for the conversion to start. This tactic can speed up the data acquisition rate, since the S/H can be acquiring the next signal while the digitized previous sample is being transferred to the microcomputer.

The S/H can be thought of as a pair of FET switches, a capacitor, and an amplifier, as indicated in Fig. 10-12. The "sample" signal opens the switches, leaving the acquired input voltage value at that moment stored in the capacitor. The S/H amplifier provides drive for the conversion devices but no net gain.

When digitization is complete, the A/D control logic provides a "conversion complete" output signal, which may be monitored by the microcomputer or used to interrupt it. If the interface provides for automatic stepping to the next channel, this signal may also be used to switch the multiplexer to the next channel.

In the meantime, the digitized value is held in registers in the converter. When selected by the microcomputer, the digitized value is transferred onto the internal bus, then to the microcomputer data bus via the bidirectional drivers. The number of steps required to transfer this value depends on the converter resolution and the microcomputer word size. If the computer is an 8-bit type and the converter resolution is greater than eight bits, more than one transfer step is required.

The A/D converters typically used with microcomputers have 8-, 10-, or 12-bit resolutions, with 16-bit converters available but relatively rare. Converters with 12-bit resolution are, perhaps, most common among general-purpose board-level interface products because they represent the best compromise between performance and cost and are suitable for most applications. Converters of less resolution are more often found in custom designs in which low cost is important and it is known beforehand that their resolution will be adequate.

If greater than eight bits, the digitized value may be distributed over a pair of bytes in several ways. The output of a 12-bit converter must be divided into two parts, one of eight bits and the other four bits. Which part holds the most significant bits is the question. In some designs, it is "4 most significant and 8 least significant." In others, it is "8 most significant and 4 least significant." The best choice from the software viewpoint depends on how converted values will be handled in the computations.

The "number system" employed by the converter may also be one of several. "Offset binary" and "2's complement" are two popular methods of representing analog values that may be positive or negative. In the jargon of analog interfacing, signals that may be both positive and negative are called *bipolar* signals (not to be confused with the use of the term when forms of logic are discussed). If a signal is not bipolar, it is said to be *unipolar*. Some designs allow the user to configure the interface for different types of signals and different number systems by means of on-board jumpers.

The last major item in Fig. 10-11 is timer logic. This is sometimes called a *pacer clock* and is not found on all designs. It consists of a crystal-controlled or RC-type oscillator that can be enabled by the microcomputer. When enabled, it provides converter start commands at a regular frequency, causing automatic sampling and conversion at this frequency. At each "tick" of this clock, a conversion is performed and the microcomputer is interrupted. Channels may be stepped automatically, if the multiplexer has been placed in this mode of operation, or all of the time-based readings may be taken from the same channel.

Although interface designers can assemble subsystems from discrete components, op amps, and FETs, it is usually more practical to engineer them from the large collection of commercially available analog function modules, such as complete S/H circuits, programmable gain amplifiers, A/D converters, and CMOS FET switch IC packages. These are relatively easy to use and help designers avoid tricky circuit design problems when dealing with complex analog functions. For some applications, it is possible to obtain multiplexing, the A/D converter, and everything that goes between them, in a single metal case module. For some transducers, the same may be said for signal conditioning circuit modules. These still must be combined with some care, but a sufficient amount of engineering cost may be saved for small volume designs to more than offset their somewhat greater cost.

Analog-to-digital converters are also available in IC packages. The most common types are successive-approximation and integrating converters. Their operation is explained later in this chapter. The IC converters most amenable to use with microprocessors (and usually advertised as being "microprocessor-compatible") have three-state digital outputs with separate controls for transferring converted values in byte-sized sections.

Digital-to-analog Interfaces

At the block diagram level, digital-to-analog interfaces are much simpler than A/D interfaces. Figure 10-14 is a block diagram of a multichannel D/A interface in which three channels are shown. Depending on board size, one may find from two to eight analog output channels on a typical board-level product interface. General signal flow from board edge to board edge is like that for the A/D interface, but in the opposite direction. The D/As are "pure output."

Because they are output-only devices, the D/A interface microcomputer boundary area is simpler. The data bus interface may consist of simple receivers and these only if needed to minimize bus loading. Subsystem recognition and control logic need only recognize the interface address block and enable the device select logic.

An ideal D/A converter can be thought of as a "black box" that instantly converts the binary number presented at its input pins to an analog voltage on its output. The moment one bit in this number changes, the output voltage changes correspondingly. This characteristic presents a problem when the number of bits converted by the D/A is greater than the number of bits in the microcomputer word. The purpose of the latches shown in the block diagram is to address this problem.

A typical D/A input latch arrangement is shown in more detail in the inset in Fig. 10-14. Latch A holds as many bits as are converted by the D/A. It holds this value (and, therefore, causes the output analog voltage to be held) while other activities are going on. If the word size of the microcomputer and the D/A are exactly the same, only latch A

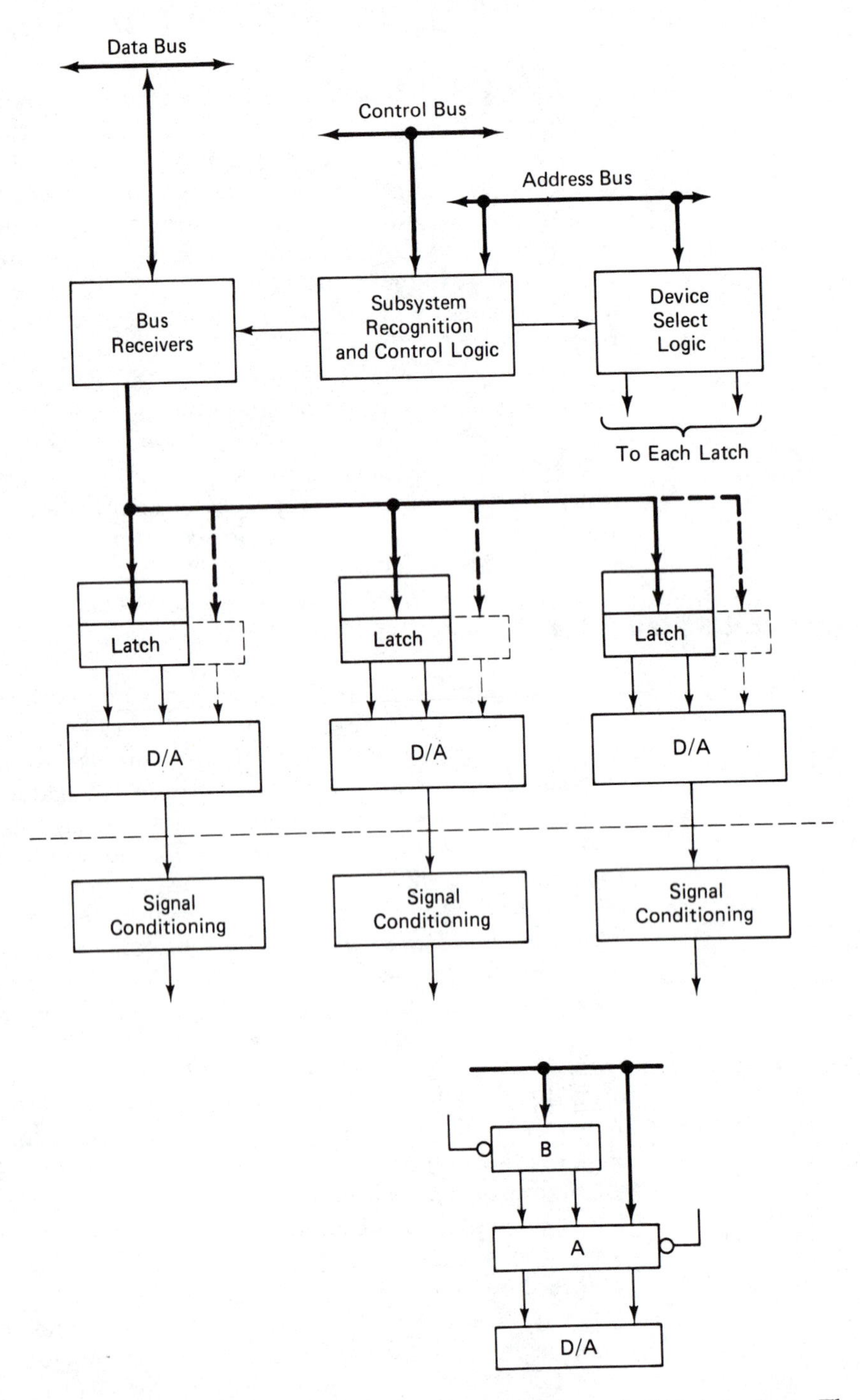

FIGURE 10-14. Block diagram of a multichannel D/A interface. The inset shows detail of the D/A latch arrangement that provides for transfer of the complete binary number to the converter.

is needed. If the D/A is "wider" than the microcomputer word, more is needed.

To see why this is so, suppose latch B is not present and that one output transfer (device select signal) loads, say, the most significant eight bits of latch A, while a second output transfer loads the least significant bits of latch A. Since the microcomputer can transfer only eight bits in a single step, time elapses between the loading of one part of latch A and the loading of the rest of latch A. During this time, the input pins of the D/A will be presented with part of the new number and part of the old number. The effect can be a pronounced "glitch" in the output voltage; clearly undesirable.

Placing latch B between part of latch A and the internal data bus solves this problem. As before, two device select signals are used, acting on the latches in the following way. The first device select signal loads the most significant part of the new number into latch B. The contents of latch A are not changed by this, so the converter output voltage remains unchanged for the moment. The second device select is now used for *two* operations. First, it transfers the least significant part of the new number into latch A. *Simultaneously* with this, it *also* transfers the contents of latch B into the most significant part of latch A. The result is that *all* bits in latch A change at the same time. With the complete new number presented at one time, the voltage output of the ideal D/A swings cleanly to the prescribed value, without a glitch.

The most common D/A converter resolution is 12 bits, with substantial numbers of 8- and 10-bit types represented in the population. Sixteen-bit D/A converters are relatively rare, but available. The D/As were among the first linear devices to receive a substantial integration effort. As a result, most types that you might need can be found in IC packages that require only a few external components in most applications. Some D/A ICs contain the equivalent of latch A. Others (which may be advertised as "microprocessor-compatible") also include the equivalent of latch B.

Most board-level product D/A interfaces provide simple voltage output signals of moderate drive capability, only sufficient for operating control level circuits and rarely enough to drive analog actuators directly. Many industrial applications call for 4–20 mA transmitter outputs. Although provided by some interfaces, they are relatively rare. When it comes to analog output signal conditioning, the user must often take responsibility for finding and assembling what is needed. Few board-level products include analog output signal conditioning because of the variety of potential requirements and the substantial amount of printed wiring board area needed by these circuits. This makes it difficult to arrive at a suitably general-purpose configuration that includes an adequate number of output channels.

Output connectors provided for D/A interfaces are generally similar to those on A/D interfaces: card edge or receptacle connectors for flat or braided cable located at the upper or outer edge of the printed wiring board. Signal conditioning circuits are interposed externally between the D/A interface and analog actuators and are usually supplied by the user.

Digital-to-analog Converters

A digital-to-analog converter is a device that multiplies an input reference voltage by an n-bit binary fraction. This is depicted in Fig. 10-15 and is represented mathematically by the following equation:

$$V_o = V_R \left(\frac{D}{2^n}\right) \tag{10.1}$$

where V_o is the output voltage, V_R the reference voltage, and D an n-bit binary integer.

Most common D/A converters use a network of switched resistors to step the reference down to the value implied by the number D. Figure 10-16 shows the circuit diagram for a 4-bit converter that uses the "R-2R" resistor network form. Although other network forms are sometimes used, R-2R networks are preferred because they can be more easily manufactured to the exacting tolerances required for converters having many bits. The resistors are switched by means of switches controlled by individual bits of the input number D. There are several forms of internal D/A design which have different output and performance characteristics. Some operate by acting as current, rather than voltage, dividers. Some are available without output amplifiers.

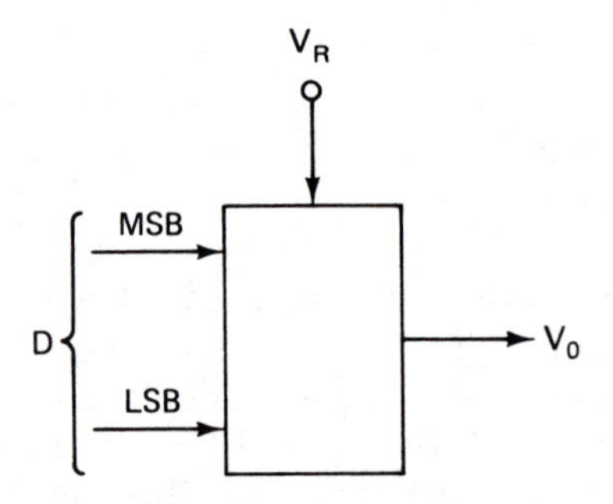

FIGURE 10-15. A D/A converter multiplies an input reference voltage by an n-bit binary fraction.

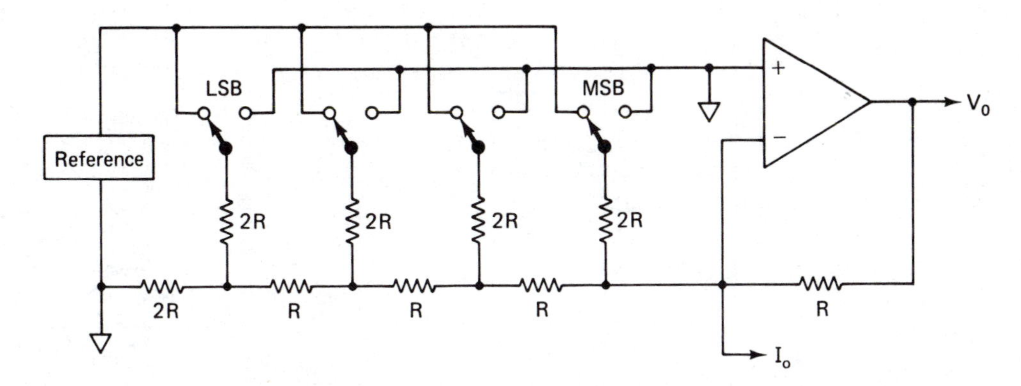

FIGURE 10-16. A 4-bit D/A converter that uses an R-2R network of switched resistors.

Although D/As are relatively simple as idealized circuits, the effects of variations in R-2R network resistance values and resistances in the switches can combine to produce less than ideal results. Figure 10-17(a) shows the ideal quantized output of an error-free converter. The light 45-deg line represents the one-to-one analog transfer ratio. In an ideal D/A, midpoints of the horizontal segments are coincident with the one-to-one transfer ratio line. There are several ways in which the output of an actual D/A may deviate from the ideal, and they may be present individually or in combination. Three are shown in parts (b), (c), and (d) of Fig. 10-17.

Gain error is manifested by slope deviations in the line which intersects the actual midpoints of the horizontal segments in the quantized output. *Offset error* is manifested by a translation of this line to one side or the other of the ideal. *Nonlinearity* is manifested by failure of the midpoints to lie on any straight line. Gain and offset errors may be corrected in most cases by trimming components on the output amplifier. Nonlinearity is generally due to failure to meet tolerances in the internal components, if other obvious causes (such as a poor output amplifier or a faulty reference voltage source) have been eliminated. A fourth deviation, *nonmonotonicity,* is manifested by reversals of the slope, drops in the voltage output when the code input is increased. This is usually due to a defect in the converter.

It becomes more difficult to meet required tolerances as the number of bits increases. Twelve-bit converters are quite practical; 14-bit and 16-bit converters are achievable, but relatively expensive. Converters with more bits are outstanding engineering and manufacturing achievements for which a healthy price is asked.

The usual manufacturing procedure is to use lasers to trim re-

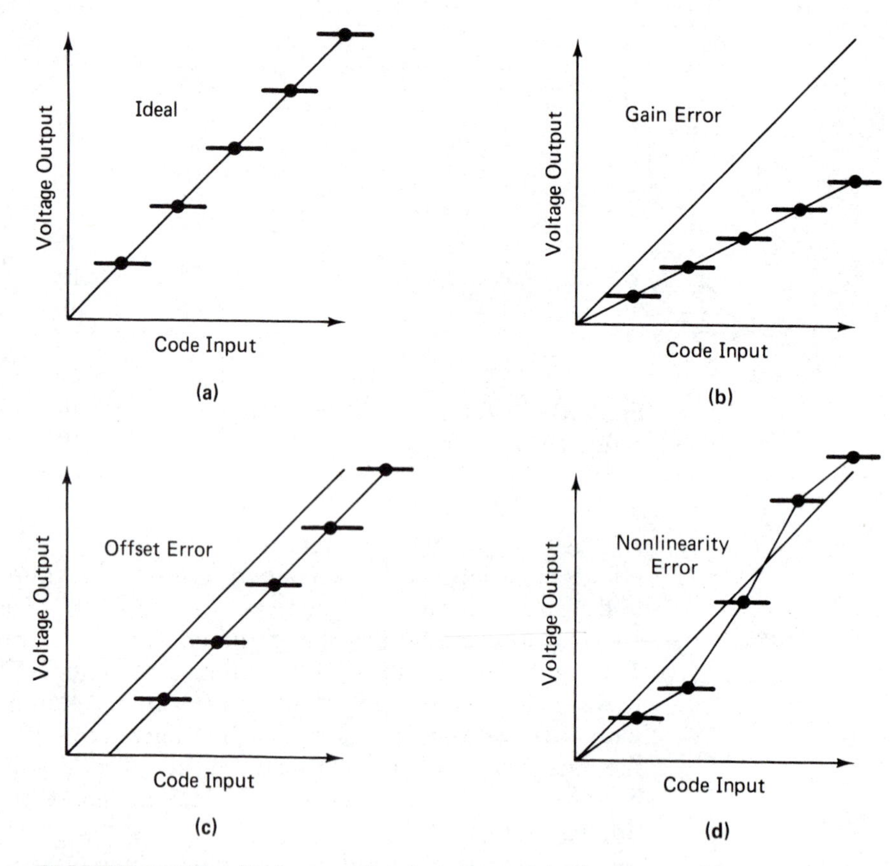

FIGURE 10-17. D/A quantized output: (a) ideal error-free converter, (b) gain error effect, (c) offset error effect, (d) nonlinearity error effect.

sistors in the R-2R networks in order to balance their values to produce the required linearity. This is a relatively slow process requiring costly equipment. A new approach, represented in the design of the ICL7134 D/A converter manufactured by Intersil Corporation, is to use EPROMs on the integrated circuit. Correction codes are written into the EPROMs as the device is tested. These codes correct linearity errors in lower-order bits. This approach holds promise for improving high-resolution D/A converter cost and performance and should also bring improvements in A/D converters, as we shall see in a moment.

Analog-to-digital Converters

An analog-to-digital converter is a device that produces an n-bit binary fraction equal to the ratio of a measured input voltage and an input reference voltage. This is represented mathematically by the following equation:

$$\frac{D}{2^n} = \frac{V_i}{V_R} \qquad (10.2)$$

where V_i is the input voltage to be measured, V_R the reference voltage, and D an n-bit binary integer. Note the similarity with Eq. (10.1).

In working with analog converters, it is usual to speak of D as a "count," with little attention to its fractional relation to the analog signal and reference. This is because treating it as an integer is most natural when one is dealing with A/D and D/A hardware. The nature of D as a component of a fraction is noted here because it is more natural to think of it this way when one is writing computer programs that deal with analog inputs and outputs.

There are many ways of producing a binary number that is proportional to an analog signal, but the two most popular methods are based on successive approximation and integration techniques, respectively.

Successive Approximation Converters. Figure 10-18 shows the organization of an A/D converter that uses the successive approximation method. Its principal components are a D/A converter, a comparator, a successive approximation register, a clock, and control and status logic.

The *successive approximation register* (abbreviated SAR) is essentially an n-bit latch designed for convenience in setting and complementing its individual flip-flops in most significant to least significant order. The output bits of this register are the inputs to an n-bit D/A converter and also hold the final quantized value D. The input signal V_i and the D/A output V_o are compared by the comparator. The output of the comparator indicates whether V_o is greater than or less than V_i.

At the start of a conversion, the SAR is cleared to "0s" and its most significant bit is set to a "1." This results in a V_o value that is one-half of full scale. The output of the comparator is then tested to see whether V_i is greater than or less than V_o. If V_i is greater, the most significant bit is left on; otherwise, it is turned off (complemented).

In the next step, the next most significant bit of the SAR is turned on. At this stage, V_o will become either three-quarters or one-quarter of full scale, depending on whether V_i was greater than or less than V_o, respectively, in the first step. Again, the comparator is tested and, if V_i is greater than the new V_o, the next most significant bit is left on. Otherwise, it is turned off.

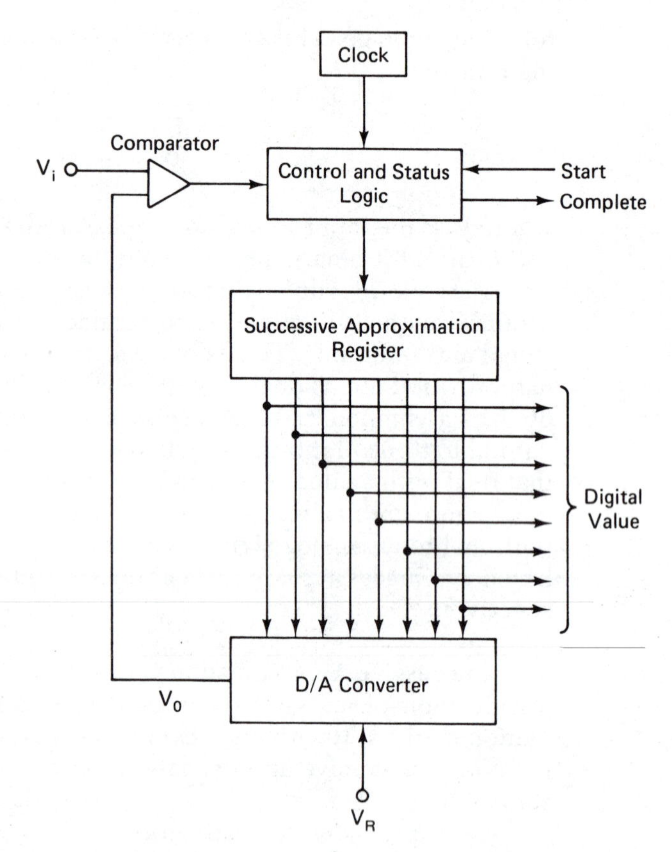

FIGURE 10-18. Organization of a successive approximation A/D converter.

This process is repeated for each remaining SAR bit. The effect is a "bracketing" and "closing in on" the input signal level V_i by the D/A output V_o, with the result of each such successive approximation recorded in the register for which it is named. When this process has been carried out for each bit, the SAR contains the binary number D that is proportional to V_i.

It is clear that a high-quality D/A converter is very important to a high-quality successive approximation A/D converter. This is the reason for the remark about the EPROM-based D/A converter in the previous section. Successive approximation converters are valued principally for their conversion speed at moderate cost. Because most of the cost of a high-resolution successive approximation converter is invested in the D/A, and because the cost of a high-resolution D/A grows dramatically for each additional bit, high-resolution A/Ds are expensive.

The functions of the control logic and SAR can be carried out by a computer program. Although not so fast as conversions done with hardwired logic, this approach can be used to reduce cost in small, low throughput systems. The essential ingredients are the D/A and the comparator. If the measured signals vary slowly enough, it is even possible to eliminate the S/H function.

When the microcomputer is embedded in the conversion loop in this way, other conversion techniques may also be used (e.g., by counting or ramping in a way that is similar to the method used by the integrating converters described below). For simple monitoring of signals to see if they have exceeded preset limits, the computer need only output the limit value to the D/A and examine the comparator output to determine the condition of the signal.

Integrating Converters. There are several integrative conversion techniques, but one of the most used is the *dual-slope converter.* Figure 10-19 illustrates its organization. Its principal components are a current source referred to a reference voltage V_R, a voltage-to-current converter that sinks a current i_i proportional to the input voltage V_i, a switch that applies one of these currents to an integrator, a comparator, a clock, control and status logic, and an n-bit up/down counter.

At the start of a conversion, the voltage-to-current converter is switched to the integrator, causing it to start ramping up at a slope proportional to V_i. The integrator is allowed to ramp up for a fixed time (as shown in Fig. 10-20) which is terminated by switching its input over to the reference current source. At this instant, the integrator output voltage is proportional to V_i. As the integrator is switched

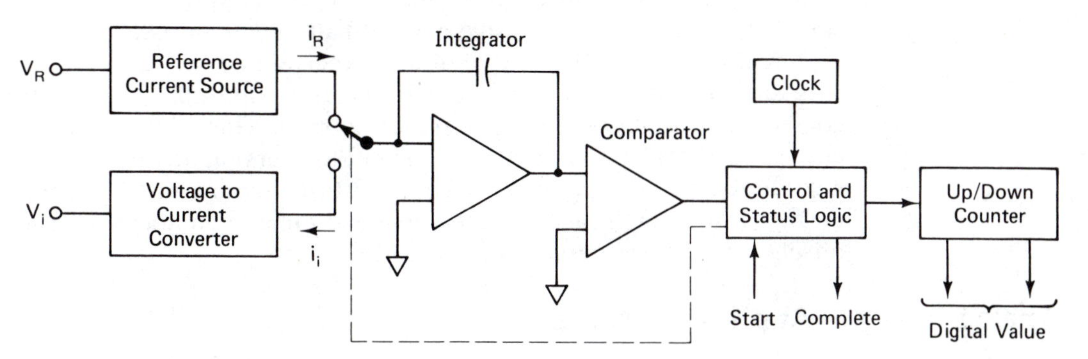

FIGURE 10-19. Organization of an integrating dual-slope A/D converter.

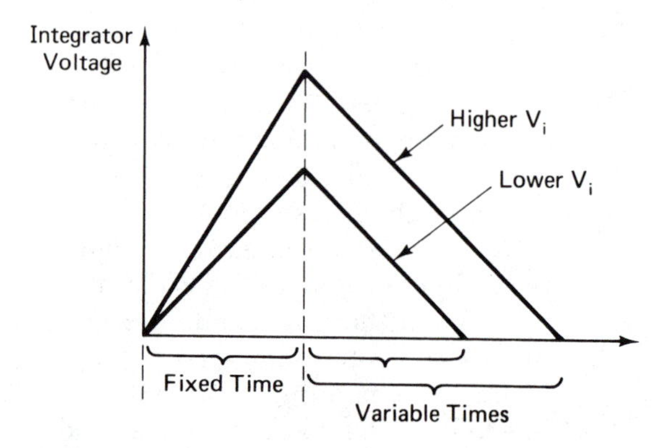

FIGURE 10-20. Graph of integrator voltage during conversion for two different input voltages.

to the reference, the counter is enabled and begins counting at a rate set by the clock. Meanwhile, the reference current causes the integrator to ramp down at a slope proportional to V_R. When its output reaches analog ground, the comparator switches and stops the counter. At that instant, the counter contains the quantized value D.

Integrating converters are valued for their relative immunity to noise, achieved by virtue of the fact that their outputs are proportional to the average value of V_i during the initial integration period and, therefore, tend to reject small noise signals imposed on V_i. This characteristic can be optimized by using clock frequencies integrally related to the frequencies of the most likely sources of interference (e.g., 60 Hz power lines).

The time required for integration and the common practice of matching it to interfering frequencies results in slower conversion speeds than those of successive approximation converters. This is its principal drawback. It enjoys cost and precision advantages because it is more easily made highly linear and is composed of less costly components. BCD readouts are easy to obtain. This, along with low cost, makes dual-slope converters useful as components of inexpensive digital panel meters and hand-held test instruments. Variations on the design of these converters make automatic compensation for offset and drift errors possible, as well.

Analog Interface Application Notes

In dealing with analog interfaces, it is useful to have a feeling for the relative magnitudes of the numbers involved and the variables being

TABLE 10-1
Analog Converter Resolution

n	*Fraction*	*μvolts at* V_R *= 5 volts*	*% F.S.*	*ppm*	*dB w/r F.S.*
8	0.003906	19531	0.3906	3906	−48.2
10	0.0009766	4882	0.0977	977	−60.2
12	0.0002441	1220	0.0244	244	−73.3
14	0.00006104	305	0.0061	61	−84.3
16	0.00001526	76	0.0015	15	−96.3

measured and controlled. Tables 10-1 and 10-2 list some of these for several common converter sizes.

Table 10-1 lists the following items:

1. The value of the least significant bit (LSB) as a fraction (2^{-n}).
2. The number of microvolts corresponding to the LSB value when the reference voltage V_R is 5 volts.
3. The LSB value as a percentage of the full-scale range of the converter.
4. The LSB value in parts-per-million (ppm) with respect to the full-scale range.
5. The range in decibels from full scale to the value of the LSB.

The second, third, and fourth columns are useful to have at hand because system resolution and accuracy specifications are often stated in one of these units.

Note that a 12-bit converter can resolve about 1.22 millivolts (0.02 percent or 244 ppm) at a 5-volt full-scale range. To obtain finer resolution with this converter size, the input signal must be amplified.

TABLE 10-2
Approximate Temperature Resolution (°C) For Typical Thermocouples

℄ type: *Range:*	*J* *(−100 to +760)*	*K* *(−100 to +1230)*	*S* *(+300 to +1760)*	*T* *(−200 to +400)*
n				
8	305–475	458–640	1609–2137	316–1240
10	76–119	115–160	402–534	79–310
12	19–30	29–40	101–134	20–77
14	4.8–7.4	7.2–10	25–33	4.9–19
16	1.2–1.9	1.8–2.5	6.3–8.3	1.2–4.8

But this, of course, means a corresponding reduction in the full-scale range.

The usual (and technically correct) accuracy claimed for a converter is $\pm\frac{1}{2}$LSB. That is, the true value of the input voltage, referred to the *actual* reference voltage (remember, the reference voltage may not be exactly what you think it is), is within $\pm\frac{1}{2}$LSB of what the converter says it is. Many forget that noise is *always* present and are surprised by its magnitude. To hear that an input filter has been applied to push relevant noise down, say, 60 dB below the signal may encourage one to believe that a perfectly stable reading of a dc signal will result. However, as the right-hand column of Table 10-1 shows, the LSB of a *10-bit* converter is 60 dB below full scale. This means "−60 dB noise" would still be seen by the converter, resulting in "random" values in the two or three least significant bits of a 12-bit converter and creating the impression of general chaos in the least significant *third* of a 16-bit converter. Within the digital logic areas of many microcomputer systems, it is common to find an ample supply of "−60 dB" noise, even when reasonable care has been taken to keep noise low. Thus, although the *converter* may be accurate to $\pm\frac{1}{2}$LSB, measurements made with it in these circumstances may be far from being this accurate.

Table 10-2 lists the approximate resolution, in degrees Celsius, of n-bit converters with $V_R = 5$ volts for several thermocouple types when the thermocouple signal is applied directly, without amplification. The column entries are the range of resolutions. They vary over the reading span because of the nonlinear characteristic of the thermocouple output with respect to temperature. Column entries are obtained by dividing the A/D resolution (in microvolts) by the minimum and maximum Seebeck coefficients (which are the slopes of the thermocouple curves in units of °C/microvolt).

Note first that there is a substantial variation in resolution over the temperature measurement range. This is most pronounced for the type T thermocouple, which is the most nonlinear of the four types shown. Note further that only the 16-bit converter even begins to approach a resolution comparable to the accuracy of a typical thermocouple, and it is still far from being satisfactory. This means that extra gain on the converter input signal is absolutely necessary. For 12-bit converters, this gain should be on the order of 50 or more, depending on the thermocouple type.

Gain is best taken "early," in order to have as high a signal-to-noise ratio as possible over as much of the signal path as possible. This strategy is also useful in combating the effects of thermal drift in system components. For example, it would be wrong to imagine that one might use the programmable gain amplifier of a typical interface board at a gain of 8 with a 16-bit converter and thereby get roughly 0.25°C resolution from a J-type thermocouple input. This is because the temperature coefficients of the resistors used in the amplifier would be, at best, roughly 1 ppm/°C even for very high-quality wirewound resistors. Over a system operating temperature range of, say, 25 to 55°C (common for industrial applications), these resistors would contribute about 30 ppm to gain error, negating the effects of the two least significant bits of the converter.

We noted that resistors are an important economic issue in converters. Precision resistors for programmable gain amplifiers are also costly. To put this into perspective, the cost of a set of three precision (±0.01 percent) wirewound resistors was approximately the same as the cost of *one* single-chip microcomputer in mid-1982. That a few resistors can contribute more to the cost of a system than the computer helps to illustrate the magnitude of the achievements of large-scale integration.

SERIAL INTERFACES

Serial interfaces are used primarily for communication among computers and terminals. The word "terminal" may mean devices ranging from complex control and display stations, capable of showing process conditions in elaborate color graphics, to simple data entry devices having only a small numeric keypad. In between, we may find printers, chart recorders, and even small disk and cartridge tape recording devices. Because so many of these devices are now organized around a microcomputer, it is common to find a computer of *some* type at each end of every link in a serial communication network.

It is important to understand how serial interfaces are used at a higher level in order to appreciate the lower-level characteristics of their designs. For this reason, we begin with a discussion of protocols, network organization, and communication methods.

Protocols and Standards

The principal feature of serial communication is that bytes, words, and groups of bytes and words are disassembled and transmitted, one bit at a time. Received bits are reassembled into bytes, words, and groups of bytes and words. Clearly, both the transmitter and the receiver must agree beforehand on the rules to be used in exchanging information. These rules are generally called *protocols*.

Reaching such an agreement can be an enormously complex technical, economic, and political activity. First, a technical design of the protocol must be made. It must specify such things as data rates, electrical means, and formats by which information will be exchanged. Because there are many ways to satisfy such technical specifications, many solutions have been developed and are manifested in the hardware and software systems of many different manufacturers of computing and information systems. Each reflects, to a degree, its manufacturer's own special insights and prejudices. So long as they were required to communicate only with their own "cousins," there were few problems that could not be solved easily.

Now, systems from many different manufacturers are expected to exchange information. To accomplish this, protocols must be agreed upon. Because each manufacturer has a vested interest in having the protocols suited to the communication methods manifested by the equipment and software in which it has such a large investment, reaching such an agreement has been complicated by their economic interests. Because computer networks are now international in their scope, political considerations have also entered the picture.

As of this writing, several standards organizations are working to achieve such agreements. Although much progress has been made, no single comprehensive standard has been agreed upon. It is possible that complete agreement may never be reached. Some feel that comprehensive standards discourage innovation. Others feel that they are needed for the application work at hand to progress satisfactorily. The truth may lie somewhere in between.

Nevertheless, developments in this area are important to the application of microcomputers to industrial control problems. The distribution of control over a plant has created a high level of interest in *local area networks* (LANs). These are networks which can have a geographical dispersion of up to several thousand meters and which interconnect the maxicomputers, minicomputers, and microcomputers that are responsible for the management and control of the machines and processes that do the work of that facility.

Network Organizations

Substantial *de facto* agreement has been reached in the terminology of network organizations, describing their physical and information-handling topologies. Although it does not constitute officially required practice, it helps everyone who discusses this subject to use a reasonably common language.

Network Topologies. As we saw in Chapter 6, when all network nodes are directly connected to one another, topological complexity increases rapidly. Chapter 6 noted two alternative topologies for the example process control system discussed: a hierarchical *tree* network and a *bus* network. Figure 10-21 shows three new topologies, the *star* and two forms of the *ring*, and repeats the direct, tree, and bus topologies in slightly different form, so that they may be compared.

These can be divided into *broadcast* and *point-to-point* categories. The ring, in part (c) of the figure, and the bus belong to the broadcast category. Any transmitted message can be "heard" by any node that is listening. With point-to-point topologies, only nodes directly connected to the transmitting node hear the message. To communicate with a node to which there is no direct link, the services of other nodes must be engaged to pass the message along to its destination. Intermediate nodes may serve as switching centers to establish virtual direct links between other nodes, or they may receive messages and send them on later when the intermediate links are available. Systems employing this technique are called *store-and-forward* systems. Note that store-and-forward message-handling facilities are not always necessary even when not all nodes have direct links with one another. In many systems, there may be node pairs that simply have nothing to say to each other. The example system discussed in Chapter 6 was such a system.

Formal local area networks are usually organized as rings or busses. With these broadcast-type networks, nodes may be connected to the network or may be removed from it without disturbing its structure. This is important for system reliability and allows users to modify it by adding new equipment with less trouble.

There are many examples of the other topologies in existing process and factory automation systems, for the most part designed to solve particular facility problems with the technology that was available at the time. For example, star topologies are characteristic of older centralized computer systems, and tree topologies are found in many early systems in which minicomputers were first used for distributed control.

Open Systems Interconnection Reference Model. Much of the terminology is embodied in the "Reference Model of Open Systems Interconnection" developed by the International Standards Organization (ISO). The ISO Reference Model depicts communication processes as consisting of seven "layers" at each node of the network, as illustrated in Fig. 10-22. Protocols are organized to relate to these layers.

The objective of the communication process is for one "user" to carry on a "conversation" with another "user." The users may be humans or computer programs. The protocol is layered as depicted in the figure in order to insulate higher levels from a need to be inti-

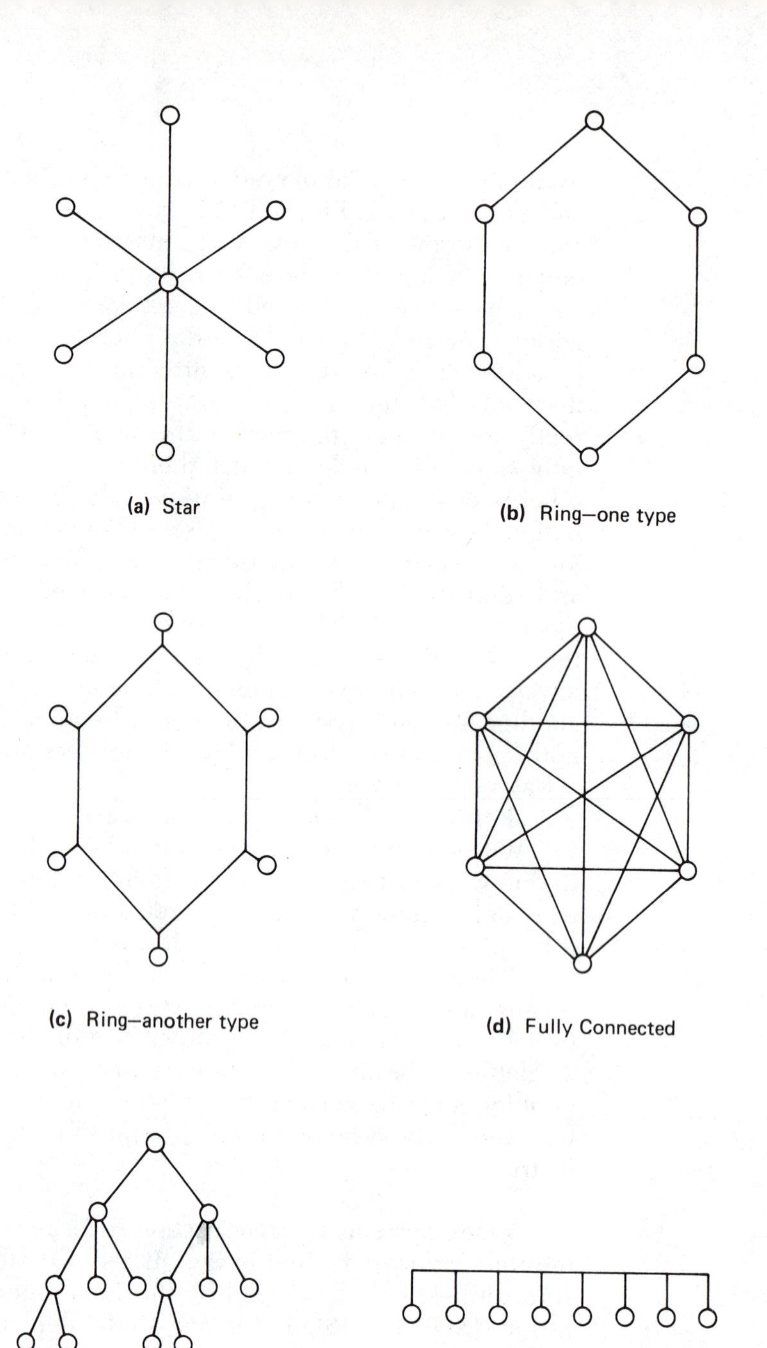

FIGURE 10-21. Network topologies: (a) star, (b) one type of ring, (c) a different type of ring, (d) fully connected, (e) tree, (f) bus.

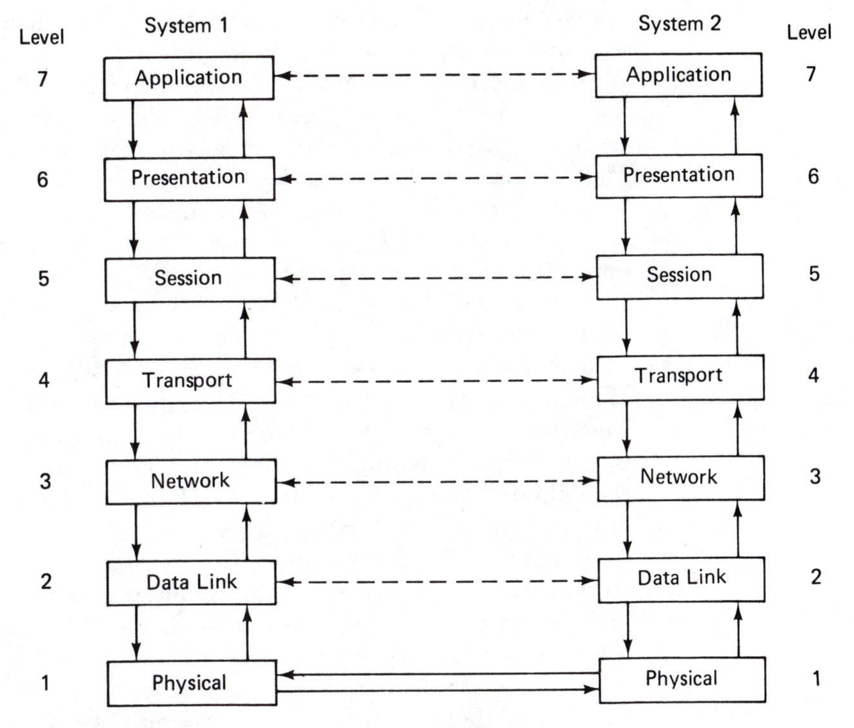

FIGURE 10-22. Open Systems Interconnection Reference Model—"layered" protocols.

mately involved in the details by which information is actually exchanged, in much the same way that you are insulated from needing to know the details of the phone company's central switching office when you place a telephone call. Most of the layers are usually composed of programs (more formally, "processes"—the software term, not the manufacturing term—which we explore in Part III). At the lowest layers, hardware becomes increasingly dominant.

A consistent overall protocol includes "subprotocols" for each layer. Processes at each layer communicate with processes at their corresponding layer in the other system (*peer processes*) by using the services of the layers below and are therefore said to be carrying on *virtual communication* (dotted lines in the figure). Actual *physical communication* takes place between layers along the solid vertical lines and between systems along the solid horizontal lines at the physical layer.

The *physical layer* deals with mechanical and electrical connections between the systems. Its rules specify the type of mating connectors used, which signals are on which connector pins, how many volts or milliamperes constitute a "1" and a "0," how many bits are transmitted per second, how multiple streams of data are multiplexed, and how they are physically switched if more than one physical channel is available (as might be the case, for example, in a network provided with redundant wiring to be used in the event of a malfunction). The physical layer design does not assume any format or meaning in the information it handles.

The *data link layer* is responsible for structuring messages and controlling low-level aspects of using the physical channel over which the message will be sent. This work involves dividing messages into tractable lengths and appending error detecting and synchronization code patterns to these subdivisions (generally called *frames*). This layer is also responsible for managing acknowledgment procedures. Together, these help to ensure error-free physical transmission of the message. If the channel can carry data in only one direction at a time, the data link layer protocol specifies how the channel will be "turned around" so that each party can speak in turn. Some form of handshaking is needed for this and also to regulate the rate of transmission so that data do not overflow the receiver's room (buffers) for it. Remember that the receiver must pass received messages up to higher-level processes that may not be always ready to accept them.

Packets of data are exchanged at the *network layer*. Packets are simply a higher level of message structure. A packet may contain several messages for the same destination, or one long message may be spread over several packets. At this level, the communication system may have several routes from which it may choose. To transmit a long multipacket message more quickly, separate physical channels may be used for different packets. The network layer decides how each packet is to be routed and is responsible for reassembling received packets into proper order. If the network is shared by many systems, the network layer is also responsible for managing "contention" for access to the network. (Network contention is described later in this section.)

A particular host system may have multiple copies of the lowest three layers in its architecture, either as part of its own integral software and hardware or available to it in the form of one or more communications computer systems. Any given set of these may handle messages from its host system and may also be responsible for passing through messages from other systems.

At the *transport layer*, we reach what amounts to message handling in true "end-to-end" fashion. Transport layer processes accept messages from the next higher layer (the session layer) and establish a network connection for their transmission, much as a shipping manager decides whether a shipment will be made by the "air network" or the "surface network" (the choice of staging points and specific routes being left to lower levels—the network layer). The transport layer may offer several services: end-to-end delivery of messages in order, broadcast of messages to all or a subset of destinations, and so on. If there are many independent programs operating in its host, the transport layer must establish and maintain many virtual channels and must be responsible for multiplexing messages onto a single network from multiple processes within its host and for demultiplexing received messages.

The *session layer* manages the initiation, maintenance, and termination of the virtual connection between two user processes. A user process must specify the "address" of the other process with which it wishes to communicate. The session layer selects the transport virtual connections that accomplish the formation of this link between user processes. If something goes wrong during a session, the session layer may be responsible for recovering in a way that is "transparent" to the user or for managing the session in such a way that no harm results from a broken transport connection. For example, it may be a requirement of the system application that no part of a message be delivered to its user until the complete message has been correctly received at the destination station. This might be the case when the message is part of a complex operation that cannot be interrupted at certain points. The session layer would be responsible for ensuring that this rule is observed by holding parts of the message until it had been completely received and validated.

The *presentation layer* is often a collection of "convenience" services which includes such functions as converting characters from one type of coding (e.g., ASCII) to another, compressing or abbreviating data so that they can be transmitted more efficiently, reformatting data so that they are compatible with the types of interactive terminals and printers used, and so on.

Finally, the *application layer* consists of the actual user processes themselves—the parts of a program on one computer that are actually communicating with parts of another program on another computer.

Local Area Networks. In distributed control systems, where local area networks are of great interest, the principal involvement is with the physical, data link, and network layers. Local area networks are presently being subjected to standardization efforts by the ISA and the IEEE (Institute of Electrical and Electronics Engineers) and to many proprietary network product introductions from manufacturers of process and machine control systems. In the trade press, these are commonly called *data highways*.

Much of the controversy in data highway standardization centers on the choice between two prominent methods of resolving *contention*. In a bus or broadcast ring, only one node may transmit at a time, yet more than one node may have messages to send. Contention methods deal with this. The principal methods of contention resolution are the *carrier sense* method and the *token* method. There are several variations on these methods, but they work basically as follows.

The most prominent variation on the carrier sense method is called *carrier sense multiple access/collision detection* (abbreviated CSMA/CD). It works in a way that has been likened to the way conversation is controlled in a group of polite people. When one node wishes to communicate with another, it must first see if the communication channel is free. It does this be sensing the channel to see if the carrier is on. The *carrier* is a signal that is modulated by the physical layer hardware when it transmits data. For example, if frequency shift keying is being used to send data, the hardware will be switching between two closely spaced frequencies, one for a "1" and one for a "0." If either frequency is present on the channel, it means that the channel is busy and the node must wait (just as a person should when someone else is speaking—it is not polite to interrupt).

If the node detects no carrier, it may start its transmission, first by turning on its own carrier generator. In most cases, this act will now keep other nodes waiting. However, there is a chance that some other node may *also* be waiting for the channel and attempt to seize it simultaneously. Both nodes then transmit at the same time and their messages are garbled—a *collision* is said to have taken place.

In CSMA/CD protocols, collisions are detected by requiring each node to monitor the channel for a certain time after it starts transmitting. If it observes an exact copy of its transmission, it may assume that there has been no collision and continue sending. If not, it stops transmitting—in other words, it "backs off." This is much like what happens when more than one person in a group begins speaking at the same time.

The next issue is how to ensure that both nodes will not keep colliding by continuing to attempt to transmit simultaneously. (Remember, they both have messages waiting to be sent and, being computers, they are operating on very fast clocks which virtually assure that they will try again the instant they sense that the channel is free.) This is done by having each node wait for a randomly selected time before trying again. Each node computes a new pseudorandom number at each collision and waits for a time that is a function of that number. To keep nodes from generating the same sequence of pseudorandom numbers, each uses a different initial number.

The token method of resolving contention is used with ring topologies. Messages travel circularly in a token ring network, from the transmitting station around the ring, *through* the network interfaces of all the other nodes, and back eventually to the transmitting station. A *token* is used to decide which node may transmit. The token is a special character or message packet, a unique pattern of bits. When the network is started up, the "master" station transmits the token character to the next node. If that node has no message to send, it passes the token on to the next node beyond it. In this way, the token circulates in the ring until it arrives at a node that does have a message to send. That node "holds" the token. Instead of immediately retransmitting the token, it sends its message. When it has finished sending its message, it then passes on "permission to speak" (in the form of the token) to the next node.

In a sense, the two methods are complementary in some characteristics. For example, while the carrier sense method is random (because one cannot say precisely how long a station may wait for access), the token method is deterministic because, if there is a maximum permitted packet length and a node is allowed to send only one packet before passing the token, there is a fixed maximum time between channel accesses that is proportional to the number of nodes on the ring. Carrier sense interfaces are passive in that they handle only their own messages. Token ring interfaces are active in that they receive and retransmit every message that passes by them.

Each method has advantages and disadvantages. Lack of determinacy is criticized in the carrier sense method. Some feel that an industrial communication method must be highly deterministic. On the other hand, token ring hardware is more complex, and the network will fail if this hardware malfunctions or is turned off (unless elaborate precautionary procedures and fail-safe hardware are included).

Conclusion. Computer networking is a complex subject, but very important to modern industrial control systems, because effective control is increasingly coming to be synonymous with distributed control. Distributed control depends heavily on communications networks—it cannot exist without them.

Common Physical and Data Link Layer Methods

Historical origins of computer serial links are based on early work in connecting interactive terminals to computers. Early widely used computer terminals were teletypewriters manufactured by Teletype Corporation. These electromechanical terminals included a keyboard and printer and operated on switched currents. Succeeding terminals incorporated cathode ray tubes for displays and operated on switched voltages according to Electronic Industries Association (EIA) standards, of which the most widely used is EIA RS-232-C, revision C of a standard formidably titled *Interface Between Data Terminal Equipment and Data Communication Equipment Employing Serial Binary Interchange.* ("Interchange" is another way of saying "communication.") Much of the practice in this area was influenced by the use of public telephone and telegraph switching networks for data communications, and its terminology is widely used, even when an actual telephonic network may not be involved (e.g., in directly connecting an interactive terminal to a computer for programming).

Modems. The term "data communication equipment" in the RS-232-C standard title refers to the equipment that is closest to the physical channel. Although the term (abbreviated "DCE") predates the OSI Reference Model, it can be thought of as a physical layer element, which also includes the actual communication "medium" (the "wire," as it were). The term "modem" is more generally used and is a synonym for DCE in informal technical conversation.

Properly speaking, *modem* is short for "*mo*dulator-*dem*odulator"—the hardware that transmits data by modulating a signal and retrieves transmitted data from the signal by demodulating it. There are many modulation-demodulation techniques, and equipment that performs this function must often include logic for other related purposes. From the viewpoint of the RS-232-C standard, the equipment to which the DCE is connected is the "data terminal equipment" (abbreviated "DTE"). Technically, DTE can be anything from a simple interactive terminal to the largest supercomputer imaginable.

Figure 10-23 illustrates these relationships. However, other interpretations are common. For example, modems are becoming more

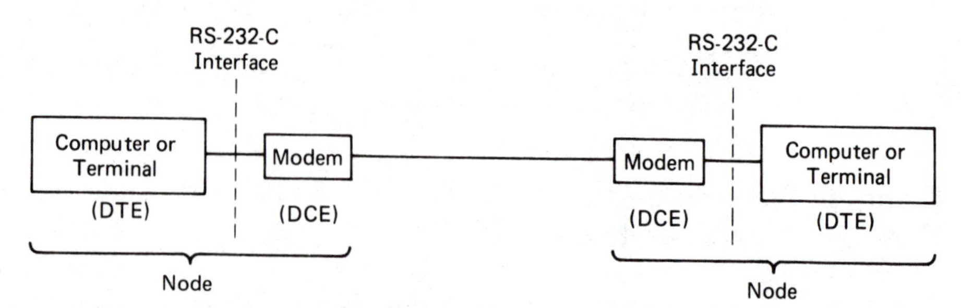

FIGURE 10-23. Relationship between data terminal equipment (DTE) and data communication equipment (DCE) in the EIA RS-232-C standard.

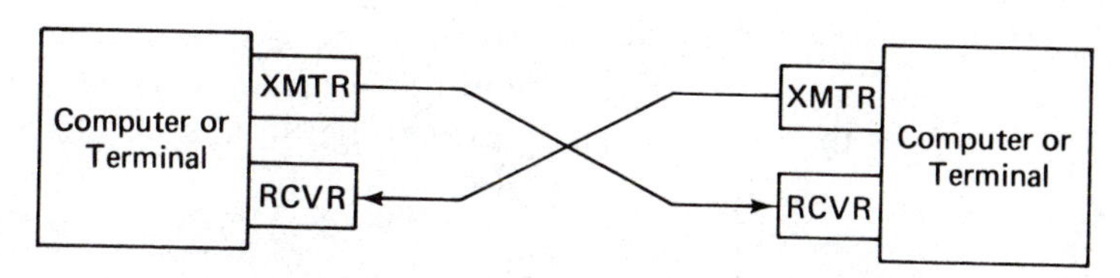

FIGURE 10-24. A full-duplex link.

"intelligent." This intelligence is obtained by using LSI components that do not always respect their place in the scheme of things imagined by creators of standards. Having equipped them with more intelligence, their designers have also increased the work they do to aid the communications process—that is only common sense and to be expected. But it results in modems that also perform sophisticated networking functions such as multiplexing, automatic dialing, signal quality analysis, diagnostic checking of their own operation and the operation of the network, and so on. Hardly a simple modulator-demodulator, but still called by the name "modem" in much general use.

Full and Half Duplex. Of interest to the parties on either end of a computer link is whether both can "talk" simultaneously. If they can, the link is said to be a *full-duplex* link. As indicated in Fig. 10-24, this means there are two channels available, each with a virtual transmitter/receiver pair of its own. The physical link between two nodes may be capable of supporting many simultaneous transmissions. This can be done, for example, by having each virtual link use different frequency bands for frequency modulation (FM) signaling. There may be one physical wire or two used for this. They may even be optical fibers or radio waves. Their physical nature does not matter logically to the communicating parties; what does matter is the ability for both parties to transmit at the same time without having their messages interfere with one another.

If one party must wait for the other to finish before it can begin its own transmission, the link is said to be a *half-duplex* link. This is shown in Fig. 10-25. Even though the wire (or optical fiber or "air-

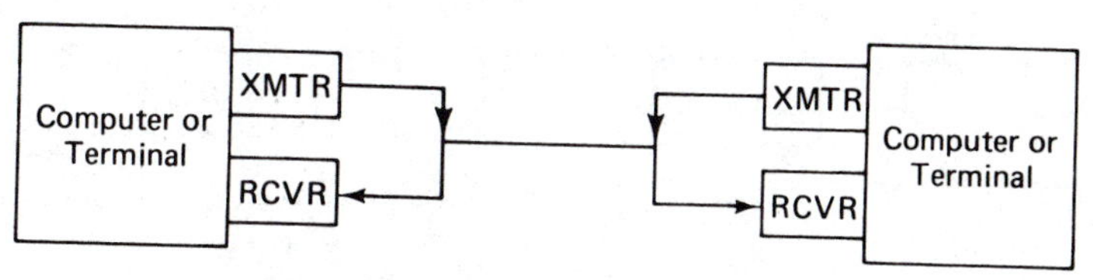

FIGURE 10-25. A half-duplex link.

waves") may be physically capable of carrying simultaneous messages, the data communication equipment is not designed to handle them. A strictly unidirectional link is said to be a *simplex* link. Traffic goes in only one direction. When there are more than two nodes on a link, the link may be said to be a *party line*, although the more modern term is bus.

Similarities to our experiences with radio communications (if we have found our way into citizens' band or amateur radio activities) must come to mind. The need to press the "talk" button on a "walkie-talkie" and its effect if the other party happens to be talking at the moment (for those who have never used one, the other party is shut off) is a sure sign of a half-duplex link at work. When you listen to the radio, you are on the receiving end of a simplex broadcast channel. And the Morse code you may have heard while turning the short wave dial trying to find something intelligible is the ancestor of the way in which "raw data" are transmitted among computers and terminals.

Asynchronous and Synchronous Methods. Samuel F. B. Morse invented the telegraph in 1837. Morse encoded the characters of the alphabet, numerals, and punctuation marks into a series of long and short contact closures and pauses—a binary code that substantially predated digital computers. The serial transmission of binary data is based on the same idea, enhanced with new capabilities that depend on the logic and speed of modern electronics.

Morse telegraphy is essentially *asynchronous*. Although the transmission of each character, as it is keyed, must be neither too slow nor too fast to be intelligible, substantial latitude is permitted in the pauses between characters. This is the essence of asynchronous communication among computers and terminals. As you might expect, there is another method, which is *synchronous*, but we come to that in a moment.

Figure 10-26 diagrams the transmission of a character using the asynchronous method. The line is initially in an "idle" condition. In a system that operates by switching current loops, this is usually defined as the condition in which the circuit is completed and current

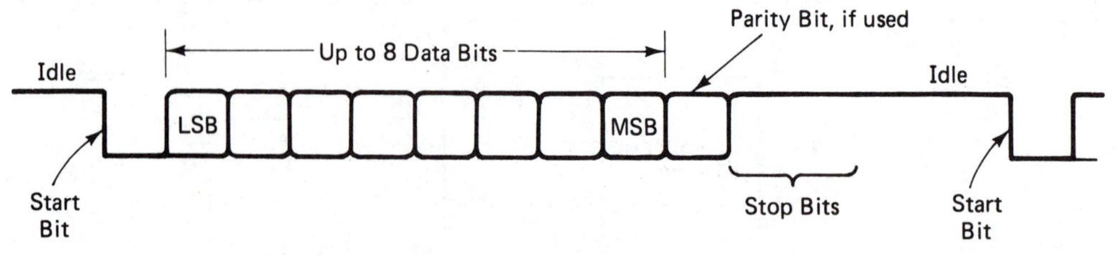

FIGURE 10-26. Asynchronous character transmission.

is flowing. In RS-232-C interfaces, which switch voltages, it is the more negative voltage condition (from −3 to −15 volts). In general, no matter what electrical method is used, it is considered to be a logical "1." This state is called the *MARK* state. An idle line is said to be *marking.*

The start of a character transmission is signaled by switching the line to the opposite state—opening the circuit to interrupt current flow in current loop systems, switching to the more positive voltage in RS-232-C systems (from +3 to +15 volts, the region from −3 to +3 volts being a "transition" region in which the signal is undefined.) This is considered to be a logical "0" and is called the *SPACE* state. A line in this state is said to be *spacing.*

The line is held in this state for one *unit interval.* The reciprocal of this time is called one *baud* (after Emile Baudot, one of the pioneers of telegraph printers). Baud is equal to the number of signaling events per second. There are other forms of signaling in which several transitions ("signaling elements") are used to represent a single bit of data, but in typical asynchronous links (which are also called *start-stop*), baud is equal to the number of bits transmitted per second.

This initial SPACE state represents the *start bit.* It is followed by *data bits,* in which the line marks for one unit interval to indicate a logical "1" and spaces to indicate a logical "0." The usual convention is to transmit the least significant bit of the character first. Most codes used for transmitting characters represent the alphabet, numerals, and punctuation marks with from five to eight bits per character. The Baudot code (an early telegraphy code still in use) uses five bits. The ASCII code uses seven. Data may also have meanings not represented by a standard code. A character of eight bits may be a simple binary number, half of a 16-bit binary number, a pair of BCD digits, and so on. The data bits may also be followed by a parity bit.

Following the most significant data bit or the parity bit, the line is returned to the marking state for a certain minimum time, after which the start bit of the succeeding character *may* be transmitted. This minimum time at the end of the character is made up of *stop bits.* The protocol may establish 1, 1.5, or 2 stop bits as a minimum requirement. There is no requirement that the next character follow immediately. An indefinite amount of time may elapse between characters in asynchronous communication. This is characteristic of the rate at which characters are typed at the keyboard of an interactive terminal. Most such terminals send the typed character out on the interface as it is typed, so the rate (in characters per second) varies from painfully slow (in the case of a hunt-and-peck typist) to rapid bursts (in the case of a skilled typist). Computers, on the other hand, usually transmit in relatively uniform bursts.

Although there is a great deal of latitude in asynchronous communication, it does depend on several points of agreement among the parties. The number of data bits and stop bits and the use of parity must be compatible with the expectations of the systems on both ends of the link. The baud rate must also be agreed upon. The most commonly used rates are 110, 150, 300, 600, 1200, 1800, 2400, 3600, 4800, 9600, and 19200 baud. Slower rates are used more often with terminals that include printers and correspond to effective "instantaneous" character rates of about 10 to 30 characters per second with most common character formats. Rates of 9600 and 19200 baud are commonly used with directly connected interactive CRT terminals and provide character rates of from about 870 to about 1920 characters per second, depending on the character format. Intermediate rates are used when the physical link is moderately long, since slower rates provide somewhat more reliable communication.

The stop bits delineate characters. Older electromechanical terminals require more time to "recover" (to reset and prepare to accept the next character) than modern fully electronic terminals and, therefore, usually need both stop bits. With a modern terminal, a 9 percent improvement in the character transmission rate can be had by using only a one-stop bit when transmitting ASCII characters. Most terminals have a CRT screen format of 24 lines of 80 characters each. At 9600 baud, it is possible to transmit a full-screen image to such a terminal in about 2 seconds.

The desire to transmit long bursts of data more efficiently with computers and electronic terminals is the motivation behind synchronous communication methods. There are several forms of synchronous communication, but we limit our discussion to some of their general characteristics.

The typical character format for ASCII asynchronous communication is only about 72 percent true data. The remainder is "formatting." In synchronous communication, the idea is to increase the true data proportional content in a message. Figure 10-27 illustrates

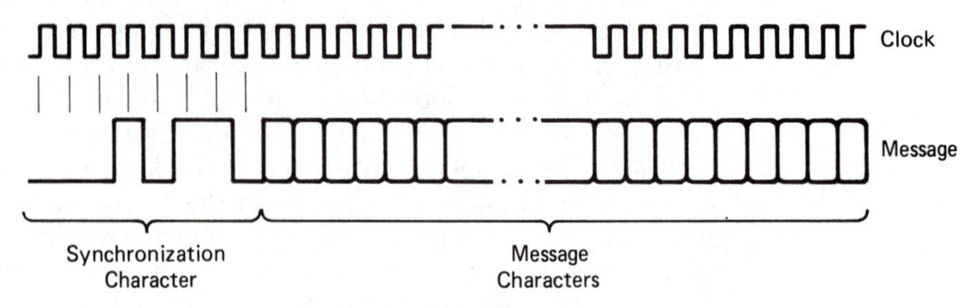

FIGURE 10-27. Synchronous data transmission.

SYN	SYN	SOH	Header	STX	Text	ETX	CRC

FIGURE 10-28. General form of a synchronous message packet.

the general idea. Since we are not dealing with electromechanical terminals, a faster clock can be used, and synchronism can be maintained between the communicating parties for a much longer time. The clock may be generated by the transmitting station and sent over the physical link to the receiving station, which must then synchronize itself to the clock, or both may operate from their own clocks, which must have the same agreed frequency within certain close tolerances. In either case, the clock frequency establishes the duration of the unit interval and, therefore, bit synchronization.

So the receiving station knows when each character begins; character synchronization must be established also. To do this, the transmitting station sends one or more *synchronization* (abbreviated SYN) *characters*. This character has a unique bit pattern and is reserved for this purpose. It cannot be used in transmitted messages as a data character unless protocol procedures are followed to keep it from being confused with SYN characters. The receiving station monitors the channel for this character. When it observes the SYN character pattern, it knows that the next unit interval following the SYN character contains first bit of the first message character. To improve synchronization reliability, the protocol may call for two or more SYN characters. To improve it still more, once a physical link is established between nodes, the transmitting station may even fill any idle time between message packets with SYN characters.

Because synchronous messages may be directed to any of several destinations within a node (that is, any of several programs in a single computer may be communicating with any of several other programs in another single computer at the application layer), it is customary to add certain control information to each message packet in the form of a *header*. This information may identify the source and destination of the message, the message type, and the number of the message packet if it is part of a longer message or sequence of messages. It may also include a running count of the number of messages sent by this station to the receiving station. This is useful if it becomes necessary to retransmit messages that have been lost or garbled.

The general form of a synchronous packet is shown in Fig. 10-28. Synchronization characters are used to establish synchronism, and the beginning of the header is delimited by a "start-of-header" (SOH) character, followed by the header itself. The beginning of the message data itself (often called "text" though it may actually be binary data) is delimited by a "start-of-text" (STX) character, followed by the message data. The end of the message data is delimited by an "end-of-text" (ETX) character. At the end is one or more check characters, formed from the header and message data characters according to whatever error-checking method the protocol calls for. *Cyclic-redundancy-check* (CRC) codes are the most widely used, usually requiring 12 or 16 bits of the packet.

The efficiency of synchronous communication is due primarily to the ability of a single packet to carry hundreds and even thousands of message characters under the fixed overhead of the relatively small number of characters required for synchronization, formatting, header data, and error checking. In theory, the length of the text section is indefinite and practically limited primarily by the amount of memory available in buffers in the communicating computers. Synchronous communication efficiency suffers when messages are small, for the overhead then becomes a larger percentage of the transmission.

Serial Interface Hardware

General Serial Interface Organization. Figure 10-29 is a block diagram of a typical board-level product serial interface subsystem. Like digital and analog interfaces, its microcomputer boundary area consists of (1) bidirectional bus drivers that mate with the microcomputer data bus, and (2) subsystem recognition and control logic which recognizes the subsystem address and responds to read and write operations by enabling and controlling bus driver directionality and by enabling the device select logic, which is the third part of the boundary area.

The figure shows only one serial interface channel, the internal logic of which is represented by a baud rate generator and a Universal Asynchronous Receiver/Transmitter (UART) or Universal Synchronous/Asynchronous Receiver/Transmitter (USART). In some board-level products, there may be four or eight such sets of internal logic. Such multichannel interfaces are found in moderately large microcomputer system configurations used in distributed control applications. The elements shown in the figure are also often found as part of the logic on "single-board microcomputers," where a single serial interface, parallel interfaces, and memory are combined with the microprocessor on a single printed wiring board.

The electrical interface is analogous to the signal conditioning and isolation areas of digital and analog interface subsystems. Signals handled by the electrical interface include serially transmitted and received data, output signals used to control modem operation, and input signals that indicate the status of the modem and the physical

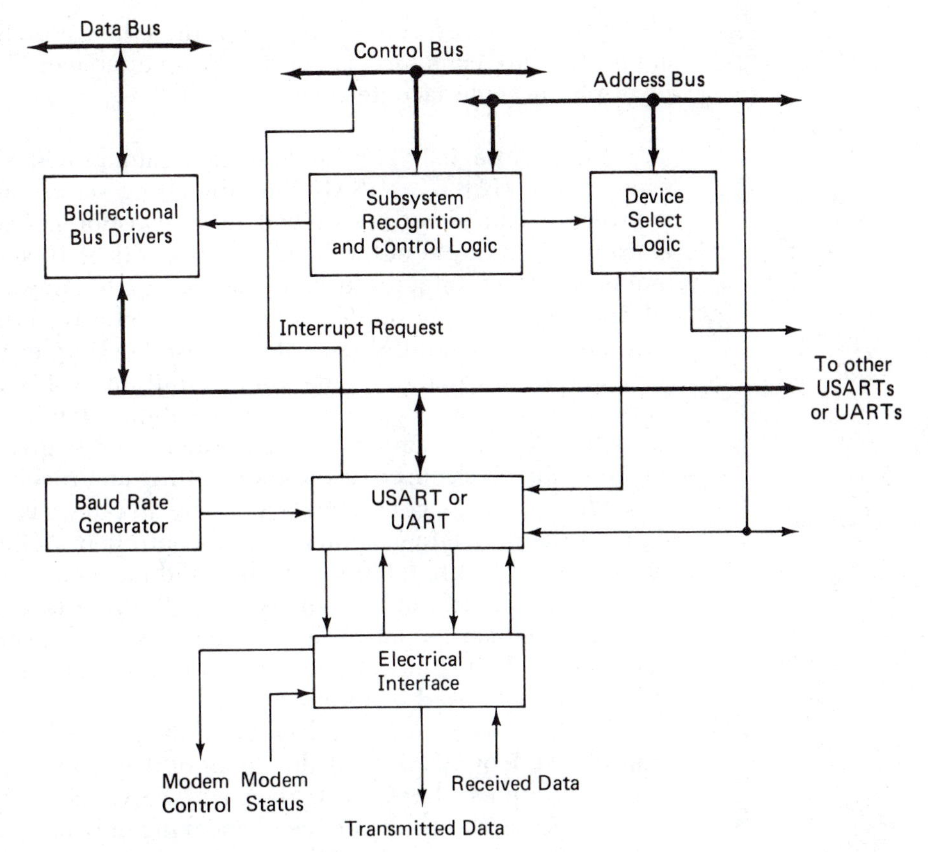

FIGURE 10-29. Block diagram of a typical board-level serial interface subsystem.

channel to which it is connected. Internal logic usually operates at TTL levels which cannot handle the currents and voltages required for serial communication over long distances and are not compatible with RS-232-C voltage levels or switched currents. The electrical interface is, therefore, responsible for converting between TTL levels and RS-232-C and switched current signals.

The electrical interface provides some measure of isolation between the microcomputer system and the physical wiring, which may be exposed to noise. In RS-232-C interfaces, the modem isolates the computer from the longer physical wiring. In current loop interfaces, optical isolation is often used and made a part of the electrical interface on the subsystem board. Because different types of modems are often used, the modem itself is not usually included on the interface subsystem board. Some interfaces include both RS-232-C and current loop electrical interfaces, some only one or the other. RS-232-C interfaces are much more common than current loop interfaces in recent board-level product interface designs.

Internal Logic. The baud rate generator provides a clock signal used by the UART or USART in generating serial data output and recovering data from the received serial waveform. The clock signal is usually a multiple of the baud rate. Factors of 16 and 64 are commonly used. Use of a multiple of the baud rate allows the UART or USART to take many samples of the received waveform within each unit interval. The multiple samples are used in simple "majority vote" logic to decide whether the datum in a unit interval is a "1" or a "0." This enhances asynchronous communication reliability by reducing the effects of small amounts of noise and permits greater tolerances between the clocks used at the transmitting and receiving stations.

Most baud rate generators operate by dividing a faster basic clock signal, usually obtained from a crystal-controlled oscillator. Division factors are set by digital inputs to the baud rate generator. These may come from switches or jumper wires on the interface board or from control words set by software. A commonly used basic clock signal frequency is 1.8432 megahertz, since it has factors that correspond to most commonly used baud rates.

Figure 10-30 is a block diagram of the internal organization of a typical UART or USART. At this level of diagramming, there are no visible differences between UARTs and USARTS, so from this point on we exercise the convenience of speaking only of USARTs, though what is said may apply to both types.

The USART, like the interface board on which it resides, is organized around an internal bus. Note the structural similarities between this device and others at the same component level (such as parallel input/output support chips) and similarities to things we have seen at higher levels. It is a useful exercise to find analogies between devices at this level and at levels even far above. For example, in what ways are the internal organization of a USART like the organization of a plantwide process control system?

The USART is interfaced to the internal bus (or the microcomputer data bus in a single-board computer) by the data bus buffer, which is essentially a set of bidirectional bus drivers (but of more limited electrical drive capability than those used at the higher level). The general operation of the USART is controlled by its control logic, which has read/write (R/W), chip select (CS), address (A_0), and clock inputs. The address input distinguishes control and data words and is usually called the "Control/Data" (C/D) input on pin assignment

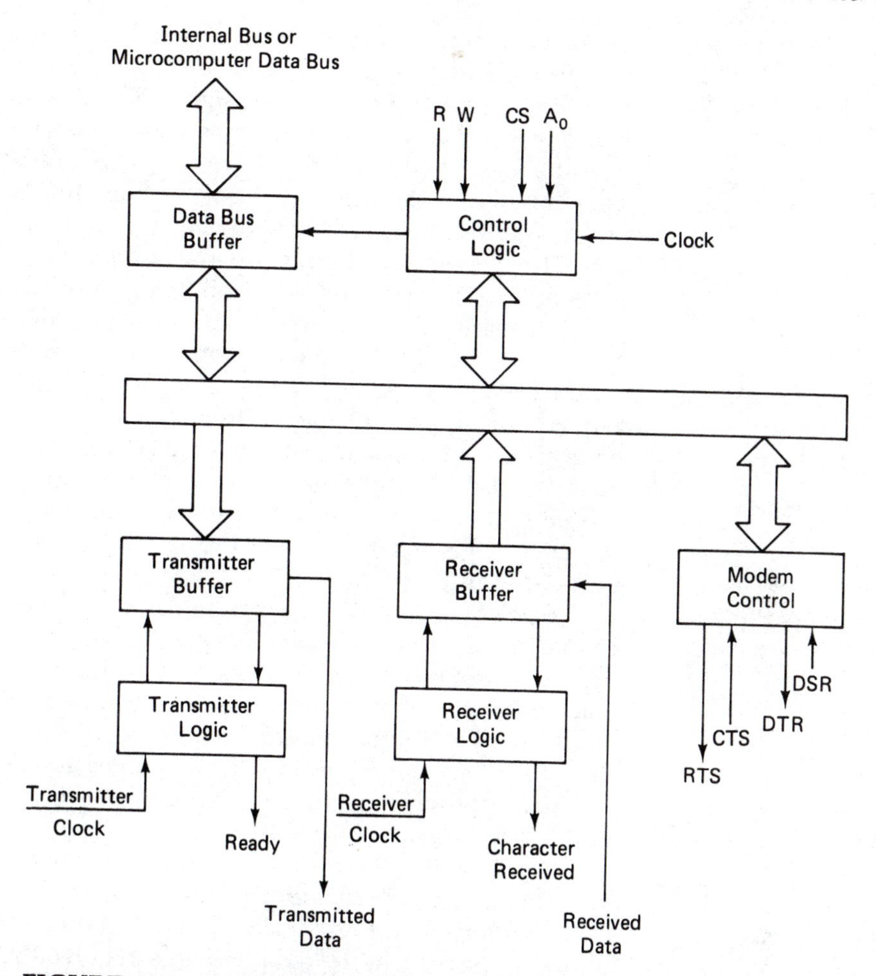

FIGURE 10-30. Internal organization of a typical UART or USART.

diagrams. Because it is common practice to use the least significant bit of the microcomputer address bus to control this, we use "A_0."

The clock paces the general USART logic and is unrelated to the baud rate clock, except that it must be sufficiently faster to provide enough periods for the logic to complete its operations within a unit interval. Usually it must be 30 or more times the frequency of the baud rate clocking signal.

The transmitter buffer consists of a parallel-to-serial shift register preceded by an octal latch. This arrangement is said to be "double-buffered." One character can be written to it from the microcomputer while the shift register is still in the process of transmitting the previous character, so that no delay results between transmissions. The transmitter buffer is controlled by the transmitter logic. Both operate from the transmitter clock input signal. This is provided by the baud rate generator and in some designs may be separate from the signal used for the receiver baud rate, so they may operate at different frequencies, if necessary (for example, in using a microcomputer to buffer fast data bursts on their way to a slow terminal).

"Ready" outputs from the transmitter logic indicate USART transmitter readiness. Two are commonly provided: the first to indicate that both the shift register and the latch are empty, the second to indicate only that the latch is empty. The "latch empty" (sometimes called TBMT, for "Transmitter Buffer Empty") means that the USART can accept another character from the microcomputer. The "both empty" condition (sometimes called EOT for "End of Transmission") means that the USART has transmitted all of the characters given to it by the microcomputer. These status conditions can be sampled by the microcomputer by reading the USART status word, or they may be used to interrupt it.

The serial output of the shift register is, of course, the transmitted character. The transmitter logic appends start, stop, and parity bits (in the asynchronous mode of operation) to the data as they are shifted out. The format of the transmitted character is specified by the control word from the microcomputer. Control words are usually loaded into the USART as part of an initialization operation. The microcomputer need only provide data for transmission once the USART has been initialized.

The receiver buffer consists of a serial-to-parallel shift register followed by an octal latch, providing double buffering for received data. Without double buffering, it would be necessary for the microcomputer to read in the most recently received full character before the first bit of the next character arrived. The burden of having to respond so quickly is relieved by the following latch. As soon as a fully formed character has been shifted in from the serial channel, it is transferred to the latch. This frees the shift register to begin accepting the next character right away and gives the microcomputer an amount of time equal to the single character transmission time in which to get around to reading the latch.

This operation is controlled by the receiver logic, which may operate, as noted earlier, from a different baud rate clock than the transmitter section. The receiver logic also checks received character parity, if the USART is so directed by its control word, and warns the microcomputer if an *overrun* occurs, or if a *framing* error occurs. An overrun error occurs if the microcomputer does not read the latch *before* the next character is transferred to the latch. It means that one or more characters have been lost because the microcomputer did not accept them quickly enough. A framing error occurs when there are too few stop bits between received characters. Parity, overrun, and framing errors are reported by the status word which the microcomputer can read from the USART. A "Character Received" signal from the receiver logic indicates that a newly received character has been transferred to the latch. As with the transmitter ready signals, this status may be read by the microcomputer or used to interrupt it.

Some USARTs include modem control logic. Although many modems have a substantial number of control and status lines (as we later see in this section), most simple applications can be handled with four signals: two outputs and two inputs. The DTR ("Data Terminal Ready") and RTS ("Request To Send") outputs inform the modem that the computer is ready and has a transmission to make, respectively. The DSR ("Data Set Ready") input indicates that the modem is in operational readiness and the CTS ("Clear to Send") input indicates that the computer may begin or continue to transmit. In some USARTs the CTS signal is interlocked with the transmitter logic and inhibits transmission if the modem is not indicating permission to send. Other USART functions include automatic recognition of SYN characters and automatic filling of idle transmission time with SYN characters when in the synchronous mode of operation.

Modems, Terminals, and Electrical Interfaces

Modems and Switched Networks. The geographical dispersion of a system network affects its operation in many ways. The data capacity of physical links, the costs of using them, their reliability, and their flexibility all enter into the overall "operational equation" of a communication system. The dispersion of an industrial control system communication network may vary widely, from a simple 2-node local link, to a multinode LAN, to continental and even intercontinental network links.

Most data communication traffic is carried over switched networks maintained by telephone utilities responsible for different geographic areas. In many cases, equipment for internal plant systems is supplied by these utilities. These in-plant systems are, in effect, small scale telephone networks. Although these are being replaced by high-capacity LANs, specialized for handling industrial control system communications requirements, they are still widely used, and much of the technology and terminology of LANs is based on precedents set by older local and plant-to-plant communication networks. For the most part, these networks are based on the EIA RS-232-C standard and use modems as electrical interfaces between computers and remote terminals and the telephone network.

So we have, for any pair of nodes in such a network, the situation depicted in Fig. 10-31: a switched network over which the actual link will be established, modems that connect the nodes to the switched network, and the nodes themselves. Nodes may be computers or terminals. Systems that use such networks are likely to use modems that are compatible with telephone networks and are most likely to conform to EIA RS-232-C standard requirements.

The RS-232-C standard defines 21 "interchange circuits" for the interface between "data communication equipment" (DCE—for most purposes, read "modem") and "data terminal equipment" (DTE—for most purposes, read "computer" or "interactive terminal"). These circuits provide for transmitted and received data on both a primary and a "secondary" channel, as well as modem control/status and signal control/status lines. The standard gives letter designations and names to each circuit. These are listed in Table 10-3. In some cases, circuit names are self-explanatory. We noted that use of circuits CA, CB, CC, and CD are sufficiently common that one often finds logic for them in USART designs. These, grounds (AA and AB), and the Transmitted (BA) and Received (BB) Data signals are provided in most RS-232-C interfaces.

The standard lists 14 interface types that use these circuits in various combinations for full- and half-duplex arrangements with provisions for automatic use of switched network features and secondary channels. Although this illustrates the variety of "official" arrangements one may find, there are also many that conform to the standard only in spirit, bending it for various crude, but effective, practical considerations where the full features of the standard are not truly required. In some cases, for example, it is not necessary for actual logic to be associated with these control and status signals. One might simply "strap" Request To Send (CA) and Data Terminal Ready (CD) to a logically ON (SPACE) voltage at the interactive terminal interface

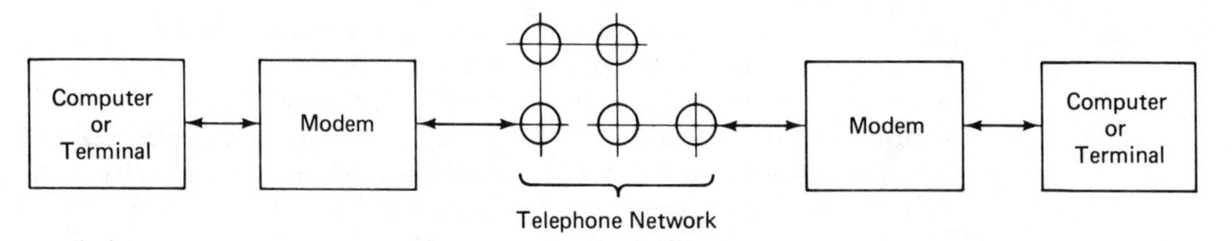

FIGURE 10-31. Communication over a switched network using modems.

TABLE 10-3
EIA RS-232-C
Interchange Circuits

AA	Protective Ground
AB	Signal Ground
BA	Transmitted Data
BB	Received Data
CA	Request To Send
CB	Clear To Send
CC	Data Set Ready
CD	Data Terminal Ready
CK	Ring Indicator
CF	Received Line Signal Detector
CG	Signal Quality Detector
CH	Data Signaling Rate Selector (DTE Source)
CI	Data Signaling Rate Selector (DCE Source)
DA	Transmitter Signal Element Timing (DTE Source)
DS	Transmitter Signal Element Timing (DCE Source)
DD	Receiver Signal Element Timing (DCE Source)
SBA	Secondary Transmitted Data
SBB	Secondary Received Data
SCA	Secondary Request To Send
SCB	Secondary Clear To Send
SCF	Secondary Received Line Signal Detector

and assume that Data Set Ready (CC) and Clear to Send (CB) are present, leaving it up to the terminal operator to notice whether the modem is turned on and whether there are any problems with lack of permission to send (indicated, perhaps, by a lack of response from the computer).

When a computer and a terminal are in such close proximity that modulation methods (intended to provide reliable communication over long telephone wires) are not needed, a formal modem is not required. Most computers can be equipped with RS-232-C interfaces. Most interactive terminals are similarly equipped. They are, therefore, electrically compatible and may be connected to one another directly.

When the terminal is connected directly to the computer, the meanings of the signal names become confused because both the computer *and* the terminal are "data terminal equipment" in terms of shorthand. Both are "DTE"—the standard is for DTE/DCE interfaces, not DTE/DTE interfaces. What is needed is a rearrangement of these signals that makes sense of them in terms of the direct connection of two items of DTE. But rather than rewire the computer or the terminal, a separate "black box" can be placed in the cable between them in which the required rewiring is done. As you might guess, this black box (inside, little more than a relatively simple terminal strip) is commonly called a *null modem.*

Notice that the standard provides for "secondary" Transmitted (SBA) and Received (SBB) Data and Request (SCA) and Clear (SCB) To Send circuits. These are generally used as low baud rate "back channels" to exchange other information about the transmission going on on the primary channel. For example, the receiving station might use secondary channels to request retransmission of garbled messages or as handshaking signals to pace the transmission of long packets of data. Circuits CF and SCF indicate that the modem has a signal on the line. Circuit CG indicates that the quality of the signal is acceptable. Circuit CK allows the modem to tell the terminal or computer that it is being called (its "phone is ringing"). Circuits DA, DS, and DD are used to provide signal element timing ("clocks"), the source of which may be in the terminal/computer or in the modem. Circuits CH and CI provide for selecting between two different rates.

To get a better picture of what goes on in a modem, we look briefly at the operation of a simple low-speed modem for "voice grade" lines. *Voice grade lines* are more or less ordinary telephone channels which have a frequency response of roughly 300 Hz to 3000 Hz, sufficient to carry human conversation with reasonable fidelity (hence their name). This bandwidth can be used to carry full-duplex binary data at rates up to 300 baud and half-duplex binary data at rates up to 1200 baud.

Full-duplex communication on such a modem uses frequency division to divide the bandwidth into two virtual channels and frequency shift keying to transmit data. The modem on one channel signals SPACE by transmitting a 1070-Hz tone and MARK by transmitting a 1270-Hz tone. The modem on the other channel signals SPACE with a 2025-Hz tone and MARK with a 2225-Hz tone. Usually, the 1070–1270 tone pair is used by the modem that "originates" the call and the 2025–2225 tone pair by the modem that "answers" the call. Some modems are equipped only for originating calls, others only for answering, and still others for both functions.

For half-duplex communication at higher baud rates, a 1200–2200-Hz tone pair is used by each modem, so they may not transmit simultaneously. At this higher modulation frequency the two tones must be widely separated, using more of the available bandwidth and leaving only enough for conversation to take place in one direction at a time. Some aspect of the protocol must be used to tell when one modem has finished transmitting and the other may begin. This changeover is called "turning the line around." There is usually room in the bandwidth for a single additional frequency which can be on-off keyed as a secondary channel. This can be used as a handshaking signal for turning the line around with somewhat more efficiency than would be had with ordinary pure half-duplex, where a character at the end of the message might be used to indicate the end of one terminal's transmission.

Figure 10-32 is a block diagram of a full-duplex FSK asynchronous low-speed modem. The elements are shown only for one set of frequencies. To handle both originate and answer frequencies, it would have to be equipped with two oscillators and two sets of filters, with logic for controlling which to use.

The modem is transformer-coupled to the telephone wires, for isolation. The oscillator responds to Transmitted Data (circuit BA) binary signals by shifting between the two tones of the tone pair being used (1070–1270 or 2025–2225). The transmitter filter removes side-band frequencies caused by oscillator frequency shifts. The receiver filter passes input frequencies in the region of the tones to be received. Amplitude variations are removed by limiting circuits, whose output signal is then converted to a frequency-proportional voltage. The carrier detect circuitry operates from two inputs: a threshold detector output, which indicates whether the received signal is of sufficient amplitude, and the frequency-proportional voltage, which is averaged over a short period in order to distinguish the signal from noise. The Received Data output (circuit BB) is the output of "slicing" circuitry, which converts the frequency-proportional voltage to ordinary MARK/SPACE form. It is usually controlled by the Carrier Detect output (circuit CF) so that the Received Data output marks when the carrier is not present.

Various means of utilizing the other control and status lines vary, depending on the application and the types of modems being used. One of the best sources to standard practice in this aspect of modem use is Industrial Electronics Bulletin No. 9, *Application Notes For EIA Standard RS-232-C*, published by the EIA. Further information is given in the bibliography.

Terminals. Interactive terminals are the principal interface between humans and computers. Their principal attractions are their compatibility with simple serial communication interfaces and their flexibility for information entry and display. Their capabilities range

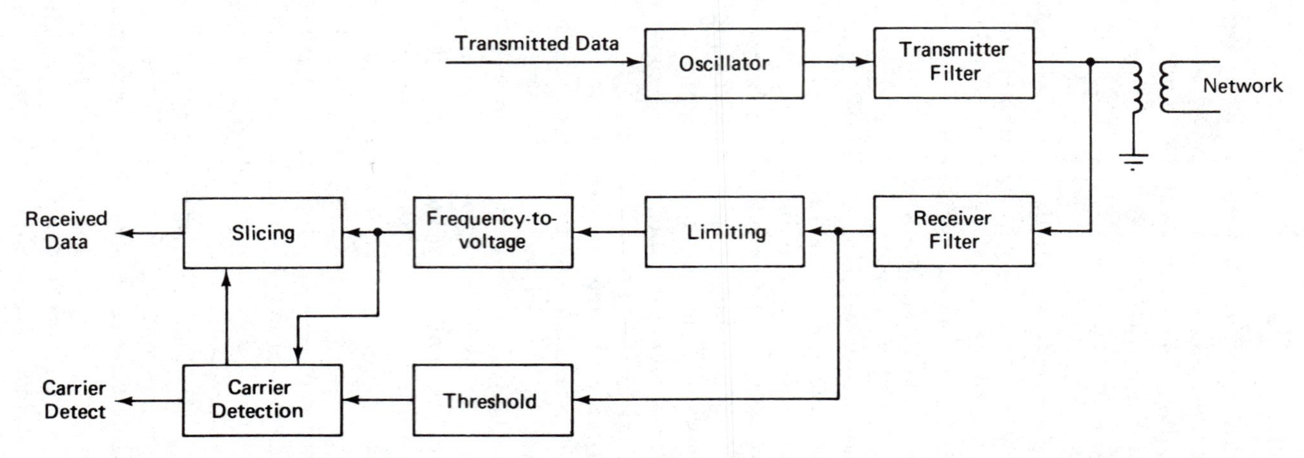

FIGURE 10-32. Simplified block diagrams of a full-duplex FSK asynchronous low-speed modem.

from simple entry and display of ASCII characters to highly sophisticated color graphics. In recent years, banks of color graphic terminals have come to be widely used in process control system consoles, replacing the familiar wall-sized process displays because they can show process status in many forms and levels of detail.

Figure 10-33 is a block diagram of a simple interactive terminal. Information is displayed on a *cathode ray tube* (CRT), the surface of which is coated with a phosphor that is stimulated by a beam of electrons. The electron beam is steered by video circuits associated with the CRT. In most terminals, the CRT and video circuits are essentially the same as those found in ordinary television receivers. In terminals used for very high resolution color graphics applications, higher-quality phosphors and more sophisticated video electronics are used for drawing fine lines. In the usual case of simple nongraphic (*alphanumeric*) terminals, the beam is steered across the CRT face in raster-scan fashion and is "blanked" (turned off) when aimed at areas where no screen illumination is desired. By blanking the beam as it scans, characters, numerals, and other symbols can be formed from the dots of light caused where the unblanked beam strikes the phosphor.

If not continuously illuminated by the electron beam, the stimulated phosphor emission decays in amplitude. Therefore, it is necessary to "refresh" the image on the CRT face periodically. The *refresh memory* holds the data used to form the picture. In most cases, data are stored in the refresh memory as ASCII-coded characters, and *character-generator* logic associated with the refresh memory is used to convert these characters into video control signals. Characters are retrieved from the refresh memory sequentially and cyclically, so that the CRT is refreshed at a rate that prevents the perception of flicker by the operator. The refresh memory may be made of ordinary static or dynamic semiconductor RAM.

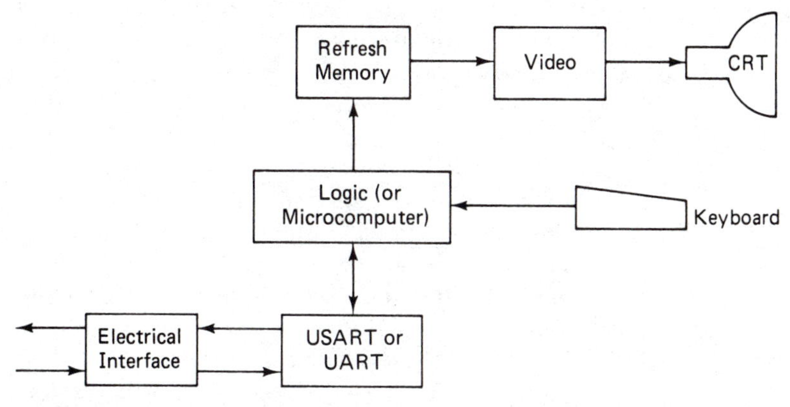

FIGURE 10-33. Block diagram of a simple interactive terminal.

The keyboard is used for data entry. Keyboards may be implemented as a matrix of contact closures scanned by logic in the terminal or by logic that is integral to the keyboard assembly. Keyboard layouts are most commonly of the "QWERTY" variety (named for the first six letters on the top rank of ordinary typewriter keys), or they may be special arrangements for unskilled typists (with letters laid out in alphabetical order in a matrix arrangement). For control applications, they may be labeled with special phrases. In some terminals, the keyboard assembly is separate from the terminal housing, connected to it by a cable so its position may be adjusted for operator comfort, or placed at some distance from the display screen (as might be required for a large multioperator control console configuration).

In modern designs, the terminal logic is implemented with microcomputers that manage the entry of data into the refresh memory and the receipt of characters from the keyboard. This work is more complex than might be imagined at first glance, even for a simple terminal. For example, it must control such things as causing lines to scroll up the CRT screen as they are entered from the keyboard or received from the computer or communication line to which the terminal is connected. In addition, certain control characters may be used to directly select the screen position at which the next character is to be written. This position is usually displayed on the screen by a blinking underscore mark or rectangle called the *cursor*. Some terminals, used to accept data to be filled into "blank forms" displayed on the screen, may protect areas of the screen from being overwritten by keyboard or computer inputs. With these, only variable parts of the data need be transmitted.

The interface to the communication line or computer uses a USART and electrical interface essentially like that on the serial interface of a microcomputer. Many terminals are equipped with electrical interfaces and connectors for both RS-232-C and current loop methods. They are also capable of handling a variety of baud rates, the choice of which is set by switches located on the rear of the terminal housing or under a covered recess on the keyboard housing.

Electrical Interfaces. We have discussed logical aspects of RS-232-C and its rules for transmitted and received character formatting and modem/terminal/computer interaction. We have also noted that a great deal of serial communication among computers and terminals is based on switched currents. Recently, the EIA has adopted new standards that address problems imposed by certain restrictions of the RS-232-C standard. The electrical characteristics of these ways of transmitting serial data are the subject of this section.

Figure 10-34 is a simplified diagram of an EIA RS-232-C DTE/

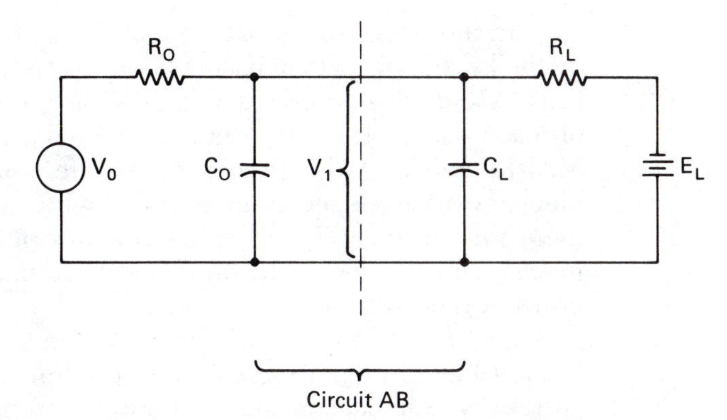

FIGURE 10-34. Diagram of an EIA RS-232-C DTE/DCE interface signal circuit.

DCE interface signal circuit. The standard specifies that the receiver will consider the signal to be a MARK when the voltage V_1 is more negative than -3 volts with respect to Signal Ground (circuit AB) and a SPACE when V_1 is more positive than $+3$ volts with respect to Signal Ground. The region between these levels is a transition region in which the signal is undefined. The driver on the circuit is required to assert a voltage of from $+5$ to $+15$ volts for a SPACE and -5 to -15 volts for a MARK, providing a 2-volt noise margin. A SPACE is considered a logic "0" or control "ON" signal and a MARK a logic "1" or control "OFF" signal. The slew rate (the rate at which the driver may change its output voltage) is limited to 30 volts per microsecond.

The standard also specifies ranges of values of circuit resistances and capacitances under certain conditions. The dc load resistance R_L must be less than 7000 ohms and more than 3000 ohms, depending on the conditions of measurement, and the load shunt capacitance C_L must be less than 2500 picofarads. Because these include the effects of the cable, there is an implied maximum cable length of 50 feet, though the standard does not state a maximum length *per se*. Other aspects of the standard call for the output impedance of an unpowered driver to be greater than 300 ohms and the driver circuit to be able to withstand short circuits to any other wire on the interface without damage, with short circuit current limited to 0.5 ampere.

The use of electromechanical terminals with early computers led to a substantial number of serial interfaces based on current switching methods. Although there are no formal standards for these, the 20-mA current loop was, and remains, widely used. Some interactive terminals and many serial interface subsystems are equipped with a 20-mA current loop transmitter/receiver.

In these early terminals, commutating brushes, activated by keys on the keyboard, switched direct current on the interface cable to form the MARK (current flowing) and SPACE (no current flowing) elements of each asynchronously transmitted character. This sequence of MARKs and SPACEs then engaged electromagnets in the printing mechanisms at the receiving terminal which selected the printing element for the received character. Each terminal included a printing mechanism connected to the other terminal's "keyboard loop"—in effect, a receiver/transmitter pair.

Figure 10-35 is a block diagram of the principal elements of one current loop. Serial TTL data from the transmitting USART directs a current switch that controls the flow of current supplied by the current source. The presence of current in the loop is detected by the current detector at the other end of the line, which provides a TTL voltage input to the receiving USART. Mechanical switches, brushes, and magnetic coils in the electromechanical terminals are replaced by solid-state circuits in modern terminals and computer interface subsystems. The location of the current source determines the type of transmitter and receiver. One or the other (but not both) may contain the current source. The figure shows an "active transmitter" and a "passive receiver." If the current source were located at the receiver, it would be called an "active receiver."

When driven by strong and properly designed sources into properly terminated receiving circuits, current loops can be used for direct communication over very long distances. Within the loop, they are relatively immune to noise (greater amounts of interfering energy being needed to disturb the current). However, switching currents

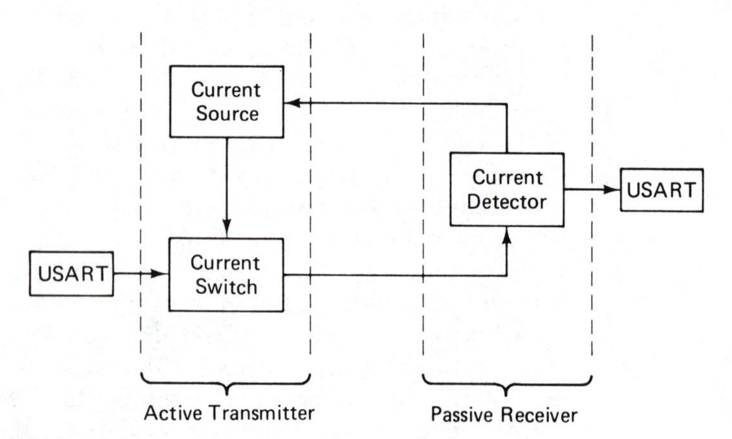

FIGURE 10-35. Principal elements of a current loop with an active transmitter and a passive receiver.

can be substantial generators of noise which can get into systems to which they are connected and into wiring adjacent to the current loop in the cable. The energy needed to switch long lines quickly generally limits the baud rates that can be used. For these reasons, current loop interfaces are often optically isolated. This isolation is best placed between the USARTs and their current switches and detectors, and isolated power supplies should be used to power the current sources.

The lack of formal standards has been a problem with current loop interfaces. Many are different in ways that prohibit simply plugging them in and expecting them to run. Threshold values on some solid-state circuits are often either exceeded or not met. Some circuits have no diodes to protect them from inductive surges, and some must be adjusted for different cable lengths. As a result, they are usually more reliable when the equipment to be placed on both ends of the line is obtained from the same manufacturer.

In 1975, the EIA released RS-422 (*Electrical Characteristics of Balanced Voltage Digital Interface Circuits*) and RS-423 (*Electrical Characteristics of Unbalanced Voltage Digital Interface Circuits*). These were followed in 1977 by RS-449 (*General Purpose 37-Position and 9-Position Interface For Data Terminal Equipment and Data Circuit-Terminating Equipment Employing Serial Binary Data Interchange*). These standards are intended to replace RS-232-C gradually and provide more accommodation to needs, practices, and advances in circuit design than was available with RS-232, Revision C, which dated from 1969. A further advance in these standards is represented in RS-485 (*Electrical Characteristics Of Generators And Receivers For Use In Balanced Digital Multipoint Systems*), which is scheduled for release in 1983 and provides for multiple transmitters ("generators") on a single physical channel—necessary in local area networks which use contention methods for access.

There is not space to comment as fully on these as on RS-232-C. Since they are relatively new, one is less likely to encounter equipment conforming to them. However, this situation will change eventually, and the reader with special interest in communication is encouraged to obtain copies of these standards from the EIA. Further information is given in the bibliography.

Figure 10-36 is a simplified diagram of an RS-422 interface circuit. Principally, RS-422 provides for differential balanced signaling in which a MARK is indicated by terminal A of the generator's being negative with respect to pin B and a SPACE by A's being positive. Provision is made for multiple receivers, and guidelines are provided for use at baud rates up to 100 kilobaud with cable lengths of up to 4000 feet, and rates up to 10 megabaud as cable lengths are shortened to 40 feet.

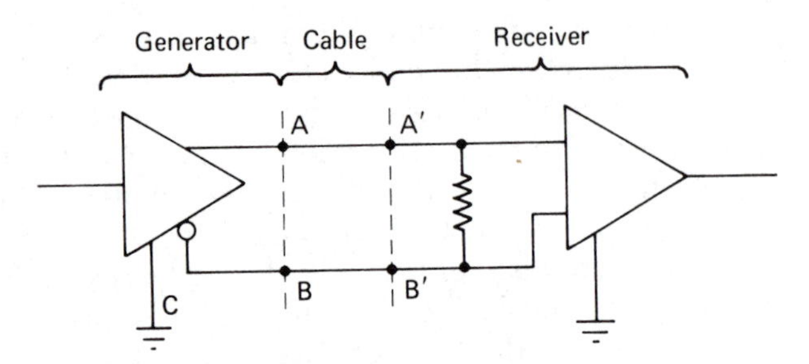

FIGURE 10-36. Simplified EIA RS-422 interface circuit.

Figure 10-37 is a simplified diagram of an RS-423 interface circuit. RS-423 follows the form of RS-422 closely, but for characteristics of an unbalanced signaling method in which MARK is indicated by terminal A of the generator's being negative with respect to a common signal return and SPACE by A's being positive. Guidelines are provided for use at baud rates up to 1000 baud with cable lengths of up to 4000 feet and rates up to 100 kilobaud as cable lengths are shortened to 40 feet.

The RS-449 standard specifies functional and mechanical aspects of these interfaces. Although the general nature of RS-232 remains evident in RS-449, new signal names are given, this time in mnemonic form, and new signals are added. Pin numbers are specified for the principal signals for a 37-pin connector, with a separate optional 9-pin connector for secondary channel signals. Table 10-4 lists the correspondence between RS-449 and RS-232-C signals.

To some degree, equipment designed to meet RS-232-C can be operated with equipment designed to meet RS-449. Guidelines for this are given in Industrial Electronics Bulletin No. 12, *Application*

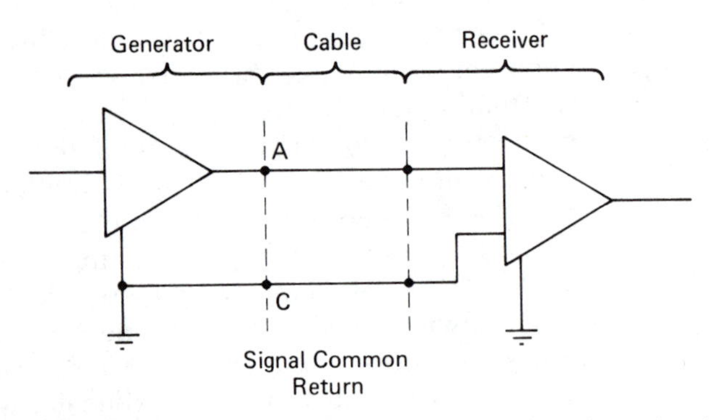

FIGURE 10-37. Simplified EIA RS-423 interface circuit.

TABLE 10-4
RS-449 and RS-232-C Interchange Circuit Correspondence

RS-449 Circuit		*RS-232-C Circuit*	
SG	Signal Ground	AB	Signal Ground
SC	Send Common	—	—
RC	Receive Common	—	—
IS	Terminal In Service	—	—
IC	Incoming Call	CE	Ring Indicator
TR	Terminal Ready	CD	Data Terminal Ready
DM	Data Mode	CC	Data Set Ready
SD	Send Data	BA	Transmitted Data
RD	Receive Data	BB	Received Data
TT	Terminal Timing	DA	Transmitter Signal Element Timing (DTE Source)
ST	Send Timing	DB	Transmitter Signal Element Timing (DCE Source)
RT	Receive Timing	DD	Receiver Signal Element Timing
RS	Request To Send	CA	Request To Send
CS	Clear To Send	CB	Clear To Send
RR	Receiver Ready	CF	Received Line Signal Detector
SQ	Signal Quality	CG	Signal Quality Detector
NS	New Signal	—	—
SF	Select Frequency	—	—
SR	Signaling Rate Selector	CH	Data Signal Rate Selector (DTE Source)
SI	Signaling Rate Indicator	CI	Data Signal Rate Selector (DCE Source)
SSD	Secondary Send Data	SBA	Secondary Transmitted Data
SRD	Secondary Receive Data	SBB	Secondary Received Data
SRS	Secondary Request To Send	SCA	Secondary Request To Send
SCS	Secondary Clear To Send	SCB	Secondary Clear To Send
SRR	Secondary Receiver Ready	SCF	Secondary Received Line Signal Detector
LL	Local Loopback	—	—
RL	Remote Loopback	—	—
TM	Test Mode	—	—
SS	Secondary Standby	—	—
SB	Standby Indicator	—	—

Notes On Interconnection Between Interface Circuits Using RS-449 And RS-232-C, published by the EIA.

CONCLUSION

In this chapter we have explored aspects of interfacing among computers and terminals that may be remotely located, and between computers and process and machine control and status signals of both the digital and analog varieties—interfacing at a relatively high level in the organization of systems. In prior chapters, we have seen how the computer and its memory are organized—system components at a relatively low level. These explorations have looked out over vistas of broad scope and also have microscopically examined the local flora and fauna, as it were. We have ranged from computing structures that are on the atomic scale to computing structures that may be intercontinental (and, perhaps someday, interplanetary) in scale. An intermediate level we have not examined closely is the interconnections within the computer itself. We return to this level in the next chapter, and find it multidimensional as well.

CHAPTER 11

Bus Systems

"Observe how system into system runs,
What other planets circle other suns."

Alexander Pope
1688–1744

Busses are shared mechanisms for transportation. In the case of transit authorities, they transport people. In the case of computers, they transport information and power.

In earlier chapters, we examined some characteristics of busses, in passing, as we discussed other subjects. In Chapter 10, we saw busses at a high level: as interconnection systems between widely dispersed computers and terminals in LANs. In that same chapter, and in Chapter 8, we also saw busses at deep levels: busses within integrated circuits and within subsystems on individual printed wiring boards.

Relationships between bus levels can be visualized as in Fig. 11-1, nested within each other. To observe the lowest levels, a powerful microscope is needed. To observe the highest levels, one must stand on a high far place, since LANs can cover many square miles of area. In this chapter, we focus on bus systems used to transport information and power among subsystems at intermediate levels: generally, between printed wiring boards.

Among the outcomes of the increased integration achieved in modern semiconductors is the appearance of "intelligence" at deeper levels of this nested structure. Components are capable of greater autonomy and relatively independent action, though they must behave cooperatively to carry out processing functions re-

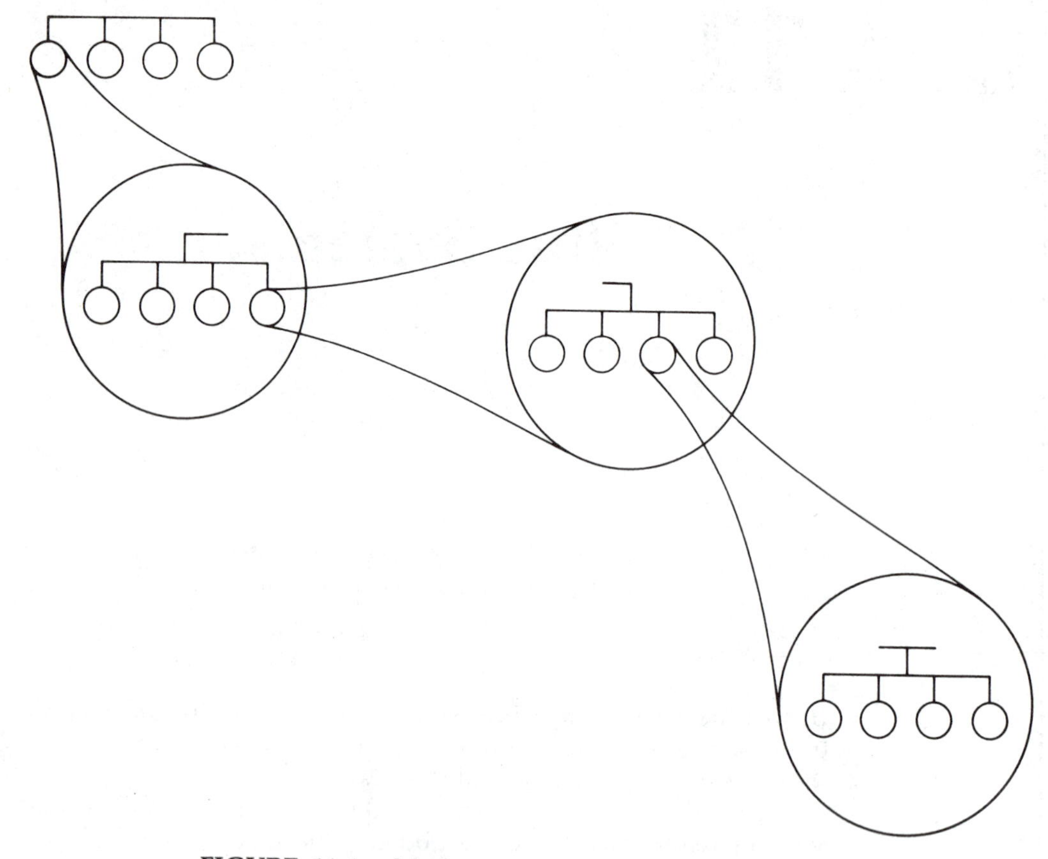

FIGURE 11-1. Modern microcomputer-based systems use busses at many levels, ranging from networks that are spread over miles to bussed elements within integrated circuits.

quiring the efforts of many components or subsystems working in concert. So we find systems running into systems, each with its own planets, each carrying out its mission.

At levels where information must be shared quickly among system components, a bus system is used. It is easy to see that, as each component becomes increasingly capable of fast autonomous action, information sharing mechanisms become more complex. The management of the MMMTA (Multiple Microcomputer Metropolitan Transit Authority), as it were, becomes more complicated.

In this chapter, we examine some of the considerations and general principles involved in bus systems, what they are like physically at this intermediate level, their logical structure, and the characteristics of a few modern bus systems.

GENERAL PRINCIPLES

Motivations

Why would one want a bus system in the first place? What is it that makes a bussed computer system better in some way? Why not just connect all of the various components and be done with it?

The first of several answers to these questions is "for flexibility." Computer systems are commonly subjected to their owners' desires for new features and capabilities, such as more memory and more input/output. Computer manufacturers have related needs. Their customers order systems in differing configurations. Manufacturers would like to build these without resorting to custom engineering and manufacturing efforts each time.

Flexibility can be achieved by designing the system so that component parts (i.e., printed wiring board subsystems) can be simply plugged in, so redesign or rewiring is not required for more memory and input/output. If the printed wiring boards can be treated like simple components, "plug-in" expandability can be obtained. By and large, bus systems have this characteristic. So long as the component printed wiring board is compatible with the bus system, it may be plugged into it virtually anywhere (although there are some exceptions, as we see later).

Microcomputer system design is easier when bus systems are used. The essence of a bus system is not so much its physical realization as its rules. These rules are carefully considered protocols for the transfer of information among subsystems. They specify timing and electrical loading. They define the location and purpose of pins on connectors between subsystems. They describe physical dimensions printed wiring boards. In short, they usually constitute a well-thought-out set of design rules.

Having to work within these rules might seem, at first, somewhat restrictive. While it is true that restrictions are sometimes inconvenient, it is also true that they often help to make design work easier because many issues have already been settled. It is easier to get started on a new design if the ground rules are already laid out. They provide a context, a standard, in which to work.

The nature of the current market for board-level products offers evidence that such standards encourage new designs. We find many small and large manufacturers developing board-level products that work with the bus systems of major microcomputer manufacturers which are either *de facto* or officially recognized standards. This offers users of these products many more choices of equipment than could possibly have been developed by a single manufacturer. Competition among these manufacturers has resulted in lower prices and has motivated manufacturers to develop enhanced products more rapidly.

Bus system design is not static. Although some bus systems have been adopted as IEEE standards, several new bus systems have been introduced in recent years. As a result, system designers still have many possible choices. If the characteristics of one bus system do not meet the objectives of a proposed system, the designer may consider others. The set of choices is sufficiently wide that it is unlikely that a suitable system cannot be found among them for virtually any industrial microcomputer application.

Needs for higher system performance levels drive bus system design evolution. One way to achieve higher performance is to use multiple processor systems. This requires logical elements for controlling processor synchronization and access to shared system resources, so that the processors can operate concurrently.

Bus systems also make it easier to incorporate higher performance components into existing systems as they become available. For example, as denser semiconductor memory devices become available, they can be designed into compatible board-level products which can be simply plugged into existing systems. As higher performance peripheral devices are developed, compatible controllers for them can be developed also, so that both main and peripheral memory can be easily expanded. This extends the useful life of older systems by making it possible for them to benefit from such improvements.

Logically well-organized systems tend to be more reliable than randomly organized systems. This is partly due to having a design context to work in. The designer with a well-defined context is less likely to make an error and more likely to develop a system with inherently better reliability.

Bussed systems are easier to maintain. Partitioning into subsystems, with standard interfaces between them, makes it easier to isolate the cause of system malfunctions. In bussed systems, it is often possible to isolate the cause of a malfunction to a single subsystem board using only exterior system behavioral symptoms. Less special test equipment is required. In multiprocessor systems, one processor can monitor the "health" of other processors, with diagnostic programs, and guide repair technicians. When the defective subsystem has been determined, repair is easily effected by simple board replacement. As a result, downtime of bus-based systems is usually much less.

Modularity and Integration

A bus system is also a vehicle for physically integrating the components of a microcomputer system. The mechanical aspects of a bus system specify the physical dimensions of its printed wiring boards. Some call for relatively large board areas, others for relatively small areas. These dimensions determine the degree to which the microcomputer system is *modular*.

It is very difficult to achieve optimum modularity because some of its aspects are self-contradictory. Consider the following situation. First, it is obviously desirable to have a complete system in a single integrated package, a single module. If it could be encapsulated in a single package small enough to fit in the palm, it would be stiff and light and might have superb mechanical reliability. It might be portable and could be installed almost anywhere. Of course, it would have to be priced right and manufacturable at a reasonable cost. Let us assume that there is a sufficiently large market and well-developed manufacturing technology that makes this possible.

Assume that we have been using this system for some time when we discover that we must expand it—more memory and input/output are needed. There is no way to expand a single encapsulated module. We must either find a way to connect it to an "expansion" module, or we must replace it entirely with a different module having greater capacity built in. The costs associated with either choice depend on the cost of the original module, the cost of the expansion modules, and the cost of a new "complete system" module of greater capacity.

If we need a certain amount of extra memory, and available expansion modules all contain twice what we need, we will be paying for more than we need. Expansion cost will be greater than it would be if expansion modules were more modular. If expansion modules contain less than we need, we must add several and pay the packaging and handling costs associated with a greater number of items. The ideal would be a module with *exactly* the amount of needed extra capacity. Unless the original module was very cheap indeed, there would be a substantial cost in replacing it entirely with a new complete system module of greater capacity.

This illustrates the paradox of cost/modularity, a very important issue in bus system mechanical packaging aspects. The economics of this issue are depicted in Fig. 11-2, which plots the cost of functional capacity increments for systems based on small printed wiring boards (solid line) and the cost for systems based on large printed wiring boards (dashed line). "Function" can be thought of as a certain increment in the amount of memory or input/output lines. For example, one small board might contain only 2K bytes of memory, along with bus interface logic, while a larger board might contain 16K bytes of memory. One small I/O board might contain two ports, while a larger board might contain eight ports. The smaller boards must support a larger overhead ratio in the costs of bus interface logic and connectors

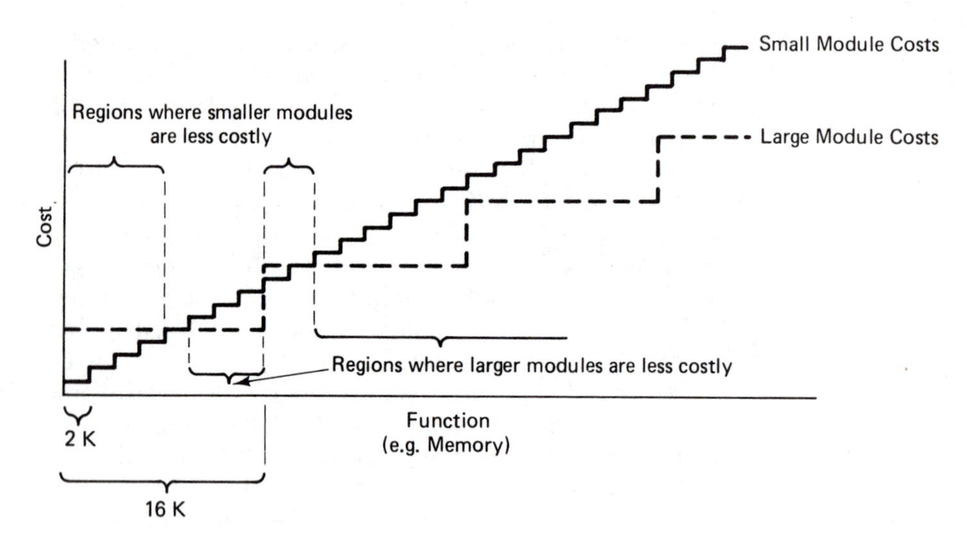

FIGURE 11-2. Costs of functional capacity increments for different degrees of modularity. Cost advantages of one over the other depend on the amount of function to be included.

(which can be surprisingly expensive compared to logical components), although they may cost less individually. The average slope of each cost plot is proportional to this overhead ratio, but actual costs increase in quantum steps because one cannot purchase part of a printed wiring board. It is all or nothing. Thus, the incremental cost of adding a certain amount of memory with small boards is smaller, but the total cost grows more rapidly than with larger boards.

It is clear from the figure that small boards enjoy an economic advantage for small systems, and larger boards have a corresponding advantage for systems that are expected to have rather more functional capacity. The most flexible packaging system would be one that permits boards of differing sizes. Some commercially available bus systems and proposed bus system standards provide for different board sizes within the same mechanical structure. In some cases, smaller functional increments are provided for by designing the individual subsystem boards to accept "add-on" modules.

Integration principles also have to do with the physical volume that a system requires. The smaller its dimensions, the wider its range of potential application. Physically large systems do not bear up as well as smaller systems in mechanically severe environments: "The bigger they are, the harder they fall." The mechanical ideal is a light, stiff structure, which is easier to achieve for a small system. Insects are good working examples of this principle. They are remarkably immune to shock and vibration and can go virtually anywhere that they are not subjected to the dangers of being crushed. Thus, one criterion for judging the mechanical aspects of a bus system is the degree to which the structures for holding the subsystem cards are light and stiff.

We end up with some conflicting objectives. Mechanical designers want to make small structures to take advantage of this principle. Computer designers want to cram a great deal of functional capacity into the system so that it will be very capable and powerful. Marketing people want to be able to offer reliable systems of great capability *and* modest incremental function cost. Users want to get what they need at reasonable cost but still have the latitude to change their minds (that is, flexibility).

Information, Energy, and Speed

Computers process information by moving it about in the form of energy. The output of a circuit provides information to other circuits by sinking or sourcing current. Current represents energy. This information-current is converted to thermal energy in resistive elements of the components and must be dissipated by conduction, convection, and radiation.

To move more information, the computer must use more energy. Using more energy creates more heat. Heat causes semiconductor failures and reduces reliability. Microcomputers use very little energy in proportion to the value we place on the information processing work they do. Nevertheless, the amount they use grows as they process information at greater and greater speeds.

For a fixed amount of input energy, the temperature rise of an object with a large surface area is less than that of an object with a smaller surface area. This is because the amount of thermal energy transferred via a surface is proportional to the product of its surface area and its temperature. Thus, there is a relationship between system size and potential operating speed. A larger system is potentially capable of faster operation because it can dissipate more thermal energy. An improved thermal design can be achieved with larger printed wiring boards and more separation between them, to promote convective dissipation.

But making systems larger entails the effects of a countervailing physical principle: the Special Theory of Relativity, which states that nothing (including energy) can travel faster than the speed of light. The farther apart the components of the system, the longer it takes for information-bearing energy signals to travel between them. This places an apparently fundamental physical limit on the information processing capability of a computer system.

The Special Theory of Relativity allows the hypothetical existence of particles which can travel faster than the speed of light. These particles have negative rest mass and are called "tachyons," but no physicist has tamed them yet, much less detected them experimentally and proved their existence, so it seems we must wait some time for the development of the "tachyonic" computer. When tachyonic computers are first introduced, it will probably turn out that they operate only in black holes, where there are as yet few operating industrial process plants to which they might be applied.

High signal speeds and long travel distances also run afoul of certain aspects of Maxwell's equations. This is one subject of the next section.

Noise

Noise interferes with the transmission of information. As we have seen in earlier chapters, a substantial effort is required to keep noise from corrupting measurements of process variables. This effort is usually manifested in isolation and filtering circuits in the signal conditioning sections of input subsystems. The design of signal conditioning for noise rejection is based on the designer's knowledge of potentially interfering signals. For example, if 60-Hz power line interference is known to be likely in the application, the designer may include notch filters tuned to reject this frequency. If noise is present at higher frequencies and the bandwidths of process variable signals are lower (as they usually are in a process control application), low pass filters are used.

Other types of noise, such as that inherent in system components and due to certain physical effects (e.g., current dumping noise), are addressed first by circuits designed to reduce its effects on their operation. Models for such noise are often stochastic, and circuit designs for its rejection are based, in part, on its assumed random nature. But noise can be more generally considered to be *any* unwanted information. This means we must consider the effects of other system components as they operate in concert. Their effects are not truly random, though they may appear so when viewed on an oscilloscope. Such noise is due primarily to reflections and cross talk. Reflection and cross-talk effects arise because wires and printed wiring board traces act as antennae and as transmission lines.

As antennae, they exchange interfering voltages and currents with neighboring wires and traces as signals are switched. These interfering signals are coupled between neighboring wires and traces by natural mutual capacitances and inductances. The more closely spaced the wires, the greater the effect. The wires become relatively efficient antennae for a particular frequency when their length exceeds one-quarter of the wavelength of that frequency. Therefore, as a bus system becomes larger (as the interconnecting wires or traces between subsystem boards are lengthened to accommodate more boards), the greater the potential becomes for cross-talk interference. Solutions to this include reducing coupling between signal carriers by providing greater separation or by providing intervening ground wires or a ground plane, in the form of a copper layer in the printed wiring board. These solutions add to the size and cost of the bus system.

As transmission lines, signal waveforms on these lines are affected by reflections caused by impedance mismatches along the wire. Impedance mismatches are discontinuities in transmission path impedances. The path includes traces and any connectors through which the signal must pass. They create boundary conditions which must be satisfied in the solution of the differential equations (Maxwell's equations) that describe the behavior of the fields and currents which represent the signal actually transmitted. The satisfaction of these boundary conditions results in reflected signals which propagate back and forth on the line following any change in its input signal. These reflections are superimposed on the applied signal and appear as interfering noise.

So long as the input signal rise time is long compared to the signal propagation time as it travels between impedance mismatches, the effect of reflections is negligible. When the signal rise time is much shorter than the travel time, substantial reflection effects appear. These can interfere with the operation of circuits on the line and can also be coupled into neighboring lines. Stubs on the line, in the form of subsystem boards and connectors plugged in along the bus, can also cause impedance mismatch. Thus, the longer the wires and the more subsystem boards installed in the system, the greater the potential for interfering reflection noise. As logic speeds are increased to provide faster system operation, signal rise times are shortened, and the importance of avoiding impedance mismatches becomes more critical.

One solution is to terminate bus lines with impedance-matching circuits. This is difficult for general-purpose system designs, since many different combinations of subsystem boards might be used in any given configuration. Each different combination could result in a system with different impedance mismatches and, therefore, different levels of performance.

In practice, a compromise is made in which line termination is provided for a reasonable range of configurations. Some control over subsystem board termination characteristics can be had by specifying them in the bus system standard. Usually, the result is a design that provides relatively good performance in terms of high-speed information transfer, with acceptable amounts of noise. But the bus system designer cannot directly control which combinations of boards are actually used. Users cannot be prevented from designing custom-boards that do not meet standards, nor can they be prevented from interconnecting several card frames to the extent that they end up with more subsystem boards than the system was designed to handle.

Reflection noise levels which are acceptable in a system that handles only digital data may be unacceptable in a system intended for precision measurements of analog process variable signals. For this reason, potential users of bussed board-level microcomputer systems must examine the noise characteristics of the system closely or must take steps in the design of analog input subsystems to protect measurements from the effects of this noise.

Synchronization

In Chapter 9, we talked of methods of dealing with electrical bus contention in memory subsystems. This is, of course, also an issue for the intermediate bus levels at which subsystems are interconnected. When a subsystem is deselected, it must "get off the bus" quickly so as not to interfere with succeeding transactions. Little more need be said about this type of contention.

Logical (as opposed to electrical) contention is a new issue at this bussing level. The earliest microprocessor systems were single CPU systems in which all I/O operations were of the "programmed," or non-DMA type. As high-speed peripheral memories and serial interface channels came to be used with these systems, it became necessary to equip them for DMA I/O operations. This meant that DMA logic had to gain access to memories in order to transfer data. To do this, the DMA had to control the bus on a temporary basis.

Direct memory access bus control was provided for by equipping the microprocessor with control signals by which the DMA could ask for, and be granted, access to the bus. A DMA with a byte to transfer would request bus access, wait for the microprocessor to grant it (the granting logic usually being provided by microprocessor logic and not a direct concern of the programmer), perform a transfer, and then release the bus to the microprocessor. The DMA could choose to transfer a complete block of data, holding the bus until the transfer was complete, or it could transfer one byte at a time, allowing the microprocessor back on the bus during intervening periods. This represented an arrangement of only slight and easily handled complexity.

Microcomputer systems have now evolved to the extent that multiple processor capability is coming to be common. Controlling and synchronizing access to the bus is much more complex when there may be several centers of intelligence. This has also become more than a hardware design issue, for now programs, operating in each processor, must intelligently and cooperatively share access to common resources. How can a program, which may be in the process of updating a table of data in shared memory, ensure that other programs, operating concurrently in different processors, will not read from this table and perhaps receive inconsistent data because they read it while it was changing? How can one program wait for another program in a different processor to complete some piece of work which must be done before it can proceed? How can it tell when the other program has finished the work?

In such systems, the bus becomes more than a vehicle for transferring data. It must also provide mechanisms for synchronization between programs operating in different processors. Few present-day commercially available bus systems provide totally adequate mechanisms for such complex systems. Nevertheless, older bus systems are being revised and new bus systems are being designed to accommodate these multiprocessor architectures because putting several processors to work concurrently is one way of obtaining higher system performance without incurring the effects of the physical limitations described earlier in this section.

Summary

In a sense, the bus system is a "matrix" that holds logical elements of the system and establishes the relationships that exist between them. Bus system designs are generally compromises between differing sets of objectives and countervailing physical principles which constrain the designer's ability simply to pack larger and larger amounts of faster and faster logic into flexible, reliable, and low-cost systems.

GENERAL ASPECTS OF BUS SYSTEM DESIGN

In the following sections, various aspects of bus system design realizations are discussed. These are to be found in different combinations in commercially available and custom designs. Because they have evolved in parallel with the microprocessors they support, some bus system designs retain the marks of their early history. Bus systems which have been around for some time have been incrementally modified and updated to support more modern requirements. Relatively new bus systems display mechanisms for modern applications incorporated more directly in their designs.

Thus, no existing bus system is a good example of all of the general design aspects with which the potential user should be familiar in order to evaluate their effectiveness for a particular purpose. Bus system design aspects are grouped into mechanical, electrical, structural, and logical aspects so that they can be addressed generally, without recourse to specific designs as examples.

MECHANICAL ASPECTS

The basic mechanical element of a bus system is its *backplane*. The backplane consists, in part, of a set of connectors into which subsystem printed wiring boards are inserted. The connectors must be held in a fixed mechanical relationship with one another, with sufficient space between them to allow for the height of components mounted on the printed wiring boards, and to provide an adequate path for convective thermal energy dissipation.

The connectors used on most backplanes are of the card edge type. The connector section of the printed wiring board is shaped to fit into connectors of this type, as illustrated in Fig. 11-3. Fingers are

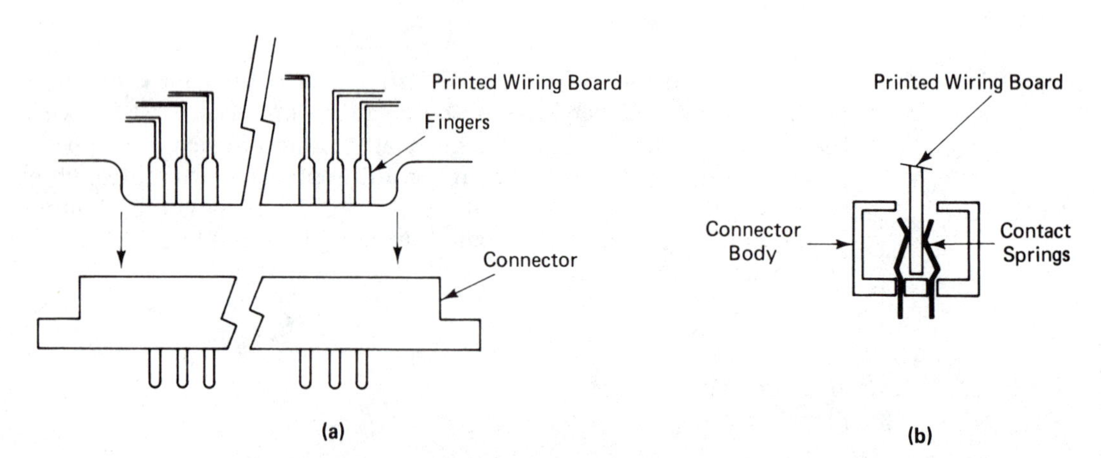

FIGURE 11-3. Card edge connectors typical of those used in bus backplanes: (a) printed wiring board edge connector section and plated fingers, (b) cutaway view of connector body showing contact springs and pins.

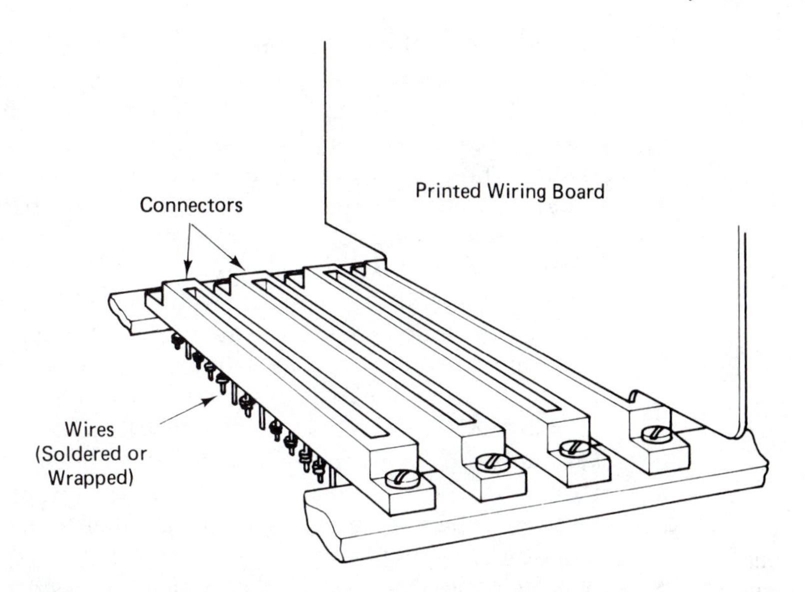

FIGURE 11-4. Wired backplane. Sometimes used for custom designs and systems that use small printed wiring boards.

plated onto the connection points in this section, generally with gold in the better quality designs. When the card is inserted in the connector, these fingers contact connecting springs installed in the connector body. The springs are connected to pins extending from the bottom of the connector body.

Figure 11-3(a) shows the shape of the printed wiring board connector section and the plated fingers. Figure 11-3(b) shows a typical interior connecting spring and pin arrangement in a cutaway end view. There are two common types of center-to-center spacing for card edge fingers: 0.100 inch and 0.156 inch. Some bus systems use pin and receptacle connector types, but these are more generally found in custom and in European designs. Although not depicted explicitly in these backplane illustrations, some use more than one connector for each subsystem board.

Conductive paths for signals and power supply wiring must be provided between these connectors. There are several means of accomplishing this.

Figure 11-4 illustrates a backplane in which bus signal and power wire connections are accomplished by individual wires which may be soldered or wire-wrapped to the connector pins. A mechanical fixture holds individual connectors in place. This may be mechanically independent or a part of the card-holding frame arrangement. (Card frame arrangements are discussed later in this section.) This fixture could be made from aluminum or steel channel, bolted together in a rectangular arrangement, punched bent metal, or formed steel wire with tabs for screws to hold individual connectors in place.

Soldered wiring is less likely to be found in a typical system than is wire-wrapped wiring. Backplanes that use individual wires are more usually found in custom designs or in systems that use relatively small printed wiring boards. Because relatively small boards have less card edge space for fingers, it is sometimes difficult to design them for completely parallel wiring. By using individual wires, designers can assign signals to card edge fingers more freely. This has the benefit of permitting more logic to be included on the printed wiring board, but it has the drawback that the backplane is no longer entirely "homogeneous." That is, complete freedom in the location of subsystem boards is given up. Individual connectors must be assigned to particular types of boards. For example, one group of connectors might be wired to accept memory cards, another to accept CPU cards, and still others for I/O cards of various types. As a result, these are somewhat less flexible than backplanes with strictly parallel wiring and are more often found in systems of relatively modest capacity dedicated to fixed applications in which changes are unlikely.

Figure 11-5 shows a more common backplane type. Here, the basic fixture is a printed wiring board into which connectors are soldered. The board generally provides mounting holes for attaching card

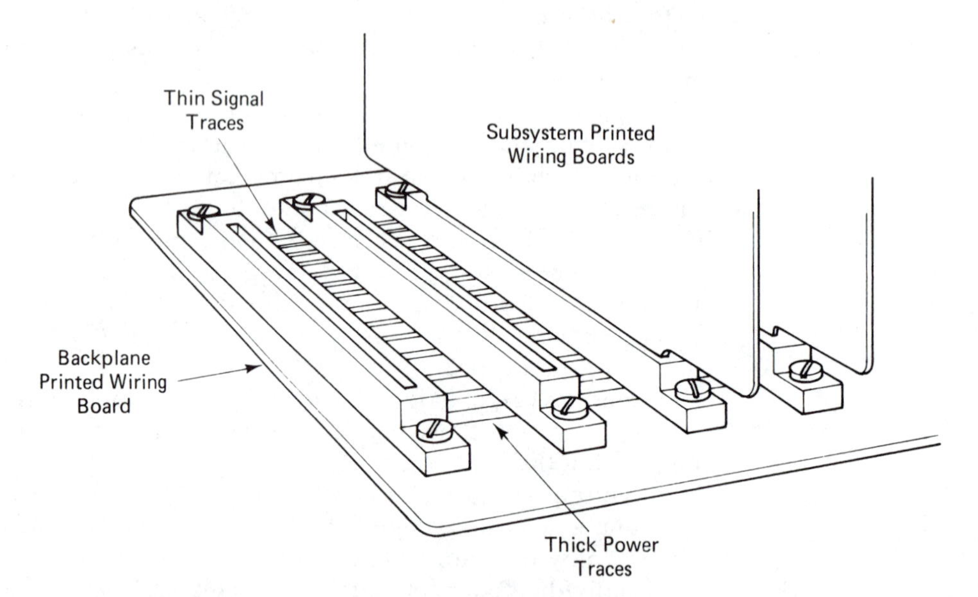

FIGURE 11-5. Printed wiring board backplane. This is the most common type for most systems based on board-level products.

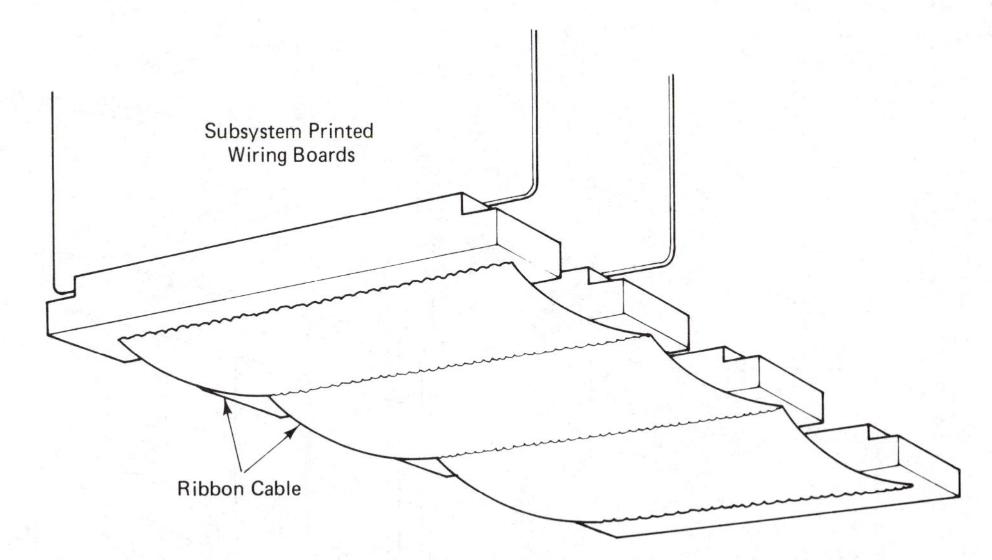

FIGURE 11-6. Ribbon cable backplane. Sometimes used for small systems and unusual enclosure interiors.

frame structures and for mounting the complete mechanical assembly on rack-mountable enclosure base plates or interior panels of NEMA-grade enclosures. Copper traces on the backplane provide electrical connections for signal and power line busses between connectors. Relatively thin traces are used for signal lines. Broader traces are used for power and ground lines that must carry larger currents. For systems that may use large amounts of current, copper bus bars may be mounted on the board. These backplanes are usually made from thicker glass-epoxy laminate material than is used for individual subsystem boards, so they can better withstand the forces involved in board insertion and removal. Impedance matching termination networks are also sometimes mounted directly on the backplane.

Figure 11-6 illustrates a third variation sometimes found in relatively small systems that can use parallel wiring and require only moderate amounts of power current. Here, ribbon cable provides electrical connections between boards. When manufactured in reasonable quantities, this approach provides a relatively inexpensive, but flexible, interconnection arrangement. This is advantageous for fitting small systems into unusual enclosure interiors, since the ribbon cable can be twisted to allow more freedom in card placement.

Figure 11-7 shows a complete bus system mechanical structure and further aspects of printed wiring backplanes. Backplanes must also provide connection points for power supply wiring. This may be

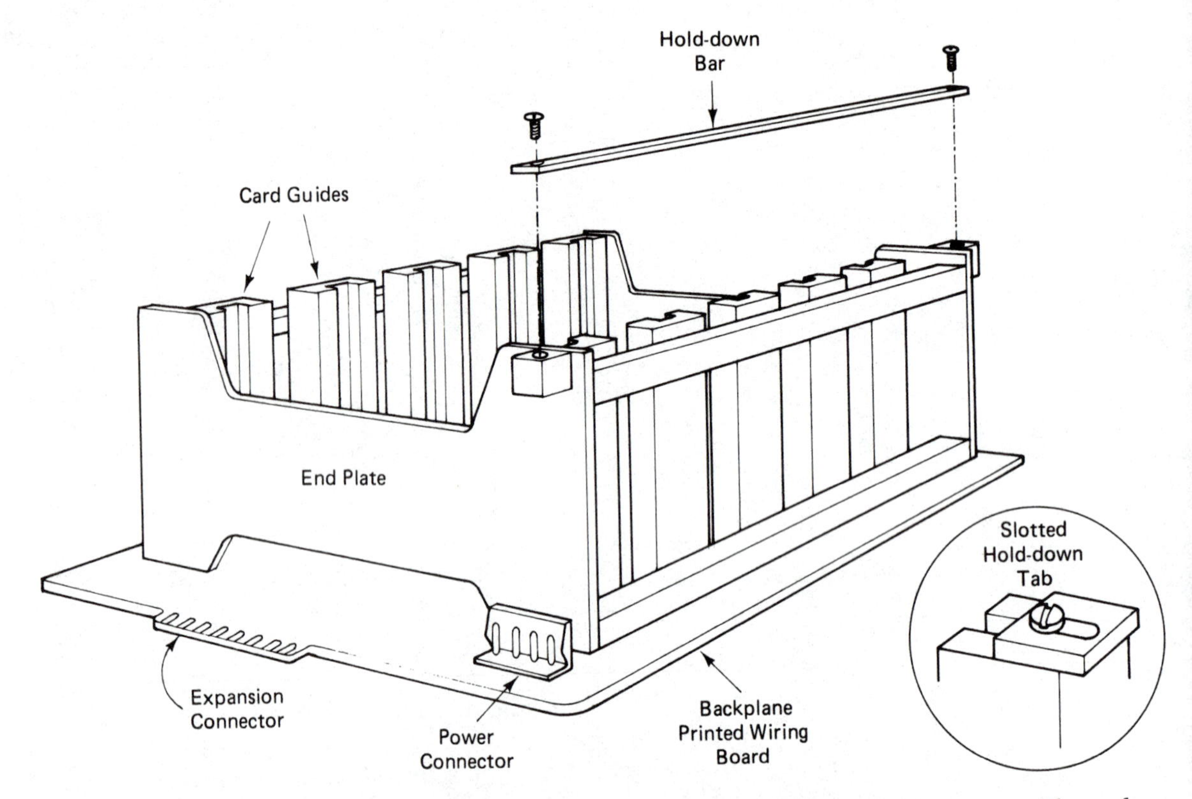

FIGURE 11-7. Complete bus system mechanical structure, with card frame, card guides, and hold-down mechanisms.

done with plug-in connectors that are adaptable to individual wires of relatively large gauge or with screw-terminal strips soldered into the printed wiring board.

Modularity is also an issue with backplanes. Therefore, some provision is usually made for expanding the card capacity of the backplane system. This can be done in several ways. The figure depicts a card edge connector section on the backplane itself. This would be inserted into a receptacle connector attached to an expansion backplane. With a receptacle at one end, and card edge connector fingers at the other end, backplanes could be connected end to end to form backplanes of greater capacity. Ribbon cables can also be used to interconnect backplanes, but in systems of that potential capacity, they may be unable to carry the required power supply currents. In that case, separate power supply wiring might be used between the connected backplanes.

A *card frame* superstructure is also required to align and hold subsystem boards if they are of moderate to large dimensions, especially if the assembly may be subjected to shock or vibration. Generally, this is an open boxlike structure formed from side bars or rods attached to end plates. Slotted *card guides* for the printed wiring boards are fastened to the side bars. These provide alignment when boards are inserted and removed and they also support the boards against shock and vibration in the horizontal plane. In some designs, a single plastic structure is used, with integrally molded card guides. This structure is fastened to the backplane with screws or bolts.

Card frame structures can also provide mechanisms to help ensure that printed wiring boards are not shaken loose if the assembly is subjected to particularly severe shock or vibration. These are often hold-down bars which are screwed to the end plates above the cards after they have been inserted in the backplane connectors, as depicted in the main part of Fig. 11-7. An alternate method (shown in the inset in the figure) is to use slotted hold-down tabs. These have the advantage of permitting easier insertion and removal of a single card but are not as secure as the heavier bars that can be used with the other method.

ELECTRICAL ASPECTS

A block diagram of the general electrical organization of a bus system is shown in Fig. 11-8. Almost all signal wires or printed wiring board traces are parallel in nature, running down the length of the backplane to each subsystem board connector. Some signals, however, run only between adjacent pairs of connectors and do not continue on to subsequent connectors. These are used for "daisy-chained" signals used to arbitrate control over access to the bus. How they work is described in a later section of this chapter.

Power bussing always provides for the +5-volt power supply level used by the most common logic families from which these systems are made. In addition, bus wiring is usually provided to accommodate other higher-voltage positive and negative supply levels that may be needed. Typically, these are for +12- and −12-volt supplies needed to power certain types of semiconductor memories. In some analog subsystem boards, the +15- and −15-volt levels needed for analog circuits are developed from these by using on-board dc-dc converters. Some older memories and microprocessors also required −5-volt supply levels. Busses that date from their era usually include power bus traces assigned to this level, as well. Some power bussing

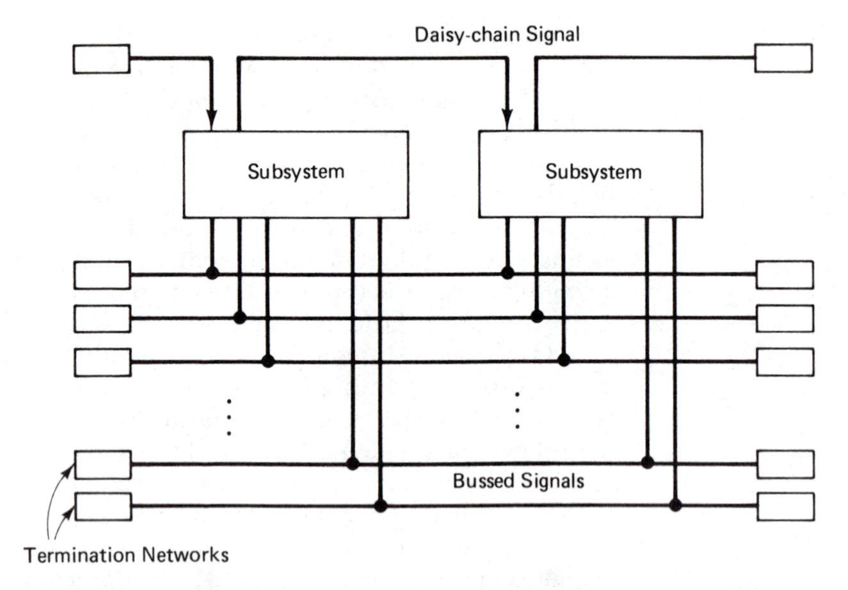

FIGURE 11-8. General electrical organization of a bus system.

schemes provide only one or two power busses for moderately higher unregulated voltages (e.g., 8 or 15 volts). Particular levels needed by each subsystem are then developed on the board itself by regulator and dc-dc converter circuits. Other variations provide uncommitted power traces to be used as required by the application. Typical uses of these are to deliver well-isolated power to analog subsystems used for measurements with particularly stringent accuracy requirements.

Termination networks may be found on one or both ends of the signal wires, depending on the approach taken by the designer. In some designs, termination networks are placed on subsystem boards rather than on the backplane. This is sometimes found on busses originally designed for single processor non-DMA applications. In these, the CPU subsystem board is placed at one end of the bus, and the termination network board is placed at the other end. In small configurations of such systems, when used with relatively slow logic, termination networks could be left out without noticeably upsetting performance. This makes it possible to reduce the cost of small configurations by simply omitting the termination board.

Termination network circuits vary from simple passive pull-up resistor arrangements to complex combinations of resistors, capacitors, and diodes that are actively switched on and off the bus wires, depending on the particular state of the bus operation at the moment. The more complex active types are not found in microcomputer bus systems, being more commonly applied to bussing systems for very high-speed minicomputers and large computers. Microcomputer bus termination circuits tend to be relatively simple. Several common microcomputer bus termination circuits are shown in Fig. 11-9.

Simple passive pull-up resistor arrangements [shown in part (a) of the figure] are quite common. They help ensure that minimum logic high levels will be maintained, and that signal rise times will be satisfactory, because they supply current to help charge the capacitance of the bus. But they do little to suppress reflection noise. The next stage of circuit complexity may include a forward-biased diode in series with the supply voltage and the pull-up resistor, as shown in part (b) of the figure. The diode serves to clamp reflection noise pulses that might otherwise momentarily exceed the supply voltage. The di-

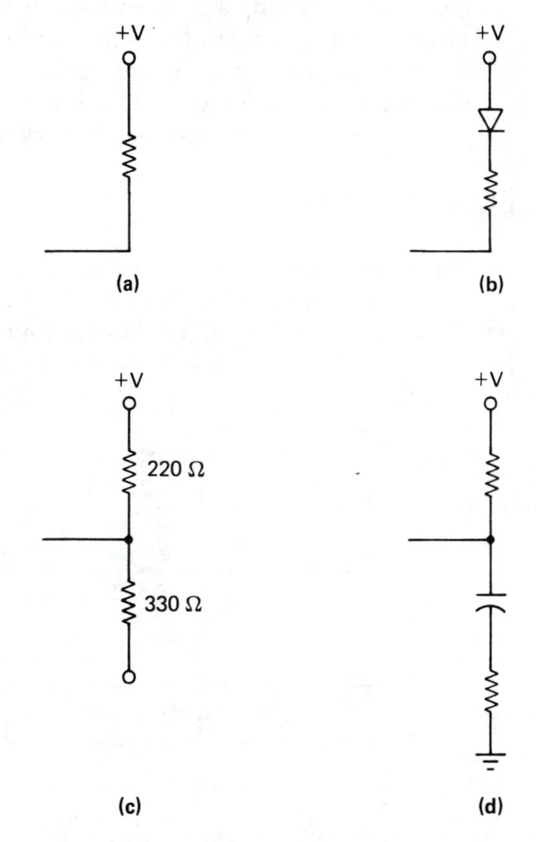

FIGURE 11-9. Common bus termination circuits: (a) passive pull-up resistor, (b) addition of a forward-biased clamp diode, (c) impedance-matching divider network, (d) addition of a filter capacitor.

vider network shown in part (c) of the figure is also a common termination circuit that can provide good impedance matching. Finally, part (d) of the figure illustrates a variation in which a capacitor between the signal line and logic ground serves to act as a low-pass filter to suppress bus line noise.

ASPECTS OF LOGICAL STRUCTURE

The logical organization of a bus system is perhaps the most important contributor to its performance. when faced with technical speed and performance limits, the designer can try an end run by putting more processors to work on the job, if the bus system will accommodate them. For some applications, this may add substantial complexity to hardware and software, even though it equips the system with more raw computing power. For other applications, it may actually simplify the work to be done. In this section, we examine the basic logical organization of microcomputer bus systems and ways in which they are designed to permit several microcomputers to work concurrently.

Time-shared Common Bus

The *time-shared common bus* is the typical logical structure for modern microcomputer bus systems. Its operation is illustrated in Fig. 11-10. Part (a) of the figure depicts four separate subsystems, all connected to the same bus. Each may be a microcomputer, a memory, or an I/O subsystem.

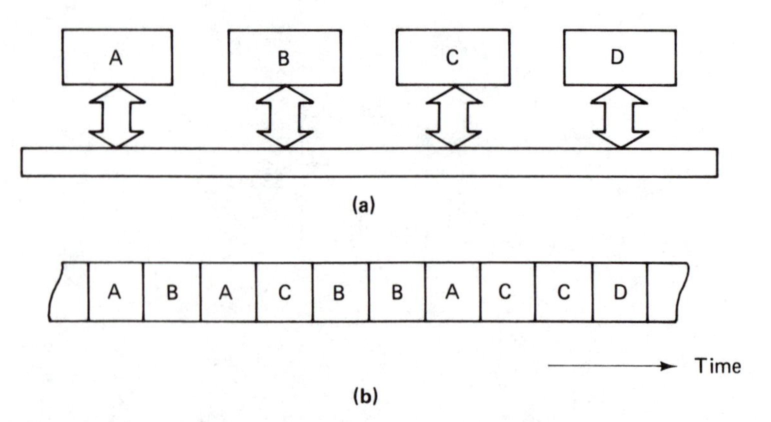

FIGURE 11-10. Most microcomputer systems use a time-shared common bus: (a) subsystems are identically connected to the bus, (b) subsystems get bus access during individual time slots.

Each subsystem is allowed access to the bus during an individual time slot, as depicted in part (b) of the figure. Which system gets access in which slot depends on what went on before. Events that can determine access include requests by one subsystem, for data from another subsystem, and the results of processing operations in any subsystems which happen to include an "intelligence center" (i.e., a microprocessor).

Subsystem A, which may be a microcomputer, may request data from subsystem B during the first slot. Subsystem B delivers the data during the next slot. Subsystem A might then read a byte of data from an input device controlled by subsystem C. Then two consecutive slots might be allotted to subsystem B to deliver two bytes of data to an output device as part of a DMA transfer. Bus transactions proceed in this fashion, each subsystem being given an opportunity to use the bus for a moment of time, then being required to give it up to some other subsystem, if there is another subsystem waiting to use it.

The first question which comes to mind is, "Who is in charge here?" How is the order of bus access decided? What if one subsystem is so garrulous that no other subsystem can get an opportunity to use the bus? The answers to these, and related questions, will be developed in this section.

One of the first considerations is the likelihood that one subsystem would be inclined to monopolize the bus. If there is only a single processor in the system, it would clearly be in charge. It would not have to compete with other processors for access to memory or I/O devices. It would have the bus all to itself, under its complete control. If it wished to give up the bus for a period of time for DMA transfers, it could do so or not, depending on the circumstances. This was the case with bus systems for many years, before multiprocessor systems came into use. When more than one processor must use the bus, the situation is quite different. If a processor needs access to the bus, it must wait for others to finish using the bus. How long it must wait, and how often it must wait, depends on the number of processors, of course, but also very much on the extent to which it and the other processors need access to the bus. The need for frequent access is a function of the degree to which the bus system is "coupled," a subject discussed in a moment.

First, it should be noted that there are other ways in which multiple processor systems may be connected. These may include virtually any of the topologies discussed in previous chapters. Hierarchical arrangements, stars, and other directly connected systems are sometimes found. In such systems, bus access control is sometimes vested in special bus controller hardware that legislates which processor can gain access. In some directly connected systems, interconnections are arranged in the form of a "cross-bar switch," a matrix of interconnecting busses which, when properly switched, provide direct independent connection paths between any two subsystems.

These arrangements are rarely found among microcomputer systems except, perhaps, as custom-designed systems for a special purpose, such as a high-performance communication system controller. They are more commonly found among special configurations of minicomputers which link other computers in complex communication systems. Because these are relatively special arrangements, we shall not consider them further.

Coupling

The extent to which processors need frequent access to the bus is related to the degree to which the bus is *coupled*. Microcomputer bus systems are most often of the *tightly coupled* variety. This means that each processor in the system has a relatively high degree of access to the resources it shares with other processors. Generally, by "resources" we mean memory and I/O device controllers. Memory and I/O subsystems on the bus are all generally available to any processor that wishes to address them. "Resources" can also mean other processors that do special work for a master processor. There must be a close degree of control and synchronization over accesses and utilization on a tightly coupled bus. In a tightly coupled bus, access is permitted only at regular times and for relatively restricted periods.

The concept of tight coupling is further illustrated by comparison to its opposite: *loose coupling*. Local Area Networks are examples of loosely coupled systems. Each node has total control over its resources and can gain access to resources in other nodes only by means of messages, exchanged according to the rules of the communication protocols. In loosely coupled systems, access timing is less regular, and nodes may use the bus for extended periods (subject, of course, to rules of protocol and courtesy to other nodes).

Single-processor systems are tightly coupled, essentially by definition. When the system includes only a single processor, it enjoys complete access to memory and input/output subsystems on its bus and is, therefore, as tightly coupled as a system can be. There would be no point in forcing it to go through an elaborate access protocol in order to get data from one of its I/O devices or memory subsystems. When more processors are added to the system, decisions must be made about how tightly the system must be coupled.

The degree to which a system should be tightly coupled depends on the latency that can be permitted in access to shared resources. If a subsystem can wait for a reasonably moderate time to gain access to a resource, then a more loosely coupled bus system may be a possible choice. As a multiple processor system becomes more widely dispersed physically, looser coupling becomes necessary. Widely dispersed multiple processor systems must be loosely coupled for physical reasons: Tightly coupled access cannot be accomplished over large physical distances because of physical limits on the speed with which bus transactions can be handled. Parallel wiring over great distances would be costly and susceptible to interference. Thus, serial interfaces are most often used as the interconnecting medium in loosely coupled bus systems.

MODERN MULTIPLE PROCESSOR SYSTEM ASPECTS

High-performance multiprocessor systems should not be very tightly coupled. When they are as tightly coupled as possible (that is, when every resource needed by a processor is a shared resource), system performance suffers. This can be seen by considering memory subsystems in which programs are stored. If all programs are stored in a single shared memory, each processor must wait for bus access in order to fetch its next instruction. The net instruction execution speed of each processor is slowed in a way that is roughly proportional to the number of processors.

For this reason, most modern multiple microcomputer systems use an arrangement in which there is a combination of shared and private resources, distributed among the microcomputers in such a way as to provide the levels of performance required by the application. Depending upon the transactions going on at any given moment, they can be considered to be somewhere between being tightly coupled and loosely coupled.

The general technique of subsystem design used to achieve this situation is shown in Fig. 11-11. Shared memory and I/O subsystems remain from the tightly coupled configuration, but each CPU is also equipped with its own private memory and, if necessary, private input/output as well. The use of *dual-port memory* is an important characteristic of this arrangement.

The dual-port memory is a memory subsystem on the same printed wiring board with each CPU and equipped with two access ports. One gives the CPU private access to the dual-port memory via the internal subsystem bus on the board. The other port provides access to the dual-port memory from the bus system. Thus, the memory can be accessed from within the processor subsystem *and* from outside the processor subsystem.

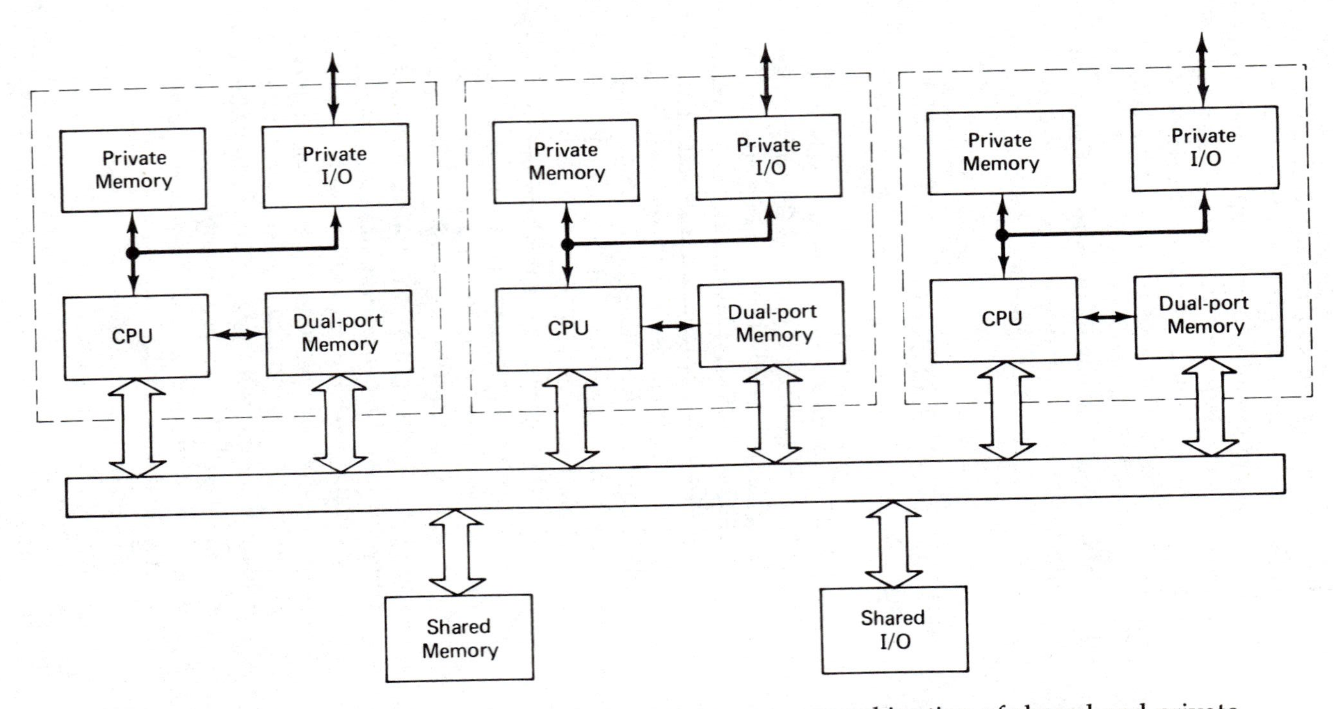

FIGURE 11-11. Most multiple microcomputer systems use a combination of shared and private resources. Dual-port memories provide for some sharing of otherwise private memory.

Programs executed by the CPU are kept in its private memory. Thus, no delays are incurred in retrieving instructions for execution. Similarly, the CPU need not wait for other processors to get out of the way for operations on its private I/O devices. Each processor may operate at full speed so long as it does not need access to any of the shared memory or shared I/O devices.

Of course, it is necessary for processors to communicate with other processors so that their work can be coordinated. This involves, among other things, the exchange of data. For this, either the shared memory or the dual-port memory can be used. Messages for other processors may be left in either place to be picked up. In a way, this is like leaving messages in someone's mailbox. In fact, "mailbox" is the name used for this interprocessor communication technique.

Generally, a CPU can expect relatively free access to its dual-port memory, since it would be tied up only when some other processor was accessing it explicitly. Simple use of the bus by other processors in addressing other subsystems has no effect on the CPU's access to its own dual-port memory. This makes the dual-port memory the preferred place to leave messages for other processors with which it wishes to communicate. In a given application, one processor tends to communicate only with a subset of the other processors. Other processors can check their mailboxes in its dual-port memory for any messages from time to time. It is also possible for the CPU to signal other processors whenever it leaves new messages for them. This can be done with interprocessor interrupts, as we see later in this chapter.

Bus Signals

The conductive paths of a bus system are divided, as depicted in Fig. 11-12, into four groups: address, data, control, and power. We discussed the power group in previous sections, so we pay it no further attention. Logical operation of the bus and its ability to provide access to certain amounts of memory and numbers of input/output devices, and to support DMA and multiple processor operations, depends on the way the address, data, and control groups work. It is to these that we direct our attention in this section.

Address Bus

The number of bits in the address bus determines the number of unique locations that can be addressed over the bus at any time. This does not necessarily equal the maximum capacity of the system for memory and I/O devices. As we saw in Chapter 9, for example, there

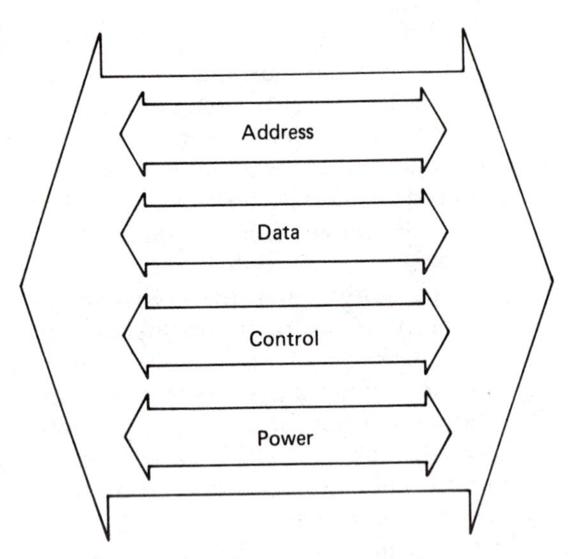

FIGURE 11-12. Bus system signals are divided into four main groups.

are methods that effectively equip the system with more memory than is directly addressable by the address bus. Because there is a certain processing overhead associated with these methods, it is best if the bus is equipped to handle a reasonable amount of direct addressability.

Until recently, most microcomputer bus systems were equipped with a 16-bit address bus, the same size as the microprocessor address bus. Since modern 16- and 32-bit microprocessors have greater addressing abilities, more modern bus systems now provide more address bus bits. Twenty-bit address busses are now common. This provides direct addressing of 1,048,576 locations (or 1M bytes). Some provide 24 bits, addressing 16M bytes.

These locations need not be the same physical addresses as in the private memories of CPU subsystem boards nor the same as those used to access dual-port memories from within the subsystem. Addresses generally may be with respect to separate base values set by switches on the subsystem boards, and accesses from the bus may be with respect to different base values. Thus, the actual amount of memory in a multiple processor system may be more than can be addressed by the address bus of the bus system.

Data Bus

Historically, the data bus in microcomputer bus systems has been eight bits wide, transferring data one byte at a time. In newer designs, provision has been added to transfer pairs of bytes, forming 16-bit words, so that 16- and 32-bit processors can read and write with fewer bus cycles.

In a multiple processor system, it is possible for 8-, 16-, and 32-bit microcomputers to be present simultaneously. There is a large number of memory subsystem designs of proven reliability and low cost that are capable of transferring only single bytes at each access. Thus, there is significant interest in bus system designs that permit mixing of processor types and the use of older, narrower memories.

To accommodate different data path widths, modern bus systems are arranged so that each CPU board can specify whether an 8- or 16-bit data item is to be transferred in each bus cycle. This is illustrated in Fig. 11-13 and operates generally as follows. Addresses are uniformly considered to refer to locations on a byte-basis. That is, each address refers to a single byte. To access a 16-bit data item, the processor uses an even-numbered address and activates a control signal that tells the addressed subsystem that a 16-bit access is taking place. The subsystem board then acts to access one byte at the even-numbered location and the next byte at the following odd-numbered lo-

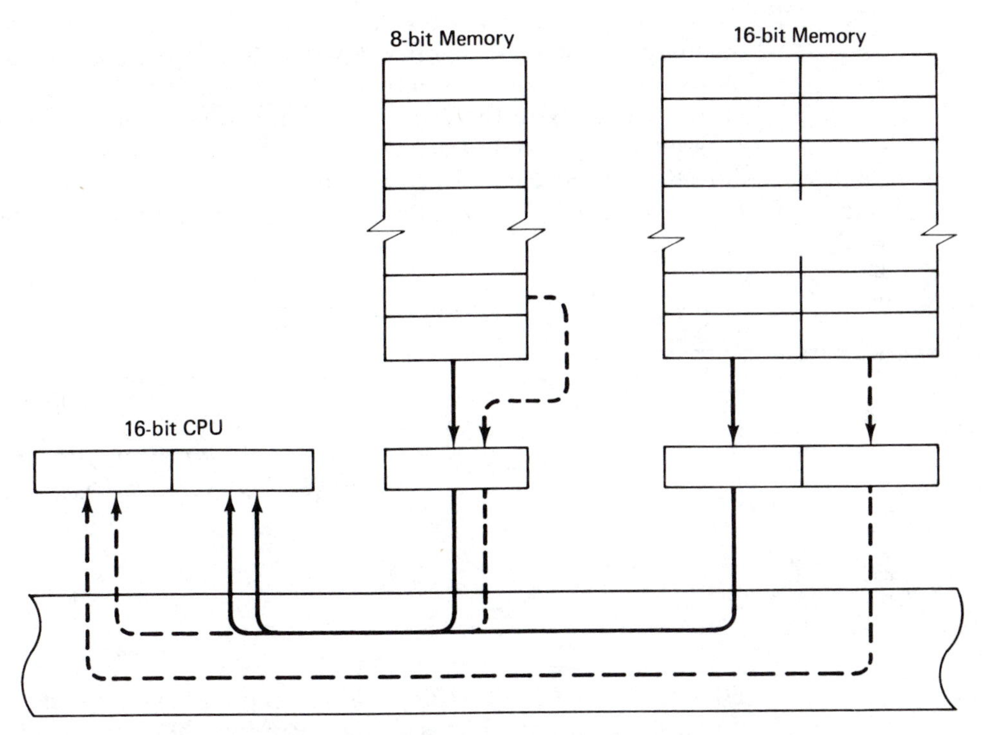

FIGURE 11-13. Bus accommodation for handling both byte and word data transfers from memory subsystems of different widths.

cation. The pair of bytes so accessed is then delivered to the processor, in the case of a read cycle, or written to from a 16-bit data item provided by the processor, in the case of a write cycle.

Most of the work of dealing with this is handled in instruction decoding logic in the microprocessor. The programmer has minimal responsibilities for ensuring that everything comes out right. Generally, this consists of not using odd-numbered addresses when 16-bit data items are retrieved. Because the logic is simpler if formal binary addition can be avoided, most implementations get the second byte by ORing a "1" into the least significant address bit. If this bit is set to begin with, the usual result is that two copies of the second byte are retrieved.

Multiplexing Address and Data. In some modern processors, the combined number of address and data bus bits needed has caught up with the number of signals available on the bus, just as it did with the number of pins which could be made available on the IC package of the microprocessor. Here, the solution is the same as was adopted for the busses most immediately surrounding the microprocessor: time-multiplexing on the same signal lines.

Two ways of doing this are illustrated in Fig. 11-14. In part (a) of the figure, the most significant part of the address is always available to the addressed subsystem, while the least significant part is placed on the bus early in the cycle to be latched or recognized by the addressed subsystem and replaced with the transferred data item at a later time in the cycle. In part (b) of the figure, the complete address is placed on the bus first and replaced with data later in the cycle.

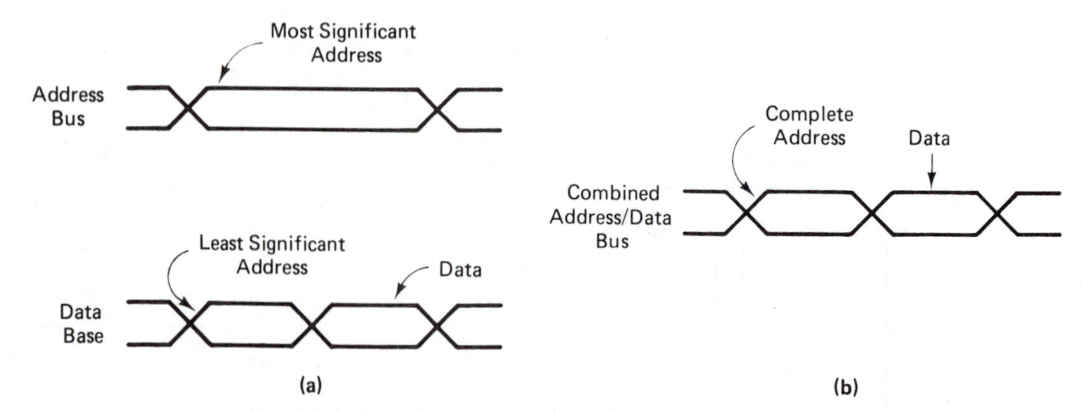

FIGURE 11-14. Multiplexing address and data on the same bus lines: (a) most significant address always available, (b) complete address followed by data.

The first method is common with larger microcomputer bus systems. The second method is found is some systems which have historically been based on 16-bit processors (e.g., the microprocessor versions of some minicomputers).

Control Signals

The group of control signals includes read and write control lines, signals which indicate when an address is valid (or what is on the bus in a time-multiplexed address/data bus system), interrupt request and response lines, control lines to manage DMA operations, and control lines used to arbitrate which processor is given control of the bus at any given time.

Generally, read/write and interrupt operations are the least complex bus functions. We discuss them first. Bus access control is somewhat more complex, and we must understand it in order to program multiple processor systems for concurrent operation. Because of its complexity, we treat it as a special topic in a separate section.

Read and Write. Once a processor has control of the bus (essentially all the time in a uniprocessor system), it may read from or write to any location that is addressable over the bus. Each bus system has specific rules for such a transaction. These include minimum times that signals must be held stable before and after the actual transfer of data and maximum times that they are allowed to remain after a transaction has taken place. Generally, these correspond to the steps involved in accessing a memory, and the considerations are essentially the same as those discussed in Chapter 9.

Although there are differences in details between each bus system, the protocols for bus system write and read operations fall into one of two general categories: synchronous and asynchronous.

General sequences for synchronous bus data transfers are shown in Fig. 11-15; part (a) depicts the sequence for a read, part (b) for a write. The essence of these transfers is that they are paced synchronously by a clock. The clock is available to all subsystems as a bus signal. This is useful in fast bus systems, but it can be done without in relatively slow bus systems.

The address is placed on the bus early, before any control signal is activated, so that subsystems may begin to decode it. It is sometimes accompanied by a control signal to indicate that the address is valid (i.e., stable) so that decoders are not confused by the slightly different times required for individual address bits to change to new states. At the next clock time, the control signal (read or write) is activated. Now the subsystem knows what direction the data are going in. In the case

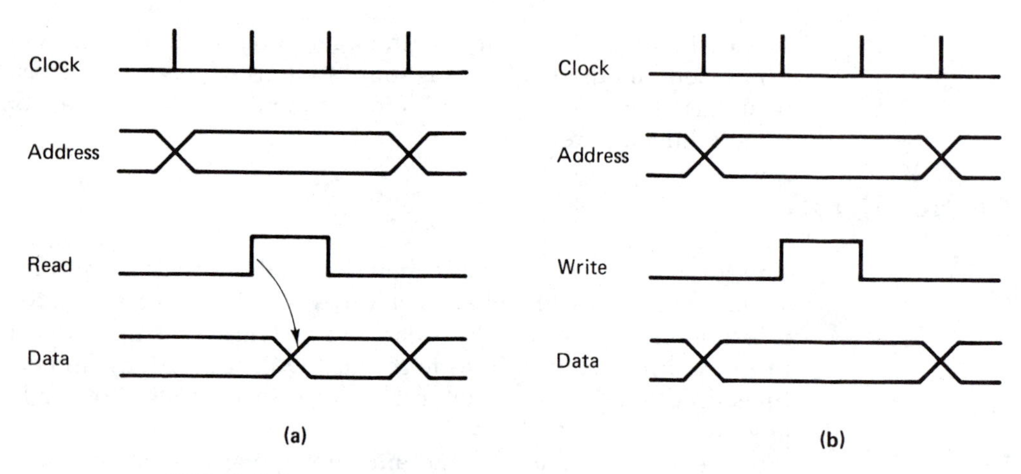

FIGURE 11-15. General sequence for synchronous bus data transfers: (a) reading, (b) writing.

of a read operation, the subsystem must retrieve the addressed item promptly and place it on the data bus. In the case of a write, data will have been placed on the data bus already, and the subsystem must transfer the item in to itself. The transfer sequence ends at the next clock time. If the subsystem has not provided the output data or accepted the input data, the transfer has not been successful.

Synchronous data transfer cycles can proceed very quickly, transferring data at speeds that are quite high. It is clear, however, that there is substantial opportunity for things to go wrong. A slow memory can deliver garbage as data. Data which the CPU may have written in good faith, and which it believes is safely ensconced in the memory at the proper location, may have been lost forever because the memory did not act quickly enough to catch it. Without logical mechanisms for interlocking one event of the sequence with the next, it is possible for the processor to push data "down the pipe" and lose it forever or to grab at "open air" and come up with nothing without knowing it.

For these reasons, synchronous transfer methods must be used judiciously. In some variations of this method, an initial stage is used in which readiness of the addressed subsystem and existence of the address is verified before the transfer cycle is started. Once this "link" is set up, data can be transferred very rapidly. But setting it up constitutes a heavy overhead for single byte transfers. It is also awkward for bus systems in which address/data multiplexing is necessary.

Asynchronous data transfer methods are more usual in microcomputer bus systems. General transfer sequences for these methods are illustrated in Fig. 11-16; part (a) shows the sequence for reading, part (b) for writing. Here, an additional control signal (often called "wait" or "ready") is used to interlock the actual transfer of the data. Again, the address is placed on the bus at the beginning, possibly with a validity signal and, in the case of address/data multiplexing systems, with a signal to indicate which kind of information is on the data bus. When it is writing, data are placed on the data bus soon after the address has been fully presented. Then the read or write control signal is activated.

At this point, there are two possible variations. In the first, it is assumed that the addressed subsystem is present and at least fast enough to know that it is being addressed but not necessarily fast enough to handle the actual transfer. If it is fast enough, it can simply accept or provide the data item corresponding to the addressed location. If it needs more time, it activates the "wait" signal, causing the processor to delay. In the case of a write, the processor leaves the data on the bus. In the case of a read, the processor defers reading. When the subsystem is ready, it deactivates the wait signal, and the remainder of the transfer sequence proceeds. In the other variation, a "ready" signal is used. The processor refrains from moving to the next step in the sequence until this signal indicates that the addressed subsystem is ready to proceed.

Because most modern semiconductor memory subsystems are fast enough to keep up with most processors and because access to I/O controllers is usually to fast registers within the controller, the "wait" variation is most efficient in terms of general simplicity. It is not necessary to equip fast subsystems with "wait" logic, and slow

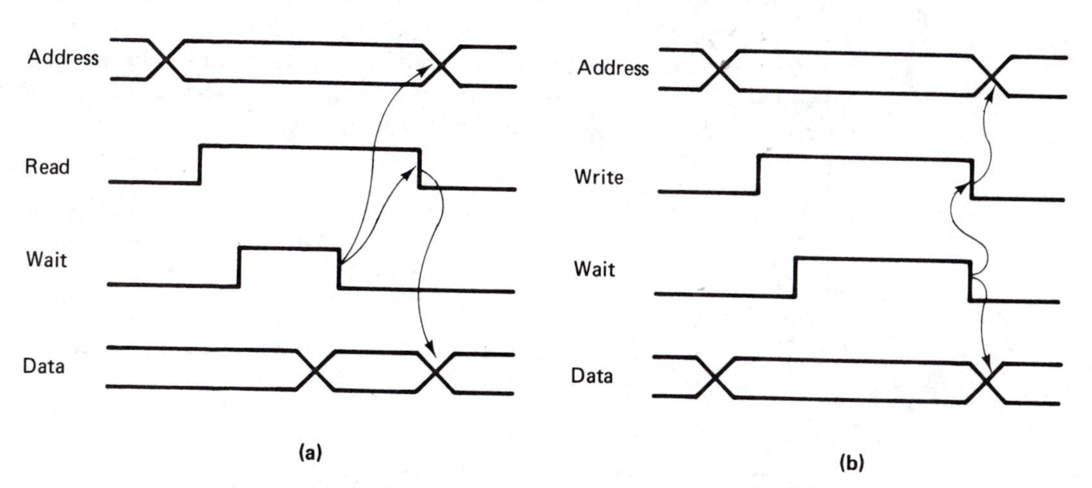

FIGURE 11-16. General sequence for asynchronous bus data transfers: (a) reading, (b) writing.

subsystems are easily equipped with it. Thus, it is commonly used in modern microcomputer bus systems.

Interrupts. Bus systems usually provide signal wires for multiple interrupt sources. That is, several separate interrupt request lines (typically eight) are provided in the control group of bus signals. The ways in which these are actually used depend on the types of microprocessors in the system and choices made by the system designer. The three principal types of interrupt protocols are "polled," "multiline," and "bus vectored." Combinations of these are also possible.

In a *polled interrupt* arrangement, each subsystem that may request an interrupt is connected to a single request line, as shown in Fig. 11-17. This line interrupts the processor whenever it is enabled and activated by one or more of the subsystems connected to it. Typically, it is arranged to serve as a "wired-OR" active-low signal which goes "active" if any subsystem pulls its interrupt request line low. To give the processor some control over when it is interrupted, the request must pass through enabling circuitry. This may be on the CPU subsystem board or in the microprocessor itself. By disabling interrupts, the processor can defer them until it is ready to accept them.

When enabled and activated by one or more subsystems, the signal interrupts the processor, and execution of an interrupt service subroutine begins, as was described in Chapter 7. Since more than one subsystem may have caused the interrupt, the subroutine must find out which was the cause. It does this by *polling* each subsystem, interrogating each in some sequence. Normally, when a subsystem requests an interrupt, it also sets a bit in one of its registers. By reading this register from each subsystem and checking the bit, the service subroutine can determine which caused the interrupt. The service subroutine must then ensure that action is taken to clear the interrupt, so that any other pending interrupts can be handled.

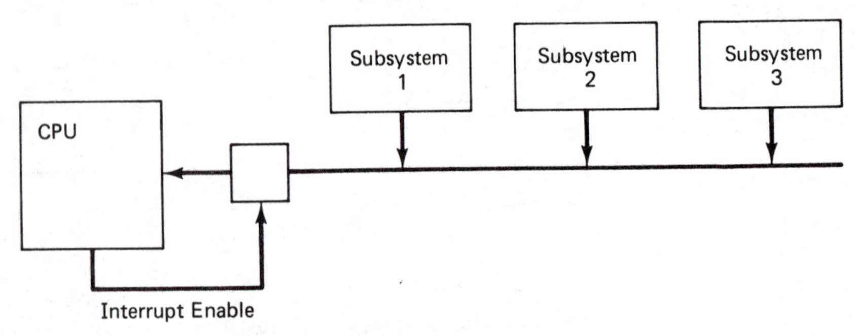

FIGURE 11-17. A polled interrupt arrangement.

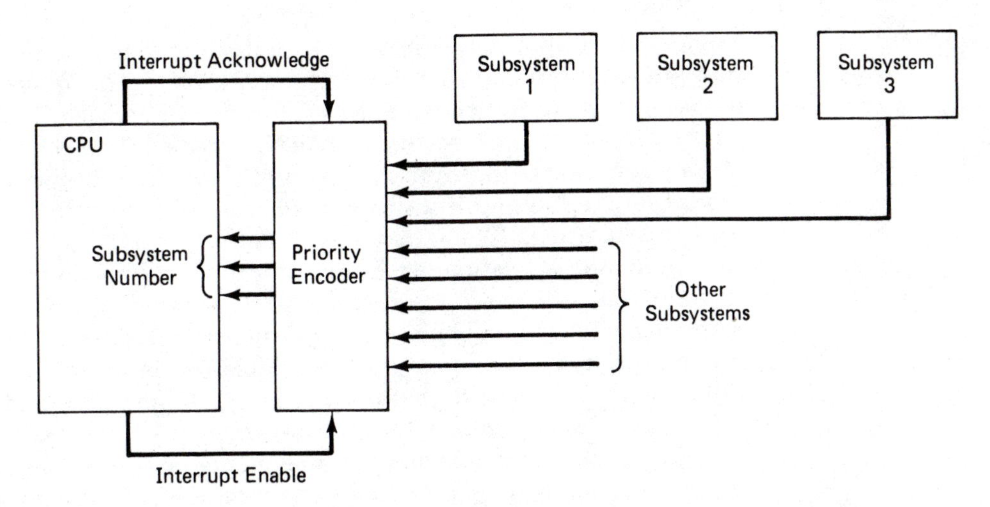

FIGURE 11-18. A multiline priority interrupt arrangement.

Since the order in which the subroutine polls subsystems determines which will be served first, priority can be dynamically modified. However, more processing time is required for polling activity. In a relatively busy system with many potential interrupt sources, this can take a substantial amount of time. However, very little hardware is needed for a polled interrupt arrangement. For this reason, it is often the preferred approach in relatively small system configurations.

Multiline priority interrupt arrangements are more commonly found in systems of intermediate scale. Their general form is shown in Fig. 11-18. Here, a single request line is allocated to each subsystem. These lines are inputs to priority encoding logic on the CPU subsystem board. The priority encoder is a common logic circuit which provides, as its output, the number of the currently active input of highest priority. If only one line is active, its number is presented as the encoder output. If more than one line is active, the number of the line having the highest priority is presented on the encoder output. Priority is determined by the encoder input pins to which the requests are wired. Priority order can be changed by rewiring these (a facility usually provided by sets of switches on some CPU subsystem printed wiring boards).

When an interrupt request is active, an interrupt signal is generated to the processor, along with the priority encoder output number. The processor uses the number to form the address of the interrupt service subroutine that is to respond to interrupts on that line. This number is thus a vector to the service subroutine. The microprocessor then interrupts any running program to that location in the program memory. Thus, polling is replaced by a direct entrance to the service subroutine dedicated to serving that interrupt request. In some cases, the priority encoding logic permits the processor to specify a priority level below which no interrupts are to be accepted. This permits high-priority requests to interrupt lower-priority service subroutines while protecting the low-priority subroutines from interrupts by still lower priority requests.

In another variation, each interrupt request line can be used as an independently polled arrangement. That is, all subsystems of a particular priority level have their interrupt requests "wire-ORed" to a single priority encoder input. This produces interrupts to service subroutines dedicated to that priority level. Polling then determines the order of service *within* that priority level.

In *bus vectored interrupt* arrangements, the subsystem that requested the interrupt is expected to provide a vector which identifies it as the source of the interrupt. Instead of generating the vector within the interrupt logic on the CPU board, it is transferred to the CPU from the interrupting subsystem via the data bus, as part of the interrupt request handling sequence. This arrangement provides the capability for literally hundreds of unique interrupt request/service routine combinations which may be dynamically altered depending on the circumstances in which the system finds itself.

Some systems also provide for *nonmaskable* interrupts. These interrupts bypass normal enabling and priority logic of the CPU subsystem and therefore constitute interrupts which are of the highest possible priority and which cannot be ignored. They are generally used for signaling such events as impending loss of power.

Bus Access Arbitration. Bus system evolution has followed a rather straightforward path to accommodate larger amounts of memory and I/O devices. Power traces are widened, to carry added current. The numbers of bits in address and data busses are increased directly or by multiplexing schemes, to provide a larger directly accessible address space. But provision for handling more than one processor is less straightforward. This affects both system hardware and system software.

To place multiple processor bus access in perspective, we can go back to consider the situation in which the system is equipped with a single CPU and an I/O device that uses DMA data transfers. In such a system, the CPU enjoyed essentially total control over the bus. Even the DMA operated at the pleasure of the CPU because the CPU initialized and started it for any transfers which used that function. This was essentially a single "master," single "slave" situation.

Now consider the situation in which there are two CPUs and a DMA-driven input/output controller in the system. This is a possible configuration for a Numerical Control system responsible for operating two identical machine tools working from a common "family of parts" data base. A single interactive terminal is used by the operator to direct both machine tools. Each CPU controls one machine. The "family of parts" data base is kept on a disk memory shared between the CPUs. Only one disk is used, so that all parts are always made according to the same data—one engineering change on one disk places the change "in effect" for all machines used to manufacture parts in this family. For speed, data transfers to and from the disk use the DMA.

The operator's interactive terminal and the disk are resources which are shared by both CPUs. When one CPU needs data from the disk, it cannot very well butt in if the other CPU is reading data from the disk. When the operator enters a directive on the terminal, how do either of the CPUs know that it is there and to which one of them it is directed? As you can see, the operation of the system can become rather complex. Oddly enough, in a configuration such as this, adding more CPUs might actually simplify things. For example, a CPU assigned to the job of communicating with the operator helps to sort things out. It could leave operator messages in mailboxes in shared or dual-port memory for the CPUs to which they were directed.

In any case, it is clear that there must be rules for deciding which CPU gets access to the bus and how long it is allowed to keep it. There must be ways to ensure that one CPU will not read data while another CPU is modifying it (for it might then be inconsistent). The process of handling these aspects of bus operation is called *arbitration*. The control signals provided in modern bus systems are upgraded versions of control signals provided in earlier designs for handling the simpler situation of transferring bus access between a single CPU and a DMA-driven I/O controller.

Early CPUs were equipped with two signals to control DMA access: "Bus Request" and "Bus Grant" (or names with similar meanings). When a DMA controller needed access to the bus, it activated its Bus Request signal, then waited. At its convenience, generally after completing execution of the current instruction, the CPU hardware switched its own bus drivers into their high-impedance state and activated the Bus Grant signal. This gave the DMA permission to proceed with its transfer. It could hold the bus for as long as it wished, the time depending on whether it was designed for interleaved transfers or for transferring entire blocks, without stopping until the transfer was complete. After the transfer, the DMA switched its bus drivers into the high-impedance state and deactivated its Bus Request. The CPU then resumed.

To go beyond this relatively simple situation and permit more CPUs to utilize the bus, the logic of the bus system must be equipped to decide which of several CPUs will be given access to the bus. DMA-type logic would certainly serve to let a running processor know that another processor needed the bus and to grant the other processor access to it. In a sense, the other processor might be considered to be rather similar to a DMA device. But if two or more other processors are asking for the bus, which should get it?

The answer is that some form of priority logic is needed. In fact, something along the lines of the logic used in interrupt systems would be a feasible approach. This is what is actually used in many modern bus systems. There are two principal forms. The first involves a serialized daisy-chain arrangement logically similar to the single polled interrupt method. The second involves a parallel priority encoding scheme logically similar to the priority encoding scheme used to generate vectors in the multiline and bus vectored interrupt methods.

The block diagram in Fig. 11-19 illustrates the daisy-chain method for a group of three independent processors. Grant of access to any processor is represented by the logically true state of its "priority in" signal. This signal comes from the "priority out" output of the next higher processor in the chain of priority, as indicated in the figure. The "Busy" and "Request" signals are wired-OR signals which can be driven and read by each processor. To see how this works, a typical sequence of bus accesses is described.

To begin, imagine that no processor is using the bus. Processor 2 then reaches a point in its program at which bus access is needed. Since Processor 1 was not using the bus, it would have enabled its "priority out" signal, thus granting access rights to Processor 2. Since Processor 2 was not in need of the bus prior to this time, its "priority out" signal to Processor 3's "priority in" would also be enabled, and so on down the chain.

Processor 2 looks to the "Busy" signal to see if any other processor is using the bus. Seeing that the bus is not busy, it lowers its "priority out" signal, which has the effect of denying access rights to any processor further down the chain, sets the "Busy" signal active, and proceeds to use the access to the bus which it has gained by this sequence.

Next, imagine that Processor 1 discovers a need to use the bus. The activity of the "Busy" signal will temporarily impede it from gaining access. It should not butt in, but it does (by system design) have a theoretically higher priority than the other processors in the figure. What if Processor 2 is in the middle of a very long, but not necessarily

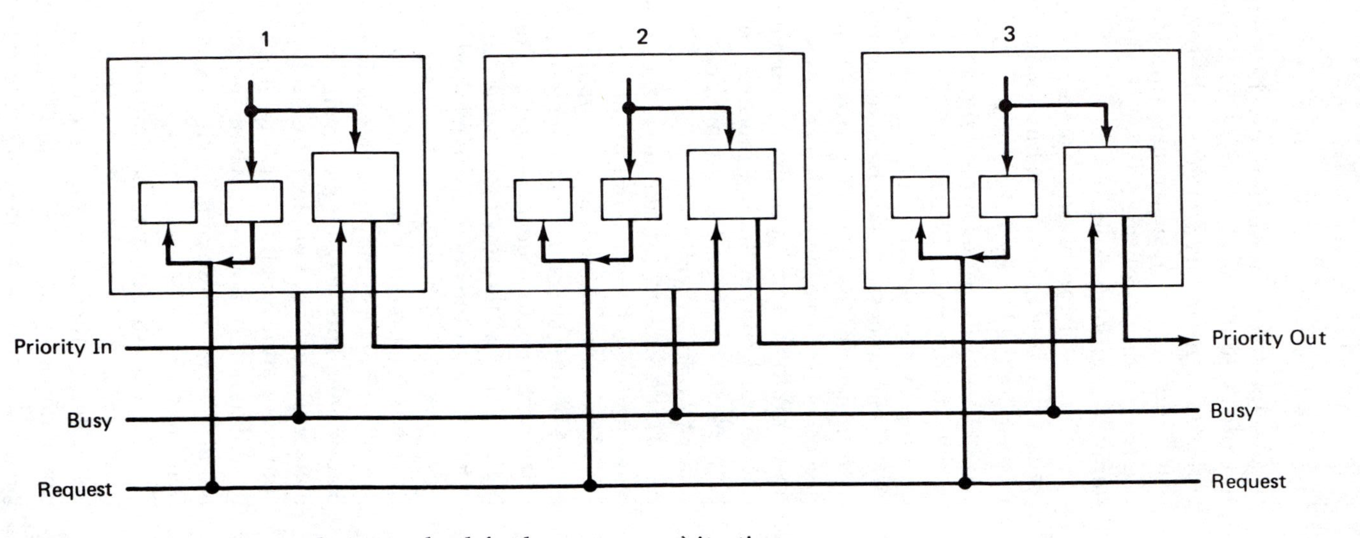

FIGURE 11-19. Daisy-chain method for bus access arbitration.

urgent, bus transaction? Similarly, how would one handle the situation in which a high-priority processor had the bus and was using it for a series of transactions which could be interrupted for lower priority work? Remember, just because a processor may be assigned a high priority in the hardware scheme, it does not mean that it is *always* doing high-priority work.

This can be resolved by proper use of the "Busy" and "Request" signals. If a processor needs bus access and finds it busy, it can signal its need for the bus by activating the "Request" line. At the same time, it deactivates its "priority out" signal. The "priority" signals then ripple down the chain. Recall that "Request" is a wired-OR signal, indicating that one or more processors want access to the bus. Then, at any given processor, the character of the requests for access can be determined by the following rule: If the "priority in" signal to the processor is active, any active requests are coming from processors of lower priority; if the "priority in" signal is not active, there is at least one request coming from a processor of higher priority. (There may be more than one, and they may include requests from processors of lower priority as well.)

If the signals indicate no higher priority request, the processor may choose to keep control of the bus by keeping its "priority out" signal inactive and its "Busy" signal active. However, it also has the option of releasing the bus to the lower priority request(s) if the work it is doing is not especially urgent. If the signals indicate requests of higher priority, the processor *must* release the bus.

Although the daisy-chain method works well logically, there are some practical difficulties with it. Because the "priority in" and "priority out" signals simply run out of one connector and into the next, the priority order of each processor is determined by its location in the card frame. This may be inconvenient and may cause difficulty with maintenance. For example, if a CPU board were returned to the wrong connector after repair, its priority would be different. This might not be immediately apparent when the system is started up again and might lead to subsequent subtle operation errors which could be difficult to trace.

Another difficulty is the time needed to decide priority order and transfer bus access from one processor to another. As more processors are added to the chain, the time required for priority information to ripple down through the chain of "priority out"/"priority in" signals gets longer, and it becomes necessary to slow down bus cycle speed in order to give the arbitration logic time to operate.

A different approach is to use a priority encode/decode scheme, as shown in Fig. 11-20. Here, the "priority out" signals become inputs to a priority encoder similar to that used to decide interrupt priority.

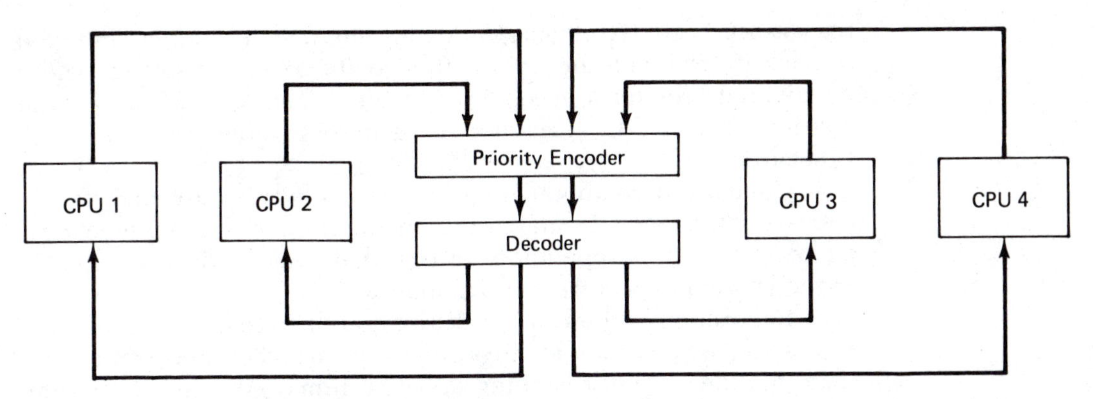

FIGURE 11-20. Priority encoding method for bus access arbitration.

The output of this encoder is the number of the requesting processor which is of highest priority. This number is then transferred to a decoder which yields one active "priority in" signal to this processor (informing it that it now has the highest priority) and inactive "priority in" signals to the other processors.

The logical behavior of these priority signals is equivalent to that of the daisy-chain priority signals, and it has the advantage of operating more quickly. However, different bus wiring is required and more bus signals are needed. A place to put this circuit is also needed.

Independent processors that are capable of controlling the bus are called "masters" in the jargon. Subsystems that cannot gain bus control are called "slaves." Most commercially available microcomputer subsystem boards that are capable of multiple processor system operation provide for using the daisy-chain method most conveniently and are quite satisfactory for systems where there may be up to four masters. The number of slaves is essentially unlimited in the logical sense, and "intelligent" slaves are freely allowed.

It is quite important that a processor be able to retain bus control and make its own decision about giving it up. This is critical to the operation of concurrent programs in different processors because concurrency requires the availability of a hardware mechanism to ensure that certain regions of program operation can have the characteristic that they are *mutually exclusive.*

The situation can be imagined figuratively to be like a section of railroad track in which there cannot be more than one train at one time. In railroading, when a train approaches such a section, the engineer observes a flag (called a *semaphore*). The railroading semaphore is a signal which is displayed at both ends of the restricted section of track and which can be turned on whenever the section of track is occupied. If the semaphore indicates that the restricted section of track is occupied, the train is then switched onto a siding, where it waits until the track becomes available. When this train enters the restricted section, the semaphore is again set to indicate that the track is occupied.

Concurrent computer programs use software/hardware mechanisms that are logically similar to semaphores in order to protect *critical regions* of their operation. In fact, they are called "semaphores" in the jargon of concurrent programming.

For example, when one processor is updating a file of information that is used by programs in other processors, no other processor should read this file. While it is being updated, it may contain inconsistent information. An example is the incrementing of a multibyte counter. Several memory accesses are needed generally to increment such a counter. If another processor were to read it while carries between bytes were being handled by the incrementing processor, data in the counter would be inconsistent. This potential inconsistency can be avoided by the use of a semaphore which is tested by a program before it accesses the file. If the semaphore indicates that the file is busy, the program waits. When the semaphore indicates that the file is not busy, the program activates the semaphore (to prevent it from being changed while it is being read), then proceeds to use the file.

Clearly, it is critical that the act of testing and activating the semaphore not be interruptible by any other processor (for example, by a master of higher priority). It is for this reason that a processor must be able to retain bus control until it can complete operations which constitute such critical regions. We examine semaphores further in the chapters on programming in Part III.

SURVEY OF BUS SYSTEMS

There are roughly a dozen or more relatively popular microcomputer system busses. Most have their origins in the pioneering efforts of system manufacturers going back to the early history of microprocessors. Although some of these manufacturers are no longer in business, their technologically successful design efforts survive in the form of modern bus systems. In some cases, designs which were not commercial successes for their originators eventually became commercial successes for others who built upon the groundwork established earlier.

In this section are some very brief comments about the characteristics of several of these bus systems. The intent is not to provide comprehensive coverage of all their important features but simply to convey a general idea of what they are like.

Some bus systems have avid adherents who feel strongly that "their" bus is the only proper one. Their strong feelings are often based on special features which are, in fact, variations on features found in other systems and are not always generally advantageous to the widest group of users. For this reason, it is best for the potential industrial user to study them carefully and make conservative choices. Remember that the system with the highest potential speed performance may not always turn out to be the most reliable.

In the case of many bus systems, compatible subsystem board-level products are commercially available from many independent manufacturers. However, many of these were developed without the "official" sanction of the companies that developed the original bus system. In some situations, the independent manufacturers took advantage of spare connector pins to provide functions not found in the original bus designs. Bus system designs are not static. Their originators continually improve them, and these improvements sometimes affect the interconnection systems for them. Because both busses and board-level products for them evolve, it is unusual to find globally assured compatibility among all of the possible combinations of board-level products that industrial systems designers might wish to use.

Unibus and Q-Bus

It is probably fair to say that the grandfather of modern microcomputer system busses is "Unibus." Unibus is the bus system used for the PDP-11 *mini*computer manufactured by Digital Equipment Corporation (DEC). The PDP-11 was the first widely successful 16-bit minicomputer and an outstanding commercial success widely used in industrial control systems. This was the first commercial unified bus system. It addresses memory and input/output devices in the same address space and provides for very high-speed data transfer between subsystems. Although microcomputer system busses are generally simpler than Unibus, to a great degree they are patterned after concepts that Unibus made popular.

The microcomputer successor to Unibus is "Q-Bus," which is the bus system used for DEC's LSI-11 microcomputer. The LSI-11 is a 16-bit microcomputer patterned after the PDP-11, but with somewhat less capability. Accordingly, the Q-Bus is a reduced-capacity model of Unibus. Mechanically, Q-Bus is designed to accept printed wiring boards of several different sizes. This permits more subsystem modularity than is typical of most microcomputer bus systems. "Unibus," "PDP-11," "Q-Bus," and "LSI-11" are trademarks of Digital Equipment Corporation.

S-100 Bus

The first popular microcomputer bus system was the S-100 Bus. This was originally developed by Micro Instrumentation and Telemetry Systems (MITS), which is no longer in existence as a separate corporation. The S-100 Bus was originally developed in a rather ad hoc fashion as an interconnection system for a microcomputer kit based on the Intel 8080 microprocessor. Nevertheless, it came to be widely used, first in the hobbyist computer market and later in the personal computer market because many other manufacturers developed subsystem boards that were generally "compatible" with it.

Because the S-100 Bus was not strictly well-defined, other manufacturers made rather free use of spare pins in its 100-pin connector. As a result, their products were not always as compatible as their advertising might lead one to believe. The outcome of the combined effects of compatibility problems and a widespread community of interested users of this bus system led to the formation of an IEEE project to develop a standard based on the S-100 Bus.

The IEEE standard for this bus system is IEEE-Std-696. In this standard, the original design has been upgraded to provide for 16-bit data transfers, 24-bit addresses, and multiple bus masters. Printed wiring board dimensions are 10 inches wide by 5.125 inches high. It provides for bus access arbitration in the form of a 4-bit encoded priority number, which theoretically allows for up to 16 bus masters. Lines are provided for eight parallel interrupt request lines. This system is also distinguished from most others by its use of unregulated power supplies in the form of three levels, which are nominally +8 volts, +16 volts, and −16 volts.

Although many functions are provided in the S-100 Bus, its assignment of signals to connectors reflects its rather uncontrolled early development. It gives the impression that the CPU board was designed first and the bus fitted to it later. Systems in which the structure of the bus is planned first generally give the appearance of being somewhat more orderly in the assignment of signals and are usually easier to work with.

STD Bus

The STD Bus is the result of a joint effort by Pro-Log Corporation and Mostek Corporation. This bus is designed to be as simple a vehicle as possible for the interconnection of relatively small systems. Its printed wiring board dimensions are 4.5 inches wide by 6.5 inches high, with a 56-pin connector. This gives it a high degree of modularity but relatively little board real estate for particularly complex subsystems.

It makes no provision for multiple bus masters but does include "Bus Request" and "Bus Grant" for DMA operations. It provides an 8-bit data bus and 16-bit address bus and uses a daisy-chain method of developing priority for bus vectored interrupts. A single line is activated to request an interrupt. The CPU acknowledgment of the request is passed down the daisy-chain, where the highest priority subsystem with an active request stops it and delivers a vector to the CPU via the data bus.

Multibus

"Multibus" is the descendant of the bus system originally developed for the "Intellec" microcomputer systems manufactured by Intel Corporation. "Multibus" and "Intellec" are trademarks of Intel Corporation. The relatively orderly arrangement of Multibus pin assignments reflects its more controlled development history. For much the same reasons as led to the development of the IEEE-Std-696 for the S-100 Bus, the IEEE also developed a standard (IEEE-Std-796) based on Multibus. Multibus provides for both 8-bit and 16-bit data transfers and for an address bus with up to 24 bits. Not all of the addressing capability of the bus is provided for on all of the board-level products which are "compatible" with Multibus. Bus arbitration is provided for in the form of a daisy-chain priority passing method. At the rated speeds of Multibus, this allows for four masters on each system. Eight parallel priority interrupt lines are provided. These may be priority decoded on a master CPU board for vector generation within the CPU, or the interrupt vector may be provided by the interrupting subsystem via the data bus.

Multibus printed wiring boards are 12 inches wide by 6.75 inches high. Two card edge connectors are provided. The principal connector is an 86-pin type with fingers on 0.156-inch centers. The IEEE-Std-796 standard defines all of the pins on the principal connector. These are the general control, address, data, and power lines.

The second connector is essentially an auxiliary connector with 60 pins on 0.1-inch centers. This is used by Intel for various special system functions, such as power fail detection status and control signals, battery power supply traces for backup power supplies, auxiliary reset switch input, and bus master status signals. A substantial number of the pins on this connector are unused and reserved for future use in Intel's descriptions of Multibus. The IEEE-Std-796 makes no specific assignments for the pins on the second connector. Backplanes for this bus often have a rectangular cutout under this connector. This permits users to mount wire-wrapped or soldered connectors in the position for customized functions.

SUMMARY

In addition to the manufacturers mentioned in this section, other principal manufacturers of microprocessors and microcomputer board-level products (such as Motorola, National Semiconductor, and Zilog) have proprietary bus systems. There is also a collection of bus systems developed by smaller microcomputer systems manufacturers who are not also semiconductor manufacturers. These present potential users of board-level microcomputer systems with a wide range of choices. For greatest future flexibility and potential for system growth and wider application, the best choices are most likely those for which the widest variety of subsystem boards are available, both from the original developer of the bus and from other manufacturers who may have climbed on to what they perceived to be a "bandwagon," provided, of course, that they first of all satisfy the technical requirements of the planned application.

PART III

SOFTWARE

"It is wonderful, when a calculation is made, how little the mind is actually employed in the discharge of any profession."

Samuel Johnson

It is wonderful, when most people think of a computer, how their minds grasp little more than its physical manifestation. Most people think of hardware when computers and automation are mentioned. They imagine a computer as a "calculating engine." They imagine programming a computer as the act of specifying a series of arithmetic operations that it is to perform. If they subscribe to Dr. Johnson's feelings about calculation, it is little wonder that it is difficult for them to feel that programming is a profession.

In fact, programming is much more than calculation, although certainly it had its beginnings in calculation. Augusta Ada Byron, Countess of Lovelace (and Lord Byron's daughter) is believed to have been the first programmer. Some notes by her, regarding Charles Babbage's Analytical Engine, describe a procedure for computing Bernoulli numbers using Babbage's invention. These notes were published in 1842 and are believed to be the first example of a computer program.

The first modern computers were developed with calculation in mind. Spurred by defense interests around the time of World War II, they were used to solve differential equations for the study of ballistics. As they evolved in the late 1940s and the 1950s, they acquired commercial purposes and broadened their military applications. But the uses to which they were put remained largely computational. During the 1950s, theories of automata were developed by such scientists as Alan M. Turing, and the first foundations of artificial intelligence were laid. Computers came to be appreciated as embodying the potential to become much more than calculating engines.

Although hardware technology improved by orders of magnitude during these years (becoming faster, less costly, and more reliable), its essence did not really change. What did change was the software. Programming is only one part of software. Software itself is very broad and, in its totality, encompasses many abstract (even abstruse) concepts which are sometimes difficult to grasp or describe clearly. Though rigorous mathematics and symbolic logic deal well with this subject, they fall short of conveying its spirit.

There is an excitement to programming that captures the minds of most of its practitioners to a degree found in few other professions. Some say it is not a profession at all, but an "art." Compared to other technological activities, it certainly offers one of the best outlets for high levels of creativity, but the creativity embodied in a program is difficult to show to one who is not a trained programmer. One can marvel at the operation of a well-designed machine because its features are easily seen. But to see the aspects of a computer program that reflect the quality of its design is difficult for one who is not already an initiate in the fraternity of programmers.

Though Augusta Ada Byron's program is over one hundred years old, true computer programming has been around, for all practical purposes, for only about one-quarter century. Although programming shares foundations in mathematics and symbolic logic with other engineering and scientific professions, it is apparent to people who spend any substantial amount of time with "real" programmers that there is something about them that is clearly different. They are a new breed, sufficiently different and intriguing to have had books written about their psychology. To my knowledge, no other profession can make this claim.

Some claim that programmers even look different from other people. By my own observation, I suspect there may be some truth to this. It is certainly true that few look like bankers. I knew a programmer who would qualify as a genius in the estimation of any who knew him, yet he was frequently mistaken for a janitor (especially since he often lurked about the building during hours when one would expect to find only janitors). Anyone who has spent a great deal of time in the company of computing equipment has probably known someone like this. Of course, I am not talking about the computing equipment found in banks and corporate headquarters. I am talking about the ones that are out where the action is: the ones being used to control equipment, to do research, to design and build things.

But times are changing. Our dependence on computers is such that we can no longer depend entirely on the "mountain men" and "trappers" of programming, the ones who first opened up the territory. We have reached the stage at which the "settlers" are needed to develop it and to build a firm foundation. It is possible that the territory of programming is now even accepting greenhorn immigrants in the form of those who are buying personal computers and struggling to understand them, both in the privacy of their homes and in executive offices.

Software is coming to be more an engineering profession and less an art form. The practice of conceiving, designing, writing, testing, and documenting software has received a great deal of scrutiny recently. Methodologies have been developed to improve the rate at which software can be prepared and to enhance its quality. Concern for "programmer productivity," reliable programs, and the shortage of programmers (qualified or otherwise) is much in evidence in the trade press. If you are new to software, you are coming upon it at a fascinating period in its history. There are still a lot of old-timers around who reflect the character it had during the pioneering days, yet it has become sufficiently civilized and settled that the danger of becoming lost in the wilderness is greatly reduced (but, please note, not entirely eliminated). In Part III, we explore the subject of software.